Physical Science
First Level

Hutchinson TECtexts

Learning by Objectives
A Teachers' Guide
A. D. Carroll, J. E. Duggan & R. Etchells

Engineering Drawing and Communication
First Level
P. Collier & R. Wilson

Physical Science
First Level
A. D. Carroll, J. E. Duggan & R. Etchells

Electronics
Second Level
G. Billups & N. T. Sampson

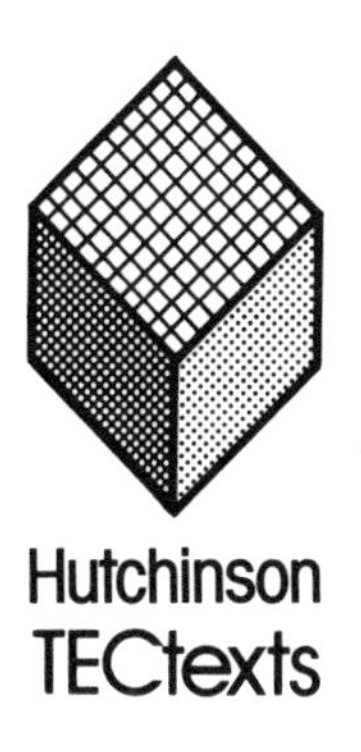

Physical Science

First Level

A. D. Carroll, J. E. Duggan
& R. Etchells

Hutchinson of London

Hutchinson & Co. (Publishers) Ltd
3 Fitzroy Square, London W1P 6JD

London Melbourne Sydney Auckland
Wellington Johannesburg and agencies
throughout the world

First published 1978

Set in Times New Roman by Oliver Burridge Filmsetting Limited

Printed in Great Britain by The Anchor Press Ltd
and bound by Wm Brendon & Son Ltd,
both of Tiptree, Essex

ISBN 0 09 133581 7

Contents

Introduction

In each of the books in this series the authors have written text material to specified objectives. Test questions are provided to enable the reader to evaluate the objectives. The solutions or answers are given to all questions.

Topic area: Materials

Section 1

After reading the following material, the reader shall:

1 Know the definitions, symbols, abbreviations and units of quantities basic to SI.
1.1 Identify the preferred units of length as the metre and the millimetre in accordance with SI.
1.2 Identify the SI unit of time as the second.
1.3 Identify the SI unit of temperature as the kelvin.
1.4 Identify the SI unit of electric current as the ampere.
1.5 Identify the SI unit of luminous intensity as the candela.
1.6 Identify a definition of mass given in everyday terms.
1.7 Define mass unit in accordance with SI.

'SI' are the first letters of two French words – *Système International*, which readers familiar with the language will translate as 'International System'. Thus, 'SI' as it is known in every language, is an international system of units for all the various technologies. (Note it is incorrect in a grammatical sense to speak of 'SI system'.) *All* the units used in the various technologies are derived from *six* basic and arbitrarily defined units shown in Table 1.

quantity	symbol	unit	abbreviation
length	l	metre	m
mass	m	kilogram	kg
time	t	second	s
temperature	T	kelvin	K
electric current	I	ampere	A
luminous intensity	I	candela	cd

Table 1: *SI units*

Note that small letters are used for both the symbol, representing the quantity, and the abbreviation of the unit, *except* for the quantities temperature and electric current and also the symbol for luminous intensity.

Note also, that the same letters, *m* and *I*, occur twice. It might appear that this leads to confusion, but in practice the context in which the abbreviations and symbols are used reduces the chance of error.

The six basic quantities listed in Table 1 will now be defined. It is not necessary for the reader to remember the details of these definitions at this stage, but they may well be useful at a later stage.

1 *Length*

The metre is defined as the length equal to 1 650 763.73 wavelengths of the orange line in the spectrum of an internationally specified krypton discharge lamp.

2 *Mass*

The kilogram is defined as the mass of a platinum–iridium cylinder, stored under controlled conditions of temperature and humidity at the International Bureau of Weights and Measures at Sèvres, near Paris.

A physicist would define the mass of a body as the quantity of matter in the body, but this is probably not a very meaningful definition, even if it is strictly correct.

If a housewife buys (say) sugar, she buys a bag labelled 1 kg, i.e. she is buying sugar by *mass*. In other words, she is buying an *amount* or *quantity* of sugar.

Suppose a doctor tells a patient he is too fat, and must follow a slimming diet. If the patient adheres to the diet, he is liable to reduce his waistline; in other words he reduces the *amount* or *quantity* of fat in his body. Thus his *mass* is reduced.

Observant readers may have noticed that no mention has been made of *weight*. This is because weight is a particular type of *force*, and force is not the same as mass. It is true that a person's weight is reduced if he follows a slimming diet, but that is only as a consequence of his mass being reduced. In fact there are other ways of reducing a person's weight: he could travel as an astronaut for example, when his weight would be reduced to zero, but he would still have essentially the same mass as if he were on earth.

Self-assessment question

1 Answer true or false to each of the three statements which follow:

(i) A person's mass can be reduced by dieting.

(ii) The weight of a spanner is constant.

(iii) The mass of a spanner is constant.

3 *Time*

The second is defined as the interval occupied by 9 192 631 770 cycles of radiation corresponding to the transition of the caesium-133 atom.

4 *Temperature*

The kelvin is based upon a scale of temperature known as a *thermodynamic temperature scale*. The reasons for the choice of the kelvin, rather than the familiar celsius scale, are beyond the scope of this book. Suffice it to say that the kelvin scale indicates that if the temperature of a body could be reduced to absolute zero, the body would not possess any thermal energy.

On the kelvin scale, the temperature at which ice melts under standard atmospheric pressure is exactly 273.15 K, i.e. 273.15 K corresponds to 0 °C.

Note that the symbol for temperature is K, not °K, although in the celsius scale it is correct to write °C.

5 *Electric current*

The rate at which electricity flows past any given point of an electric circuit is termed an *electric current*. The unit of electric current is the ampere.

6 *Luminous intensity*

The candela is defined as $\frac{1}{60\,000}$ of the luminous intensity per square metre of a black body radiator at the temperature of solidification of platinum (approximately 2 042 K) under standard atmospheric pressure.

After reading the following material, the reader shall:

1.8 Select the correct symbols and units for the quantities basic to SI.

Self-assessment question

2 Complete the table below, if possible without making reference to the text.

Write the correct symbols and units for the six basic quantities of SI. Select the symbols from the following list:

I, i, T, t, M, m, l, L, A, a

Select the units from the following list:

kilogram, ampere, gram, volt, candela, watt, metre, kilometre, second, minute, kelvin, celsius

quantity	symbol	unit
mass		
electric current		
luminous intensity		
length		
time		
temperature		

After reading the following material, the reader shall:

1.9 Distinguish between the multiples and sub-multiples of units preferred in SI.

Most readers will be familiar with terms such as kilometre, millimetre or megawatt, and may know that a kilometre is 1 000 metre, a millimetre is $\frac{1}{1\,000}$ of a metre and a megawatt is 1 000 000 watt. These are more usually written as 10^3 metre, 10^{-3} metre and 10^6 watt respectively: indices are used because the method is shorter (and therefore more convenient), and also gives less opportunity for error; e.g., when writing 200 000 000, it is not difficult to omit a nought, and it is therefore better technique to write 200×10^6.

In SI, it is recommended that 200 000 000 be written as 200×10^6 and not as (say) 2×10^8; multiples and sub-multiples should be confined to powers of 10 which are multiples of ± 3. A summary of the recommendations is given in Table 2, together with the special prefixes and abbreviations given to each multiple and sub-multiple.

power of 10	prefix	abbreviation
10^{12}	terra	T
10^9	giga	G
10^6	mega	M
10^3	kilo	k
10^{-3}	milli	m
10^{-6}	micro	μ
10^{-9}	nano	n
10^{-12}	pico	p

Table 2: *SI prefixes*

Note that the sub-multiples use small letters for the abbreviations, and the multiples, *except for kilo*, use capital letters for the abbreviations.

One or two examples may clarify the table:

(i) 20 micrometre means $\frac{20}{1\,000\,000}$ metre, or 20×10^{-6} metre, and can be abbreviated as 20 μm.

(ii) 40 gigasecond means 40 000 000 000 second, or 40×10^9 second, and can be abbreviated as 40 Gs.

(iii) 500 millimetre means $\frac{500}{1\,000}$ metre, or 500×10^{-3} metre, and can be abbreviated as 500 mm or 0.5 m.

The insistence upon powers of 10 which are multiples of ± 3, means that the centimetre (cm), a very common length measure, is not a preferred unit of length in SI. By a similar argument, neither the minute nor the hour are preferred units of time in SI.

Self-assessment question

3 Convert each of the following quantities into the preferred SI multiple or sub-multiple of 10, and write the preferred abbreviation for each one, e.g.

	1 100 metre	=	1.1×10^3 m	=	1.1 km
(i)	0.000 005 metre	=		=	
(ii)	6 320 second	=		=	
(iii)	1 654 000 candela	=		=	
(iv)	0.008 ampere	=		=	
(v)	0.000 000 009 2 metre	=		=	

After reading the following material, the reader shall:

2 Be aware of the preferred units of area and volume in SI.
2.1 Define length unit and derive units of area and volume in accordance with SI.
2.2 Identify the preferred units of area and volume in SI.
2.3 Derive areas in mm^2 and m^2.
2.4 Convert units of area from mm^2 to m^2 and vice-versa.

It has been established that the preferred units of length in SI are the metre and the millimetre. It is logical then to propose that units of area should bc m^2 or mm^2, and that units of volume should be m^3 or mm^3. Note that these units are written using indices, rather than sq. m or cu. mm; this latter method should never be used.

Solution to self-assessment question

1 The correct responses are:
(i) True, since the amount or quantity of fat in his body will be reduced.
(ii) False, since the weight of a spanner can be reduced by taking it onto the moon, for example.
(iii) True, since the only way the mass of the spanner can be changed is by removing some of the metal from which the spanner is made; in which case, it is not the same spanner.

2 The table should have been completed as follows:

quantity	symbol	unit
mass	m	kilogram, kg
electric current	I	ampere, A
luminous intensity	I	candela, cd
length	l	metre, m
time	t	second, s
temperature	T	kelvin, K

There is however one other unit of volume which may be used, in addition to m^3 and mm^3; this is the litre. The litre is defined as 1 000 cubic centimetre (1 000 c.c.) and is a unit much used in the motor industry.

Summarizing therefore:
Units of area: m^2 or mm^2
Units of volume: m^3 or mm^3 or litre

Thus the SI units of area are m^2 or mm^2.

$$\begin{aligned}\text{Now, } 1\text{ m} &= 1\,000\text{ mm} = 10^3\text{ mm}\\ \therefore\ 1\text{ m}^2 &= (1\,000 \times 1\,000)\text{ mm}^2\\ &= (10^3 \times 10^3)\text{ mm}^2 = 10^6\text{ mm}^2\\ \therefore\ \underline{1\text{ m}^2} &\underline{= 10^6\text{ mm}^2}\end{aligned}$$

$$\begin{aligned}\text{Similarly, } 1\text{ mm} &= 0.001\text{ m} = 10^{-3}\text{ m}\\ \therefore\ 1\text{ mm}^2 &= (0.001 \times 0.001)\text{ m}^2\\ &= (10^{-3} \times 10^{-3})\text{ m}^2 = 10^{-6}\text{ m}^2\\ \therefore\ \underline{1\text{ mm}^2} &\underline{= 10^{-6}\text{ m}^2}\end{aligned}$$

Until the reader is completely familiar with SI units, it is not recommended that he should remember these conversion factors, but rather how they are derived. Consider two rather simple examples:

Example 1
Calculate the area in mm^2 of a sheet of polythene, measuring 0.35 m by 0.45 m.

$$\begin{aligned}\text{Area} &= \text{length} \times \text{breadth}\\ \text{i.e. Area} &= (0.35 \times 0.45)\text{ m}^2\\ &= (350 \times 450)\text{ mm}^2\\ \therefore\ \underline{\text{Area}} &\underline{= 157\,500\text{ mm}^2}\end{aligned}$$

Alternatively:

$$\begin{aligned}\text{Area} &= (0.35 \times 0.45)\text{ m}^2\\ &= 0.157\,5\text{ m}^2 = 0.157\,5 \times 10^6\text{ mm}^2\\ \therefore\ \underline{\text{Area}} &\underline{= 157\,500\text{ mm}^2}\end{aligned}$$

Solution to self-assessment question

3 The correct multiples of 10 and abbreviations are:

(i)	0.000 005 metre	$= 5 \times 10^{-6}$ m	= 5 μm
(ii)	6 320 second	$= 6.32 \times 10^3$ s	= 6.32 ks
(iii)	1 654 000 candela	$= 1.654 \times 10^6$ cd	= 1.654 Mcd
(iv)	0.008 ampere	$= 8 \times 10^{-3}$ A	= 8 mA
(v)	0.000 000 009 2 metre	$= 9.2 \times 10^{-9}$ m	= 9.2 nm

Example 2
Calculate the area in m^2 of a hole 70 mm in diameter cut in a sheet of glass.

$$\text{Area of circle} = \frac{\pi d^2}{4}$$

$$\text{i.e. Area} = \left[\frac{\pi}{4} \times (70)^2\right] \text{mm}^2$$

$$= \left[\frac{\pi}{4} \times (0.07)^2\right] \text{m}^2$$

$$\therefore \text{Area} = \underline{0.003\,85 \text{ m}^2}$$

Alternatively:

$$\text{Area} = \left[\frac{\pi}{4} \times (70)^2\right] \text{mm}^2$$

$$= 3\,848.5 \text{ mm}^2$$

$$= 3\,848.5 \times 10^{-6} \text{ m}^2$$

$$\therefore \text{Area} = \underline{0.003\,85 \text{ m}^2} \text{ (answer rounded off)}$$

Self-assessment questions

4 Calculate the following areas, giving the answers in the units specified:

(i) A bar of mild steel has a diameter of 250 mm. Find the cross-sectional area in m^2.

(ii) A triangle has the dimensions shown in Figure 1. Find its area in mm^2.

(iii) Find the area in mm^2 and m^2 of the cross-section of aluminium alloy shown in Figure 2.

(iv) Find the area in mm^2 and m^2 of the sheet of copper shown in Figure 3.

(v) Find the curved surface area of a cylinder measuring 420 mm diameter and 1 m high. Quote the answer in both mm^2 and m^2.

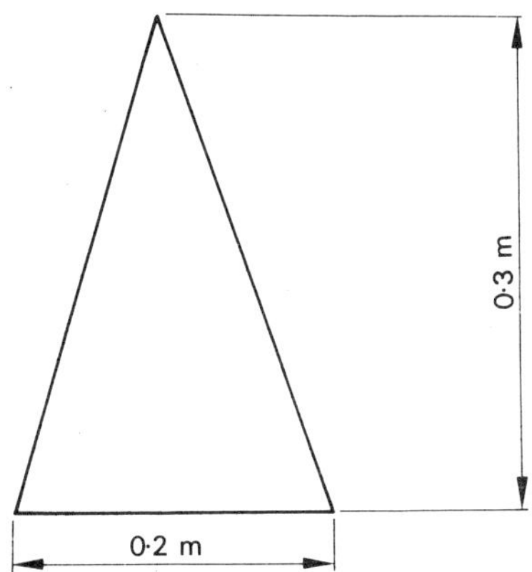

Figure 1 *Isosceles triangle*

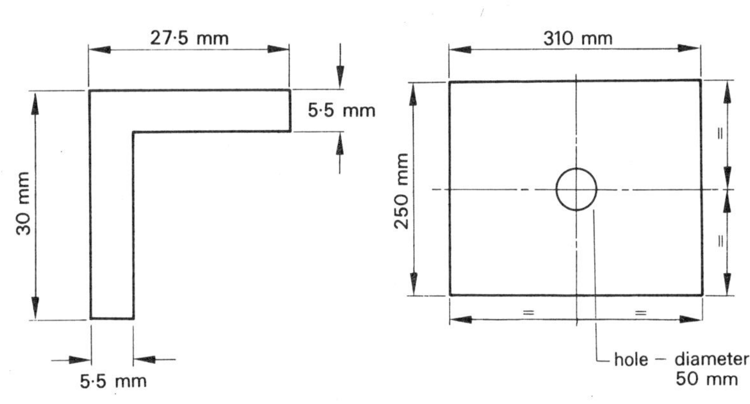

Figure 2 *Aluminium alloy angle*

Figure 3 *Copper sheet*

After reading the following material, the reader shall:

2.5 Derive volumes in mm^3, m^3 and litre.
2.6 Convert units of volume from mm^3 to m^3 or litre, and vice-versa.

It has been established that the SI units of volume are m^3, mm^3 or litre, and that a litre is defined as $1\,000\ cm^3$

$$\begin{aligned}\text{Since } 1\ m &= 1\,000\ mm\\ \text{and } 1\ m^2 &= (1\,000 \times 1\,000)\ mm^2 = 10^6\ mm^2\\ \text{then } 1\ m^3 &= (1\,000 \times 1\,000 \times 1\,000)\ mm^3 = 10^9\ mm^3\end{aligned}$$

Solutions to self-assessment questions

4 (i) Cross-sectional area of mild steel bar $= \dfrac{\pi d^2}{4}$

$$\begin{aligned}&= \left(\frac{\pi \times 250^2}{4}\right) mm^2\\ &= 49\,087.4\ mm^2\\ &= (49\,087.4 \times 10^{-6})\ m^2\end{aligned}$$

$\therefore$ Cross-sectional area $= 0.049\ m^2$ (answer rounded off)

Alternatively,

$$\text{cross-sectional area} = \left(\frac{\pi \times 0.25^2}{4}\right) m^2 = 0.049\ m^2$$

(ii) Area of triangle $= \frac{1}{2} \times \text{base} \times \text{perpendicular height}$

$$\begin{aligned}&= (\tfrac{1}{2} \times 0.2 \times 0.3)\ m^2\\ &= 0.03\ m^2 = (0.03 \times 10^6)\ mm^2\end{aligned}$$

$\therefore$ Area of triangle $= 30\,000\ mm^2$

(iii) Cross-sectional area $= [(30 \times 5.5) + (22 \times 5.5)]\ mm^2$

$$\begin{aligned}&= [(30 + 22)5.5]\ mm^2\\ &= [52 \times 5.5]\ mm^2\end{aligned}$$

$\therefore$ Cross-sectional area $= 286\ mm^2$

Also cross-sectional area $= (286 \times 10^{-6})\ m^2$
or cross-sectional area $= 0.000\,286\ m^2$

(iv) Area $= \left[(0.31 \times 0.25) - \dfrac{\pi \times 0.05^2}{4}\right] m^2$

$$= [0.0775 - 0.00\,196]\ m^2$$

$\therefore$ Area $= 0.075\,54\ mm^2$

Also, area $= (0.075\,54 \times 10^6)\ mm^2$
$\therefore$ Area $= 75\,540\ mm^2$

(v) Circumference of cylinder $= \pi \times d$

$$\begin{aligned}&= (\pi \times 0.42)\ m\\ &= 1.319\ m\end{aligned}$$

$\therefore$ Surface area of cylinder $= (1.319 \times 1)\ m^2 = 1.319\ m^2$

Also, surface area of cylinder $= (1.319 \times 10^6)\ mm^2 = 1\,319\,000\ mm^2$

$$\text{Similarly, } 1\ \text{mm}^3 = (0.001 \times 0.001 \times 0.001)\ \text{m}^3 = 10^{-9}\ \text{m}^3$$

$$\text{Also } 1\ \text{m} = 100\ \text{cm}$$

$$1\ \text{m}^2 = (100 \times 100)\ \text{cm}^2$$

$$1\ \text{m}^3 = (100 \times 100 \times 100)\ \text{cm}^3 = 10^6\ \text{cm}^3$$

Since there are 1 000 cm^3 ($10^3\ \text{cm}^3$) in 1 litre,

$$\text{then, } 1\ \text{m}^3 = \frac{10^6}{10^3}\ \text{litre} = 10^3\ \text{litre}$$

$$\text{Similarly, } 1\ \text{mm} = 0.1\ \text{cm}$$

$$1\ \text{mm}^3 = 0.001\ \text{cm}^3 = 10^{-3}\ \text{cm}^3$$

$$\text{i.e. } 1\ \text{mm}^3 = \frac{1}{10^3}\ \text{cm}^3$$

converting to litre

$$1\ \text{mm}^3 = \frac{1}{10^3 \times 10^3}\ \text{litre}$$

$$\text{or } 1\ \text{mm}^3 = 10^{-6}\ \text{litre}$$

Again, until the reader is completely familiar with SI units, it is not recommended that these conversion factors be remembered, but rather how they are derived. It may help to imagine what each measure of cubic capacity looks like. Two of them are drawn to scale in Figure 4, but it is impracticable to draw 1 mm^3 to the same scale, since it would be so small as to be almost invisible.

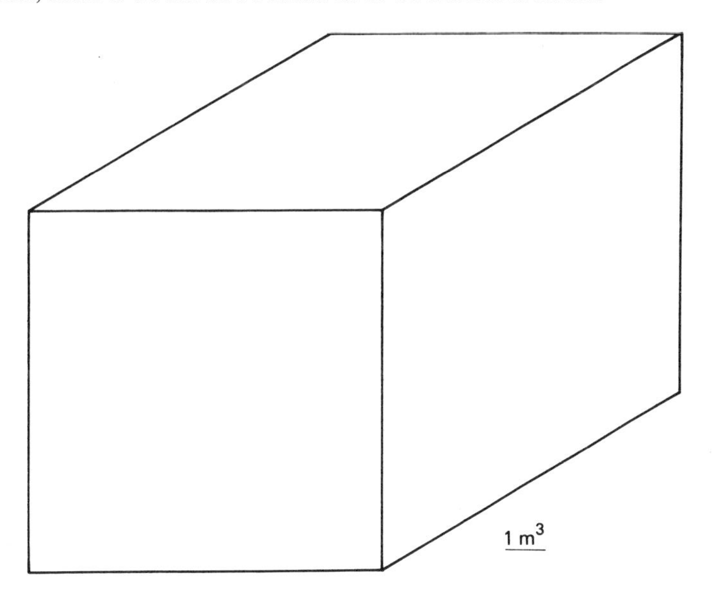

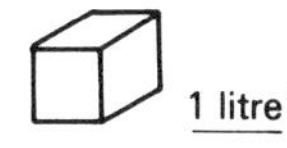

Figure 4 *Comparison of cubic capacities*

To assist in understanding conversion of the three measures of capacity, two examples will be worked out.

Example 3
A container has an internal diameter of 50 mm and is 20 mm deep. Find its volume (*a*) in mm^3, (*b*) in m^3 and (*c*) in litre.

(*a*) $$\text{Volume} = \left[\left(\frac{\pi}{4} \times 50^2\right) \times 200\right] \text{mm}^3$$
$$= \underline{392\,700 \text{ mm}^3}$$

(*b*) $$\text{Volume} = \left[\left(\frac{\pi}{4} \times 0.05^2\right) \times 0.2\right] \text{m}^3$$
$$= \underline{392.7 \times 10^{-6} \text{ m}^3}$$

(*c*) Using cm as a measure of length,
$$\text{Volume} = \left[\left(\frac{\pi}{4} \times 5^2\right) \times 20\right] \text{cm}^3$$
$$= 392.7 \text{ cm}^3$$

Dividing by 10^3 to convert cm^3 to litre,
$$\underline{\text{Volume} = 0.392\,7 \text{ litre}}$$

Example 4
A container has a capacity of 1.5 litre. Calculate its volume (*a*) in m^3 and (*b*) in mm^3.

(*a*)
$$1.5 \text{ litre} = 1\,500 \text{ cm}^3$$
$$1 \text{ cm} = 0.01 \text{ m}$$
$$1 \text{ cm}^3 = (0.01 \times 0.01 \times 0.01) \text{ m}^3 = 10^{-6} \text{ m}^3$$
$$\therefore 1\,500 \text{ cm}^3 = 1\,500 \times 10^{-6} \text{ m}^3 = 1.5 \times 10^{-3} \text{ m}^3$$
$$\text{i.e. } 1.5 \text{ litre} = 0.001\,5 \text{ m}^3$$
$$\therefore \underline{\text{Volume of container} = 0.001\,5 \text{ m}^3}$$

(*b*)
$$1 \text{ cm} = 10 \text{ mm}$$
$$1 \text{ cm}^3 = 1\,000 \text{ mm}^3 = 10^3 \text{ mm}^3$$
$$\therefore 1\,500 \text{ cm}^3 = 1\,500\,000 \text{ mm}^3$$
$$= 1.5 \times 10^6 \text{ mm}^3$$
$$\therefore \underline{\text{Volume of container} = 1.5 \times 10^6 \text{ mm}^3}$$

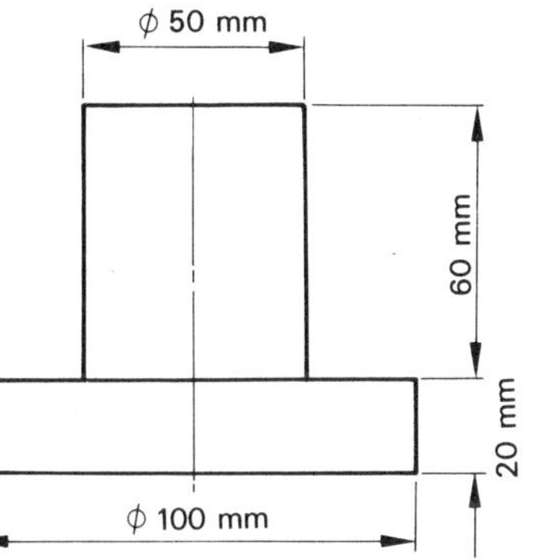

Figure 5 *Sheet metal copper container*

Self-assessment questions

5 Calculate the following volumes, giving the answers in the units specified.

(i) A rectangular storage tank has dimensions of 1.5 m by 2 m by 2 m. Find its volume in m^3 and litre.

(ii) Find the volume of a pipe 20 mm diameter and 2 m long. Give the answer in litre and m^3.

(iii) A tube has dimensions of 10 mm diameter and 100 mm long. If the tube is half full of water, find the volume of water in mm^3 and litre.

(iv) A vessel made from thin copper sheet has the dimensions shown in Figure 5. Find its volume in mm^3 and litre.

After reading the following material, the reader shall:

3 Be familiar with the concept of density.

3.1 Define density.

3.2 State that the density of a body or substance is independent of its physical size.

There is a very old schoolboy joke which, slightly modified, reads as follows: 'Which has the greater mass, 1 kg of lead or 1 kg of feathers?' It is hoped the reader knows the answer!

Although both quantities have the same mass, the lead occupies very much less space than the feathers; in other words, the lead has a higher *density*. Some readers may remember that the density of a body is defined as its *mass per unit volume*.

Representing density by ρ (Greek letter rho, pronounced ro), mass by m and volume by V,

$$\text{density, } \rho = \frac{m}{V}$$

Self-assessment question

6 Remembering that density $= \frac{\text{mass}}{\text{volume}}$, write down the units of density.

Applying the definition of density that

$$\rho = \frac{m}{V},$$

to the schoolboy joke, the volume occupied by the feathers is very much larger than that occupied by the lead, and hence the density of the feathers is very much less than the density of the lead.

Suppose the masses of the lead and feathers are increased from 1 kg to 2 kg each. The materials then occupy twice as much space as they did before, but since both the mass and the volume have been increased by the same amount, their densities remain the same, i.e.

$$\rho = \frac{m}{V} \text{ remains constant.}$$

The density of a body depends only upon the material from which it is made, and is independent of its physical size.

Solutions to self-assessment questions

5 (i) Using metre as the unit of length:

$$\text{Volume of tank} = (1.5 \times 2 \times 2)\ \text{m}^3$$
$$\therefore \underline{\text{Volume of tank} = 6\ \text{m}^3}$$

$$\text{Since } 1\ \text{m}^3 = 10^3\ \text{litre}$$
$$\underline{\text{Volume of tank} = 6\,000\ \text{litre}}$$

(ii) Using metre as the unit of length:

$$\text{Volume of pipe} = \left[\left(\frac{\pi}{4} \times 0.02^2\right) \times 2\right] \text{m}^3$$
$$\therefore \underline{\text{Volume of pipe} = 628 \times 10^{-6}\ \text{m}^3}$$

$$\text{Since } 1\ \text{m}^3 = 10^3\ \text{litre}$$
$$\text{Volume of pipe} = 628 \times 10^{-6} \times 10^3\ \text{litre}$$
$$= 628 \times 10^{-3}\ \text{litre}$$
$$\underline{\text{Volume of pipe} = 0.628\ \text{litre}}$$

(iii) Using millimetre as the unit of length:

$$\text{Volume of tube} = \left[\left(\frac{\pi}{4} \times 10^2\right) \times 100\right] \text{mm}^3$$
$$= 7\,854\ \text{mm}^3$$
$$\therefore \text{Volume of water} = \tfrac{1}{2} \times 7\,854\ \text{mm}^3$$
$$= \underline{3\,927\ \text{mm}^3}$$

$$\text{Since } 1\ \text{mm}^3 = 10^{-6}\ \text{litre,}$$
$$\text{Volume of water} = 3\,927 \times 10^{-6}\ \text{litre}$$
$$\underline{\text{Volume of water} = 0.003\,927\ \text{litre}}$$

(iv) Using millimetre as the unit of length:

$$\text{Volume of base} = \left[\left(\frac{\pi}{4} \times 100^2\right) \times 20\right] \text{mm}^3$$
$$= 157\,080\ \text{mm}^3$$
$$\text{Volume of funnel} = \left[\left(\frac{\pi}{4} \times 50^2\right) \times 60\right] \text{mm}^3$$
$$= 117\,800\ \text{mm}^3$$
$$\text{Volume of vessel} = \text{volume of base} + \text{volume of funnel}$$
$$= (157\,080 + 117\,800)\ \text{mm}^3$$
$$\therefore \underline{\text{Volume of vessel} = 274\,880\ \text{mm}^3}$$

$$\text{Since } 1\ \text{mm}^3 = 10^{-6}\ \text{litre}$$
$$\text{Volume of vessel} = 274\,880 \times 10^{-6}\ \text{litre}$$
$$\underline{\text{Volume of vessel} = 0.274\,88\ \text{litre}}$$

Self-assessment question

7 A block of steel 200 mm × 100 mm × 60 mm has a mass of 9.42 kg.
(*a*) Calculate the density of the material in kg/m^3.
(*b*) If the mass of the steel block is doubled to 18.84 kg, how will the volume change? Will the density change?

After reading the following material, the reader shall:

3.3 State the density of water.
3.4 Define relative density.
3.5 Solve simple problems relating to density, mass and volume of solids.

Precise measurement has shown that 1 kg of pure water (at maximum density and under normal atmospheric pressure) has a volume of 1.000 028 litre. Since water is rarely if ever pure, it can be assumed that for most practical purposes, *1 litre of water has a mass of 1 kg.*

$$\text{Since } 1\ m^3 = 1\,000 \text{ litre}$$
$$\text{Mass of } 1\ m^3 \text{ of water} = 1\,000 \text{ kg}$$
$$\therefore \text{Density of water} = \frac{\text{mass}}{\text{volume}} = \frac{1.000}{1}\left[\frac{kg}{m^3}\right]$$
$$\text{i.e. } \underline{\text{Density of water} = 1\,000\ kg/m^3}$$

The term *relative density* of a material is a ratio in which the density of the material is compared with the density of water. Relative density (symbol *d*) is defined as

$$\frac{\text{density of the material}}{\text{density of water}}$$

For example, considering the steel block in self-assessment question 7:

$$\text{Density of steel} = 7\,850\ kg/m^3$$
$$\text{Density of water} = 1\,000\ kg/m^3$$
$$\text{Relative density of steel} = \frac{7\,850\ kg/m^3}{1\,000\ kg/m^3}$$
$$= \underline{7.85}$$

Since relative density is a ratio of two quantities expressed in the same units, it is purely a number, and has no units.

Solution to self-assessment question

6 The units of mass are kg. The preferred units of volume are m^3 or mm^3. Hence the units of density are kg/m^3 or kg/mm^3. It is usual in SI to quote density in kg/m^3.

Self-assessment questions

8 Find the density and relative density of a copper bar 60 mm diameter, 2 m long, if its mass is 50 kg.

9 The cross-section of an aluminium alloy extrusion is shown in Figure 6. If the extrusion is 3 m long, and the density of aluminium alloy is 2 700 kg/m^3, find its mass.

10 A steel component is machined to the dimensions shown in Figure 7. If the relative density of steel is 7.85, find the mass of the component.

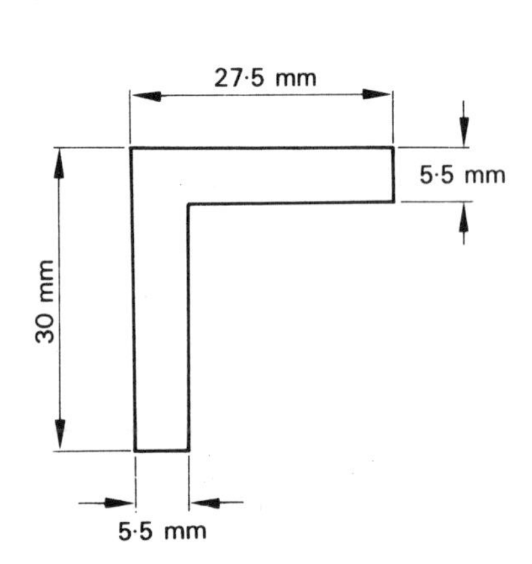

Figure 6 *Aluminium alloy extrusion*

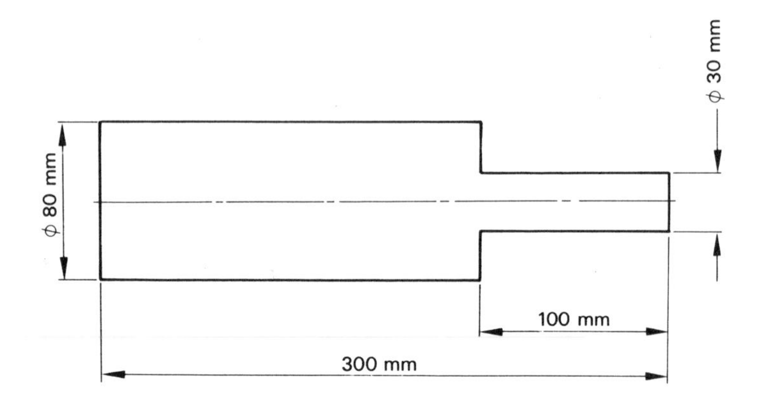

Figure 7 *Machined steel component*

After reading the following material, the reader shall:

3.6 Recognize a method of measuring the density of liquids.

The density of a solid can be derived if the mass of the solid is known or can be measured, and the volume can be calculated. In a similar manner, the density of liquids may also be found, providing that the mass and the volume of liquid are known. One method of finding the mass and volume of a given amount of liquid is outlined below:

(i) Using a physical balance, measure the mass of a clean, dry, graduated container. The graduations are normally marked with either cubic centimetre (c.c.) or fractions of a litre.

(ii) Pour a given amount of liquid into the container, and use the graduations to measure the volume.

(iii) Use the physical balance to find the mass of the container and liquid.

(iv) Find the mass of the liquid by comparing the results of (i) and (iii).

Self-assessment question

11 The results of such a procedure are shown in the table below, which contains the masses in g, and the corresponding volumes in cm^3, of four liquids: water, paraffin, petrol and beer.

Complete the table by calculating the densities and relative densities of paraffin, petrol and beer. In order to give guidance in completing the table, the appropriate calculations for water have been supplied.

liquid	volume		mass		density kg/m^3	relative density
	cm^3	m^3	g	kg		
water	50	50×10^{-6}	50.1	0.0501	1 002	1.0
paraffin	50		38.7			
petrol	50		36.5			
beer	50		49.7			

Further self-assessment questions

12 Select (by marking in pencil) from the following list, those units which are preferred units of area and volume in SI.
cm^2; m^3; litre; cm^3; m^2; mm^3; mm^2.

13 Answer true or false to the following statements:
(i) Density can be defined as mass × volume.
(ii) The preferred units of density in SI are kg/m^3.
(iii) The density of water can be written as 1 Mg/m^3.

14 Statement 1: The density of a body varies as its physical size varies.
Statement 2: The units of relative density are kg/m^3.
Underline the correct answer:
(*a*) Only statement 1 is true.
(*b*) Only statement 2 is true.
(*c*) Neither statement is true.
(*d*) Both statements are true.

15 Make a list of the measurements and calculations made in finding the density of a liquid, given the following equipment; a graduated container, a physical balance and a quantity of liquid.

Solutions to self-assessment questions

7 Using metre as the unit of length:

(*a*) Volume of steel block $= [0.2 \times 0.1 \times 0.06]\ \text{m}^3$
$= 0.001\,2\ \text{m}^3$

$$\text{density} = \frac{\text{mass}}{\text{volume}}$$

$$= \frac{9.42}{0.001\,2}\ \frac{\text{kg}}{\text{m}^3}$$

$$\therefore \text{Density} = 7\,850\ \text{kg/m}^3$$

(*b*) If the mass of the steel block is doubled, then the volume of the block also increases by a factor of 2. The density remains unaltered, since it is independent of the size of the block.

8 Using metre as the unit of length:

$$\text{Volume of copper bar} = \left[\left(\frac{\pi}{4} \times 0.06^2\right) \times 2\right]\ \text{m}^3$$
$$= 0.005\,65\ \text{m}^3$$
$$= 5.65 \times 10^{-3}\ \text{m}^3$$

$$\therefore \text{Density of copper} = \frac{\text{mass}}{\text{volume}} = \frac{50}{5.65 \times 10^{-3}} \left[\frac{\text{kg}}{\text{m}^3}\right]$$

$$= \frac{50 \times 10^3}{5.65}\ \text{kg/m}^3$$

$$\therefore \text{Density of copper} = 8\,850\ \text{kg/m}^3$$

$$\text{Relative density of copper} = \frac{8\,850\ \text{kg/m}^3}{1\,000\ \text{kg/m}^3}$$
$$= 8.85$$

9 Using, initially, millimetre as the unit of length:

Cross-sectional area $= [(30 \times 5.5) + (22 \times 5.5)]\ \text{mm}^2$
$= [52 \times 5.5]\ \text{mm}^2$
$= 286\ \text{mm}^2$

Converting to m^2,
cross-sectional area $= 286 \times 10^{-6}\ \text{m}^2$
Volume of extrusion $= [286 \times 10^{-6} \times 3]\ \text{m}^3$
$= 858 \times 10^{-6}\ \text{m}^3$

$$\text{Density} = \frac{\text{mass}}{\text{volume}}$$

$\therefore$ Mass of extrusion $=$ density $\times$ volume

$$= 2\,700 \times 858 \times 10^{-6} \left[\frac{\text{kg}}{\text{m}^3} \times \text{m}^3\right]$$

$$2.32\ \text{kg}$$

10 Using metre as the unit of length:

$$\text{Volume of larger–diameter portion} = \left[\left(\frac{\pi}{4} \times 0.08^2\right) \times 0.2\right]\ \text{m}^3$$
$$= 1.005 \times 10^{-3}\ \text{m}^3$$

$$\text{Volume of smaller–diameter portion} = \left[\left(\frac{\pi}{4} \times 0.03^2\right) \times 0.1\right]\ \text{m}^3$$
$$= 7.07 \times 10^{-5}\ \text{m}^3$$

$$\therefore \text{Volume of whole component} = [(1.005 \times 10^{-3}) + (0.070\,7 \times 10^{-3})]\ \text{m}^3$$
$$= 1.075\,7 \times 10^{-3}\ \text{m}^3$$

Since relative density of steel is 7.85,

$$\text{Density of steel} = 7.85 \times 1\,000 \text{ kg/m}^3 = 7\,850 \text{ kg/m}^3$$

$$\text{Density} = \frac{\text{mass}}{\text{volume}}$$

$$\therefore \text{Mass} = \text{density} \times \text{volume}$$

$$\therefore \text{Mass of component} = [7\,850 \times 1.075\,7 \times 10^{-3}]\,\frac{\text{kg}}{\text{m}^3} \times \text{m}^3 = \underline{8.44 \text{ kg}}$$

11 The table should have been completed as follows:

liquid	volume		mass		density kg/m³	relative density
	cm³	m³	g	kg		
water	50	50×10^{-6}	50.1	0.050 1	1 002	1.0
paraffin	50	50×10^{-6}	38.7	0.038 7	774	0.772
petrol	50	50×10^{-6}	36.5	0.036 5	730	0.729
beer	50	50×10^{-6}	49.7	0.049 7	994	0.992

Answers to self-assessment questions

12 The preferred units of area and volume in SI are:
m^2 and mm^2 for area.
m^3, mm^3 and litre for volume.
cm^2 and cm^3 are not preferred units in SI.

13 The correct responses are:

(i) False, since the correct definition of density is $\frac{\text{mass}}{\text{volume}}$.

(ii) True, since the unit of mass is kg, and a preferred unit of volume is SI is m^3.

(iii) True, since the density of pure water is approximately 1 000 kg/m³, which could be written as 1 000 000 g/m³, which in turn could be written as 1 Mg/m³.

14 Neither statement 1 nor statement 2 is true, and the correct assertion is (*c*), since:

(i) Density is defined as $\frac{\text{mass}}{\text{volume}}$, and hence the density of a body is not dependent upon its size, but only upon the material from which it is made.

(ii) Relative density is defined as $\frac{\text{density of substance}}{\text{density of water}}$, and hence is a pure number, without any units.

15 The correct order of measurement and calculation is:

(i) Measure mass of container.
(ii) Measure volume of liquid.
(iii) Measure mass of container and liquid.
(iv) Calculate the mass of the liquid from (i) and (iii).
(v) Calculate the density of the liquid from (ii) and (iv).

Section 2

After reading the following material, the reader shall:

4 Be aware of the different nature of tensile, compressive and shear forces.
4.1 Identify a tensile force as one which causes extension.
4.2 Identify the deformations resulting from the application of compressive forces.
4.3 Identify the cutting action of a shear force.
4.4 Identify tensile and compressive forces as 'normal' forces.
4.5 Identify shear forces as 'transverse' forces.

If an elastic band is pulled, it stretches. Engineers and technologists call the stretching force a *tensile force*, and refer to the elastic band as being in a state of *tension*. Suppose the elastic band is replaced with a piece of thin metal wire, and a tensile force is applied to the wire.

Self-assessment question

16 Will the tensile force cause an extension in length (no matter how small) of the wire?

A tensile force is depicted in Figure 8(*a*), with the component extending in length. A *compressive force* is depicted in Figure 8(*b*), the action being to shorten a component. Figure 8(*c*), also depicts a compressive force, but in this case the action of the force results in a more easily observable deformation than in Figure 8(*b*); the component buckles into a bow shape.

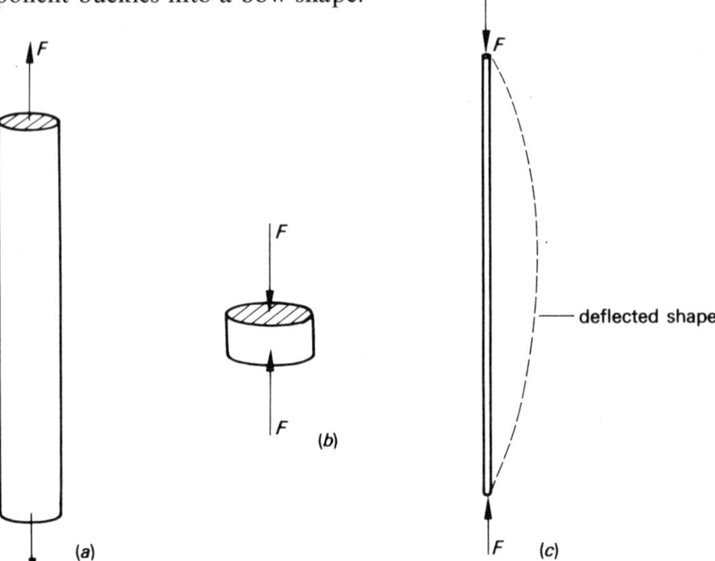

Figure 8 *Effects of tensile and compressive forces*

Thus the deformation produced by a tensile force is an extension parallel to the direction of the applied force; the deformation produced by a compressive force is either a shortening parallel to the direction of the applied force, or buckling in a direction normal to the applied force. Referring again to Figure 8, the forces are acting *normal* (i.e. at 90°) to the cross-section of the components. Thus, tensile and compressive forces are termed *normal forces*.

Consider now a gardener cutting a hedge with a pair of shears. As the name of the tool suggests, the branches of the hedge are being *sheared*, under the action of *shear forces*. Indeed, all cutting operations involve shear forces. Figure 9 illustrates the essential action of shear forces, Figure 9(*a*) representing two metal plates separated by a thin film of oil, and Figure 9(*b*) the effect on the plates of applying a shear force.

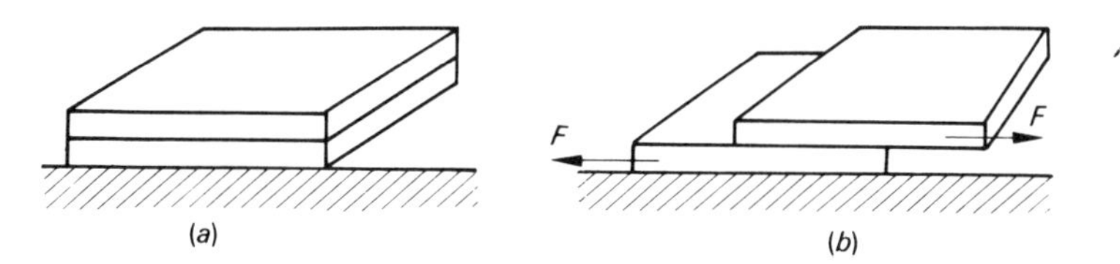

Figure 9 *Effects of shear forces*

The shear force depicted in Figure 9 is acting across the area of the plates in contact with the film of oil. In other words, the force is acting in a *transverse* direction. Thus a shear force is termed a *transverse force*.

Self-assessment questions

17 Complete the following three statements by selecting the correct phrase from the list which follows the statements:

Statements

(i) Compressive forces may produce ______

(ii) Shear forces involve ______

(iii) Tensile forces produce ______

List of phrases

(*a*) extensions in length.

(*b*) reductions in length.

(*c*) cutting actions.

18 Two of the four statements which follow are correct, the other two are incorrect. Select the two correct statements.

(i) Tensile and compressive forces are called normal forces.

(ii) Tensile and compressive forces are called transverse forces.

(iii) Shear forces are called normal forces.

(iv) Shear forces are called transverse forces.

After reading the following material, the reader shall:

4.6 Select from given examples of force diagrams those components in
(*a*) tension,
(*b*) compression,
(*c*) shear.

Figure 10 illustrates two metal plates fastened together by a bolt. Both plates are being extended, and hence the applied force F is a tensile force *when considering the effect of the force on the plates.*

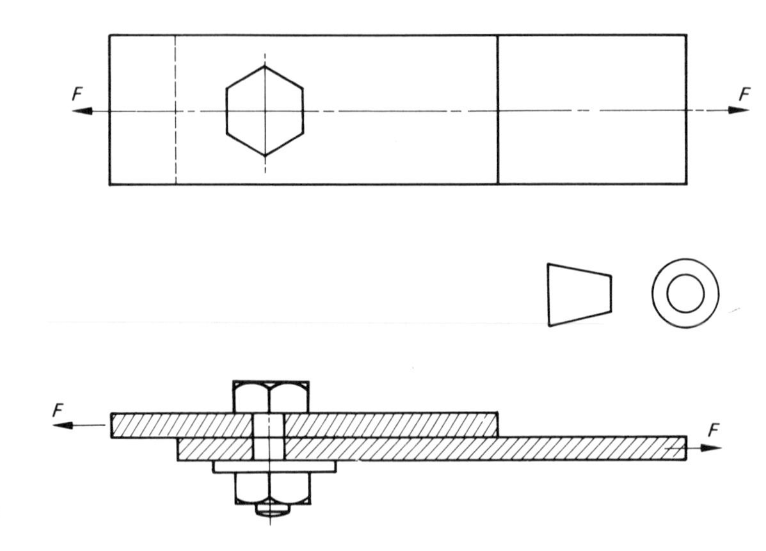

Figure 10 *Bolted joint – tensile forces*

If the direction of the force F is reversed, as in Figure 11, the plates are in compression under the action of a compressive force.

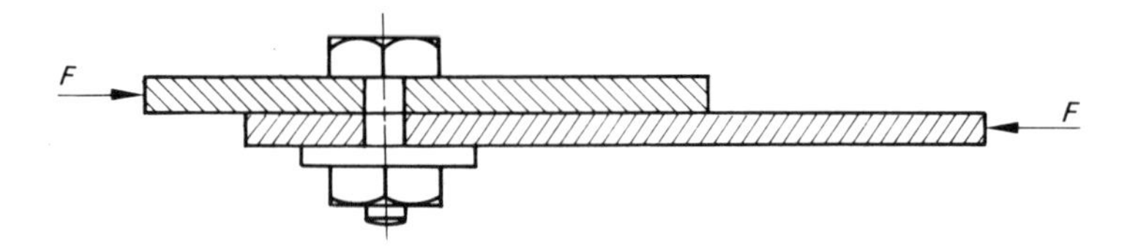

Figure 11 *Bolted joint – compressive forces*

Referring again to Figure 10, consider the effect of the force F upon the bolt. The part of the bolt passing through the top plate is tending to be displaced to the left, while the bottom part of the bolt is tending to be displaced to the right. In other words, the bolt is being sheared; using a similar argument, the bolt in Figure 11 is also subjected to a shear force.

Thus the applied force F causes tension or compression in the plates (dependent upon its direction), but causes shear in the bolt. Note that the force acts normal to the cross-sectional area of the plates (tensile or compressive forces are termed normal forces), but acts in a transverse direction to the cross-sectional area of the bolt (shear forces are termed transverse forces).

Self-assessment question

19 Write down the nature of the force (tension, compression or shear) in the components identified by the letters A, B and C, in Figure 12.

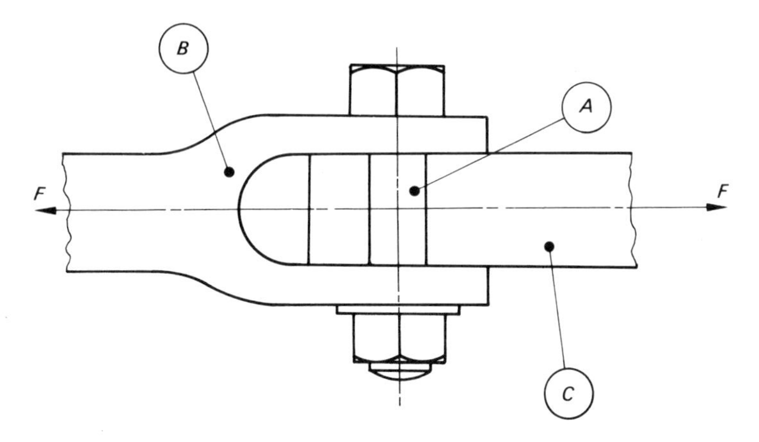

Figure 12 *Transmission joint*

Solutions to self-assessment questions

16 Yes. The wire will extend a small amount. The magnitude of the extension depends upon the size of the tensile force applied and the size of the wire.

This principle is used in the tuning of almost all stringed musical instruments, such as the piano, violin or harp. The correct pitch of note is achieved by increasing or decreasing the tension in the strings.

17 The statements should have been completed as follows:

(i) Compressive forces may produce *reductions in length.*

(ii) Shear forces involve *cutting actions.*

(iii) Tensile forces produce *extensions in length.*

18 The two correct statements are (i) and (iv). Tensile and compressive forces act normal to cross-sectional area; shear forces act parallel to the cross-sectional area.

After reading the following material, the reader shall:

5 Know the effects on a common engineering material of a gradually applied, axial tensile force.
5.1 Draw a force–extension graph for a non-ferrous material from given data.
5.2 Identify the portion of the graph of 6.1 where extension is directly proportional to force.
5.3 Identify the portion of the graph of 6.1 where extension is not directly proportional to force.
5.4 Define elasticity.
5.5 Identify the elastic and plastic parts of the graph of 6.1.

The table below contains values of axial tensile force and related extensions for a brass tensile test specimen. ('Axial' in this context refers to a force acting along axis *AA* in Figure 13.) The specimen was machined as shown in Figure 13, and placed in a tensile testing machine. Axial tensile force was applied starting from zero, increasing in the increments (i.e. intervals) shown, until the specimen fractured. Readings of the extension of the specimen were noted as the axial force was gradually increased.

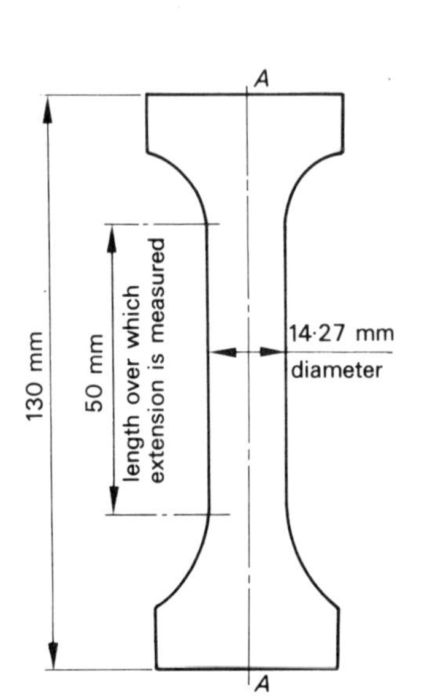

Figure 13 *Tensile test specimen*

axial force kN	0	2	4	6	8	10	12	14
extension mm	0	.0082	.0114	.0146	.0178	.0209	.024	.027

axial force kN	16	18	20	21	23	24	26	28
extension mm	.0301	.0323	.0385	.0428	.0555	.065	.088	.115

axial force kN	30	35	40	45	50	52	54	55
extension mm	.150	.255	.400	.630	1.050	1.380	1.800	2.160

axial force kN	56	55.6
extension mm	2.500	FRACTURE

Assignment

Choose appropriate scales, and draw two graphs using the data in the table. For one graph, plot force against extension up to a maximum of 30 kN and 0.150 mm; for the other graph, plot all of the data up to fracture. Plot axial force on the vertical axis, and extension on the horizontal axis. Label the axes and scales.

Self-assessment questions

20 Identify the portion of the graph of Figure 14 where extension is directly proportional to axial force. Identify the portion by the letters on Figure 14, e.g. 0 to *A*, 0 to *B*, *A* to *B*, etc.

21 Identify the portion of the graph of Figure 14 where extension is not directly proportional to axial force. Use the same notation as in the previous question.

The reader may remember from the introduction to this section that tensile force was defined by an analogy to an elastic band. What does the word *elastic* mean? It is tempting to say that it means 'easily pulled', but some elastic bands can be rather difficult to stretch, and in any event, this definition is too vague to be accepted by an engineer or a technologist. A more acceptable definition of an elastic material is that, *within limits*, after a force has caused a deformation of a specimen of the material, the specimen *returns to its original dimensions* when the force is removed. Many, but not all, materials used in the various technologies possess this property of *elasticity*.

There is an important qualification in the above definition, and that is the phrase 'within limits'. The limits are zero force and a point approximately coincident with point *A* on Figure 14 or 15. In other words, if the axial force is reduced to zero at any point between 0 and *A*, the specimen will return to its original dimensions, i.e. the material is elastic.

Suppose the specimen is subjected to an axial force greater than that corresponding to point *A*. The specimen does not return to its original dimensions, when the force is reduced to zero. The material has lost its elastic properties, and has become *plastic*.

Summarizing therefore:
Elasticity is defined as the ability of a material to return to its original dimensions, when an applied force is removed.

The *elastic region* of a force–extension curve for a metal corresponds very closely to the straight line portion of the graph.

The *plastic region* of a force–extension curve for a metal corresponds very closely to the curved portion of the graph.

Solution to self-assessment question

19 Component A, the bolt, is in shear. Components B and C are in tension.

Results of assignment

The graphs of force against extension should be as Figures 14 and 15. Note that in Figure 15 the curve does not pass through the origin whereas in Figure 14, because of the different scale chosen for extension, it appears to pass through the origin. The fact that the curve does not pass through the origin can be disregarded in this instance, since it is mainly caused by inaccurate calibration of measuring instruments.

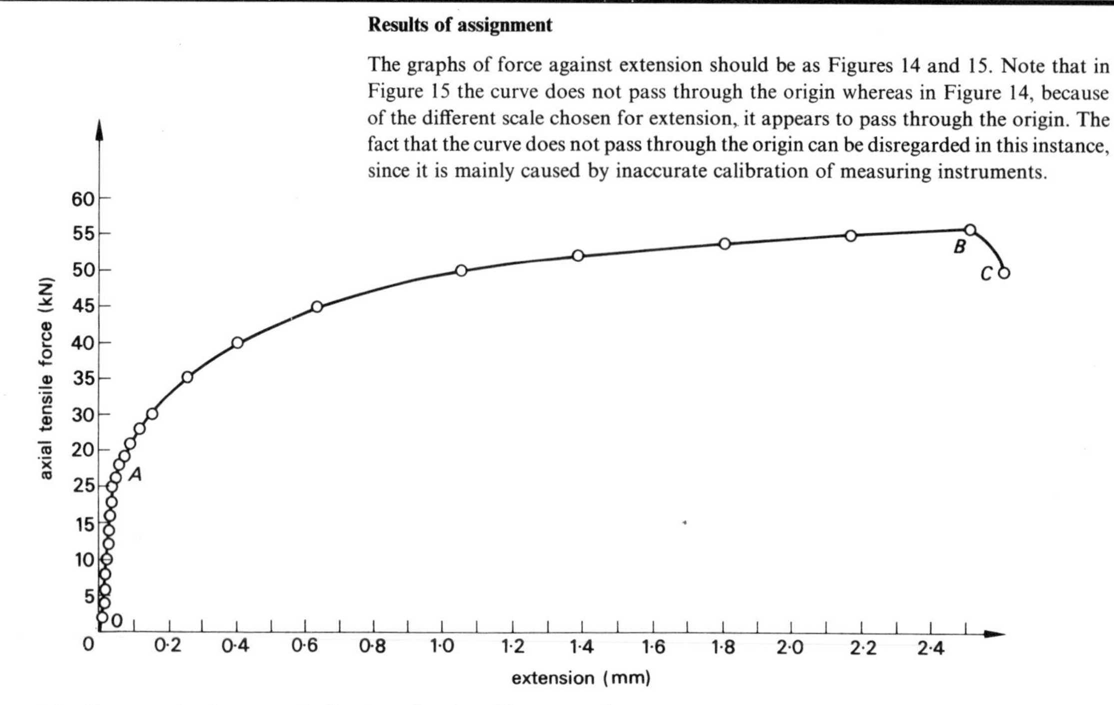

Figure 14 *Force–extension curve to fracture: hardened brass specimen*

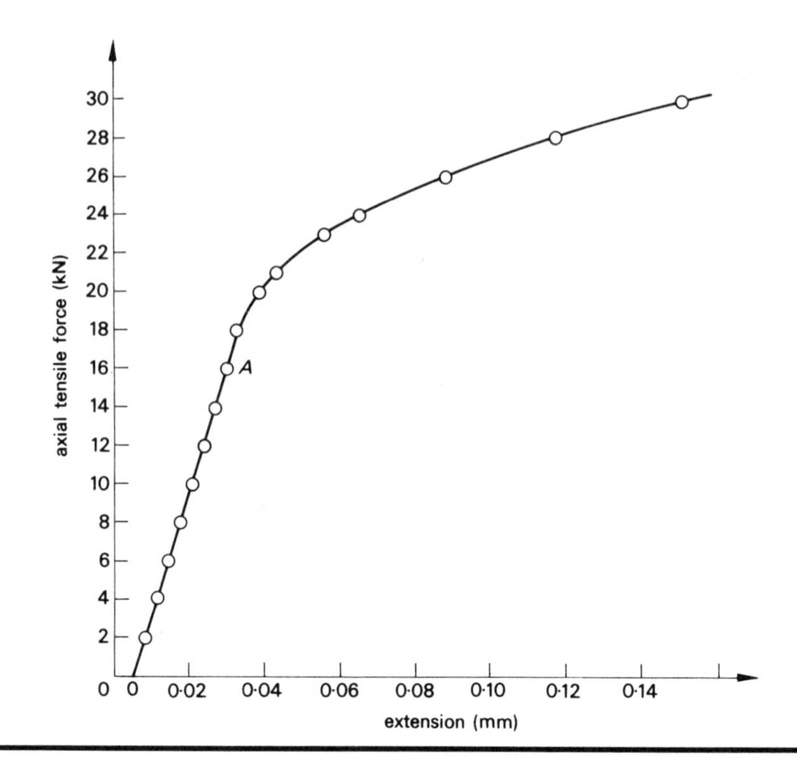

Figure 15 *Force–extension curve: hardened brass specimen*

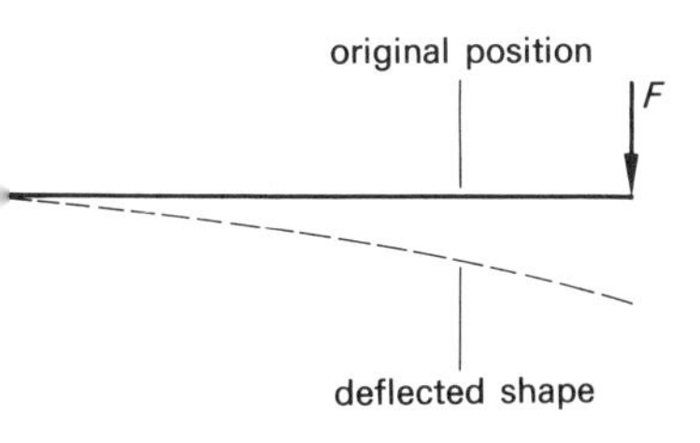

Figure 16 *Cantilever beam*

Self-assessment questions

22 A beam is held securely at one end, and subjected to a force F at the other end, as shown in Figure 16. When the force is removed, the beam returns to its original position. Is the material of the beam elastic or not?

23 The elastic part of Figure 14 is represented by:
(i) $0B$
(ii) AC
(iii) AB
(iv) $0A$
Select the correct alternative.

24 The plastic part of Figure 14 is represented by:
(i) $0B$
(ii) AC
(iii) $0C$
(iv) $0A$
Select the correct alternative.

After reading the following material, the reader shall:

5.6 State Hooke's law.
5.7 Solve simple problems involving Hooke's law.

It has been established that the elastic region of Figure 14 or 15 corresponds to the straight-line portion $0A$ of the graph, and that point A corresponds very closely to the elastic limit of the material. Also, extension is directly proportional to force in the region 0 to A. If these statements are combined, the result is *Hooke's law*, which states:

Extension is directly proportional to the force producing it, providing that the material remains within the elastic limit.

Solutions to self-assessment questions

20 The portion of the graph of Figure 14 where extension is directly proportional to axial force is the straight-line part of the graph, i.e. 0 to $A(0A)$. Note that extension ranges from 0 to approximately 0.03 mm.

21 The portion of the graph of Figure 14 where extension is not directly proportional to axial force is the curved part of the graph, i.e. A to $C(AC)$. Note that the extension in this part of the graph ranges from approximately 0.03 mm to 2.5 mm – a very much wider range than the straight-line portion. Not all materials possess this property of large extension before fracture.

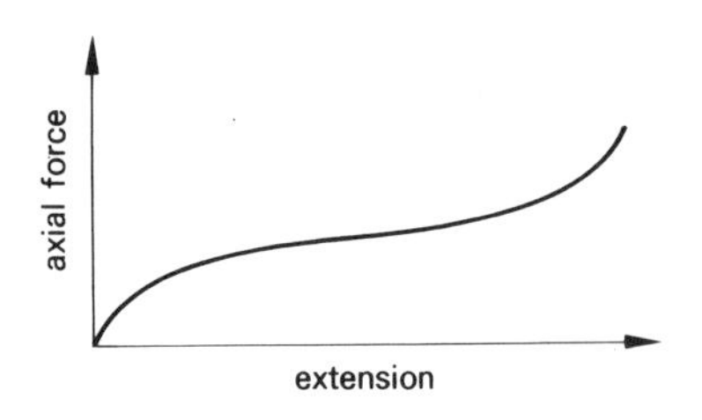

Figure 17 *Typical force–extension curves for elastic non-metallic material*

This law is directly applicable to most metals, but it does not apply to all materials. For example, soft rubber, a highly elastic material, exhibits a force–extension curve similar to that shown in Figure 17. Because there is no straight-line relationship between force and extension, Hooke's law does not apply to soft rubber. For a similar reason, many polymers do not conform to Hooke's law.

Since the behaviour of many metals is consistent with Hooke's law, it has important consequences for engineers and designers. It would be embarrassing, to say the very least, if, for example, the force applied momentarily to the connecting rod of a piston, or to the members of a suspension bridge, resulted in a permanent change in the dimensions of the components. Probably, the reader can think of many other examples where it would be extremely dangerous for such a situation to occur.

Thus, engineers and designers require to know the maximum force which may be applied in service to a particular component. They need to compare this maximum force with the force corresponding to the elastic limit of the material. Occasionally, the maximum force applied in service cannot be determined with any accuracy, but it may be possible to measure the extension or deformation of the component. By using a graph similar to Figure 14, it is possible to check whether the component remains within the elastic limit.

Example 5
A brass component, of the same dimensions as that used to derive Figure 14, is found to extend by 0.025 mm under the action of an axial tensile force. Confirm that the component has not exceeded the elastic limit of the material, and estimate the applied force.

If a line is projected vertically from 0.025 mm on the horizontal scale of Figure 14 then it meets the force–extension curve between points 0 and *A*. Thus it is confirmed that the component is within the elastic limit of the material. Also from Figure 14, the axial force corresponding to 0.025 mm is 12 kN.

In the example, confirmation that the elastic limit had not been exceeded, and determination of the axial force could only be achieved because the component had the same dimensions as the component used to derive Figure 14. If the dimensions had differed, then the figure could not be used.

The effects of varying the diameter or the cross-section of the specimen are beyond the scope of this book, but the effect of varying the length of the specimen will be examined in a simple, brief fashion.

The specimen extensions of Figures 14 and 15 were measured by comparison with an original length of 50 mm. If, for example, this length was doubled, what would be the effect on the extensions? The logical and correct answer is that the extensions would also be doubled at each increment of force. Thus the extension at a force of 16 kN would now become 0.06 mm instead of 0.03 mm. Note that this argument only applies up to the elastic limit, i.e. while extension is directly proportional to force.

Self-assessment questions

25 The maximum force applied to a brass component in service is estimated to be 23 kN. If the component has the same dimensions as that used to derive Figures 14 and 15, ascertain whether the material has been stretched beyond its elastic limit.

26 Three readings of force and extension are noted during a tensile test on an engineering material:
10 kN produces an extension of 0.003 mm,
20 kN produces an extension of 0.006 mm,
30 kN produces an extension of 0.009 mm.
If the elastic limit of the material corresponds to a force of 40 kN, does the material obey Hooke's law?

27 A force of 15 kN when applied to a particular component 100 mm long produces an extension of 0.05 mm.
(*a*) Determine the extension of a component with the same cross-sectional area made from the same material, and subjected to the same force, if its length is 500 mm.
(*b*) What would be the extension if the length were 10 mm?
Assume the elastic limit of the material is not exceeded.

After reading the following material, the reader shall:

6 Appreciate that forces may cause a number of different deformations to materials.
6.1 Identify examples of strong and weak materials from experimental data.
6.2 Describe the difference between the behaviour of
(*a*) brittle and
(*b*) malleable materials under the action of a force.

Solutions to self-assessment questions

22 The beam material *is* elastic, since elasticity has been defined as the ability of the material to return to its original dimensions when an applied force is removed.

23 The correct response is (iv), since it is only within the region $0A$ that the specimen returns to its original dimensions if the axial force is removed.

24 The correct response is (ii), since if the applied force is removed in the region AC, the specimen does not return to its original dimensions.

Ductility and brittleness

A *ductile* material is one which withstands large plastic deformations without fracture. Conversely, a *brittle* material is one in which there is very little, if any plastic deformation before fracture occurs, and thus fracture usually occurs with very little warning.

Very many materials other than metals can be classified as either ductile or brittle. Glass for example, is a brittle material at room temperature, as are many ceramics (bathroom or kitchen wall tiles, for instance). Many polymers on the other hand, are ductile, but there are exceptions to this general rule – perspex is brittle at room temperature, but becomes ductile if heated.

Strength

The strength of a material may be measured by its ability to withstand high applied forces without fracture occurring. However, this statement must be qualified by considering the character of the applied forces. A material can be weak in compression and strong in tension, or vice versa. Its strength when subjected to shear forces may well differ from its strength in either tension or compression. For example, concrete is very weak in tension, but strong in compression. Ordinary grey cast iron exhibits similar properties. Thus a designer always endeavours to avoid subjecting these materials to tensile forces.

Little reference has yet been made to polymer materials, and in particular, to their reaction to applied forces. This is because polymers are rarely if ever used to withstand large forces (unless the molecular structure is reinforced with a material such as glass or carbon fibres), since their strengths are in general very much smaller than those of the metals used in engineering. Polymers possess other useful properties such as the insulation of heat or electric current and low frictional resistance.

Solutions to self-assessment questions

25 Referring to Figure 14, an axial force of 23 kN causes an extension beyond the elastic limit. Hence the component has become plastic.

26 If the three readings of force are plotted against the relevant extensions, the result is a straight line. Since all the readings are below the elastic limit, the material conforms to Hooke's law.

27 Since the extension is proportional to the length over which it is measured, providing the material is within the elastic limit, then:

(*a*) The extension over a length of 500 mm is
$\frac{500}{100} \times 0.05 = \underline{0.25 \text{ mm}}$

(*b*) Over a length of 10 mm,
extension $= \frac{10}{100} \times 0.05 = \underline{0.005 \text{ mm}}$

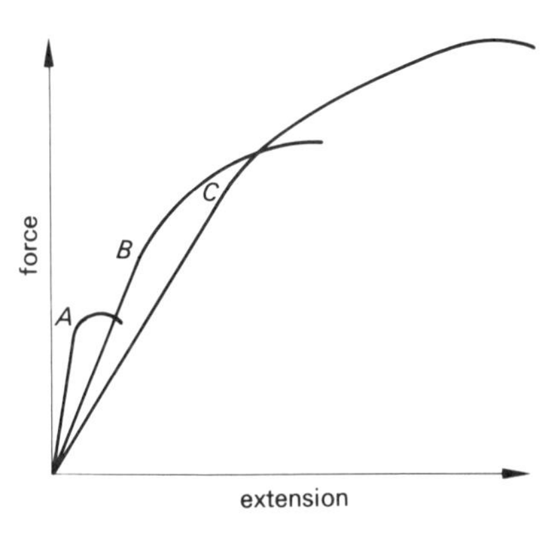

Figure 18 *Comparison of force–extension characteristics*

Malleability

Figure 18 illustrates diagrammatically the behaviour of three materials when subjected to a gradually increasing axial tensile force. The force is applied until fracture occurs.

The curves drawn in Figure 18 were derived from experimental data during tensile tests. A common feature of all tensile tests is that the axial force is applied *gradually*. The axial force does *not* increase suddenly.

Suppose a piece of mild steel is subjected to a suddenly applied force – perhaps by hitting it with a hammer. The deformation produced by the force differs from that produced by a gradually applied force of the same magnitude. However, unless the suddenly applied force is extremely high, the mild steel deforms, but does not fracture. Such a material is said to be *malleable*. Malleability can be considered as a special case of ductility, which qualitatively indicates the ability of a material to be hammered and deformed, particularly into thin sheets.

If a sudden force is applied, this time on a piece of cast iron, the results are markedly different – there is little or no deformation, just a sudden fracture. Most brittle materials behave in this fashion – as an extreme example, compare the behaviour of a sheet of glass with a sheet of aluminium alloy when hit by a cricket ball.

Self-assessment questions

28 Referring to Figure 18:

(*a*) List the materials in order of ductility, the most ductile first.

(*b*) List the materials in order of brittleness, the most brittle first.

(*c*) List the materials in order of strength, the strongest first.

29 Describe, without reference to the preceding text, the behaviour of a malleable material and a brittle material under the action of a suddenly applied force.

Further self-assessment questions

30 Define 'elasticity' as a property of a material.

31 The force–extension graph of an elastic material *must* contain a straight line portion. Answer TRUE or FALSE.

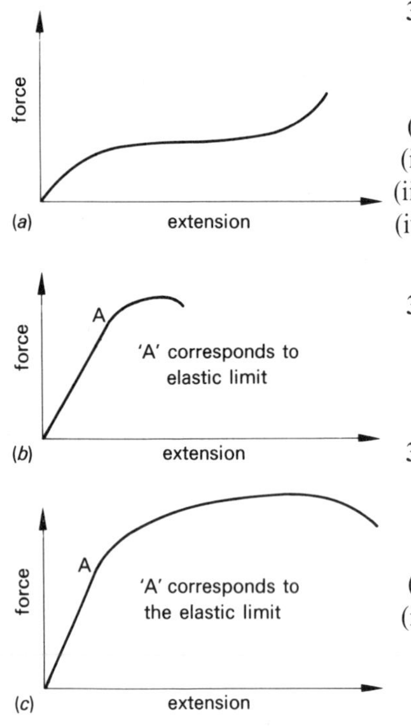

Figure 19 *Comparison of material properties*

32 The following four questions all refer to Figure 19, which contains force–extension curves for three materials, each of which was tested to destruction.

(i) Do each of the materials conform to Hooke's law?
(ii) Which material is the least ductile?
(iii) Which material is the most brittle?
(iv) Which curve represents typically the behaviour of an elastic non-metallic material?

33 Steel wire, 3 m in length, is subjected to a tensile force of 100 N, and found to extend by 5 mm. Calculate the extension of a piece of the same wire 6 m long, subjected to the same force. Assume that the elastic limit of the wire is not exceeded.

34 The slope of the elastic portion of the force–extension curve for a mild steel specimen is found to be 670 MN/m. Given that the elastic limit occurs at a force of 37 kN and a corresponding extension of 0.057 mm, calculate

(i) The extension of the specimen produced by a force of 25 kN.
(ii) The force necessary to produce an extension of 0.02 mm of the specimen.

State why it is incorrect to calculate the extension produced by a force of 45 kN, from the above information.

35 Sketch on the same axes, the force–extension curves for (*a*) a brittle material and (*b*) a ductile material.

Solutions to self-assessment questions

28 (*a*) The correct order is C, B, A, since ductility is defined as the ability to withstand large plastic deformations without fracture. Thus, material C has the largest plastic region, then B and finally A.

(*b*) The correct order is A, B, C, since brittleness can be considered as the opposite of ductility.

(*c*) The correct order is C, B, A, since strength can be defined as the ability to withstand high applied forces without fracture.

29 A malleable material, when subjected to a suddenly applied force, tends to deform, perhaps quite noticeably, but does not fracture. Malleability is usually taken to mean the ability to be hammered into thin sheets.

A brittle material, when subjected to a suddenly applied force, tends to fracture with little or no evidence of deformation before fracture.

Answers to self-assessment questions

30 Elasticity is defined as the ability of the material to return to its original dimensions, when an applied force is removed.

31 The statement is FALSE. A material such as natural rubber does not exhibit a straight-line portion in its force–extension graph, and yet it is a highly elastic material.

32 (i) The material of Figure 19(*a*) does not conform to Hooke's law ('Force is directly proportional to extension within the elastic limit'). The other two materials do conform to Hooke's law.

(ii) The material of Figure 19(*b*) is the least ductile.

(iii) The material of Figure 19(*b*) is the most brittle.

(iv) Figure 19(*a*) is typical of an elastic non-metallic material.

33 Extension = 10 mm.

34 (i) Extension $= \dfrac{25 \times 10^3}{670 \times 10^6}$ [m] = 0.038 mm

(ii) Force $= 670 \times 10^6 \times 0.02 \times 10^{-3}$ [N] = 13.25 kN.

Since a force of 45 kN exceeds the elastic limit of the material, it is not correct to use the slope of the elastic portion of the graph.

35 Typical force–extension curves would be as in Figure 20.

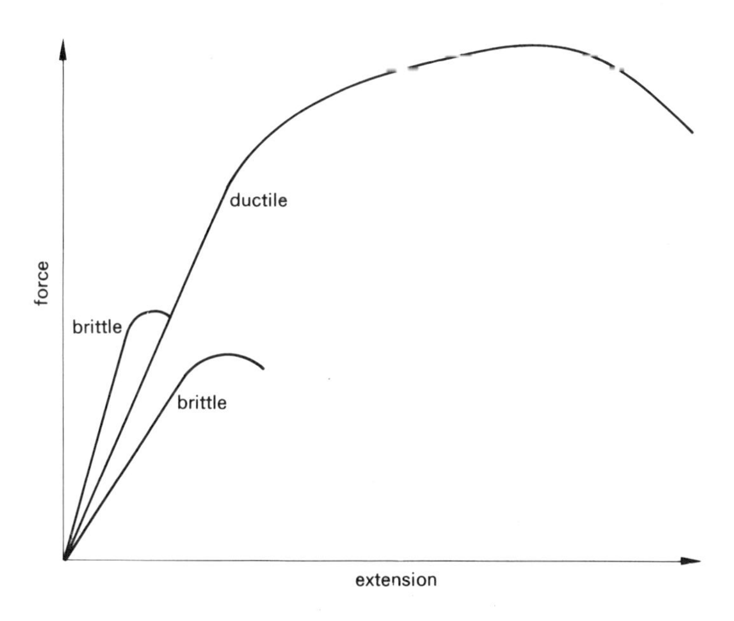

Figure 20 *Force–extension characteristics: ductile and brittle materials*

Section 3

After reading the following material, the reader shall:

- **7** Know the nature of the atomic structure of matter.
- **7.1** Describe the atom as the basic building block of matter.
- **7.2** Identify the two basic parts of the atom.
- **7.3** Describe the molecule as an independent group of atoms bonded together.
- **7.4** Explain the terms elements and compounds in terms of atomic composition, and distinguish compounds from mixtures.
- **7.5** Give three examples of each of the following:
  - (*a*) elements,
  - (*b*) compounds,
  - (*c*) mixtures.
- **7.6** Identify elements, compounds and mixtures, given a list of substances.

What does a scientist or technologist mean when he discusses *matter*? In essence, he means any material or substance which exists, whether it be solid, liquid or gas. The atomic theory of matter suggests that all matter is made up of particles, the simplest kind of particle being the *atom*. For many years, scientists thought that the atom could not be divided. Indeed, it is extremely small, and cannot possibly be seen, even with a very powerful microscope. For example, the mass of an atom of carbon is 0.000 000 000 000 000 000 000 000 02 kg (20×10^{-27} kg). A small piece of soot with a mass of 0.000 1 kg contains 5 000 000 000 000 000 000 000 000 (5×10^{24}) atoms of carbon. Even though the atom is so small, it is capable of being split. A result of this *fission* (another word for division) is that large amounts of energy are released.

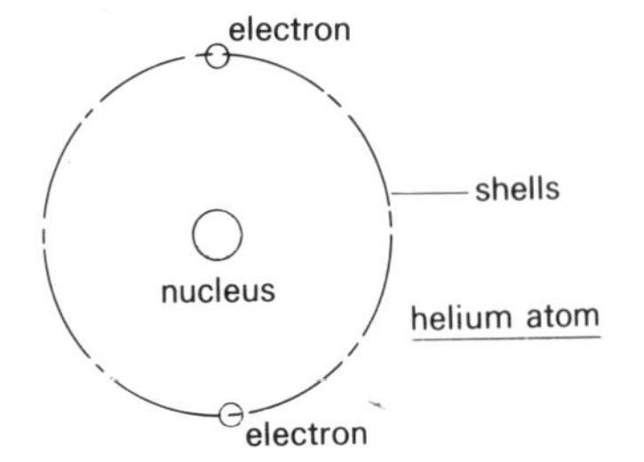

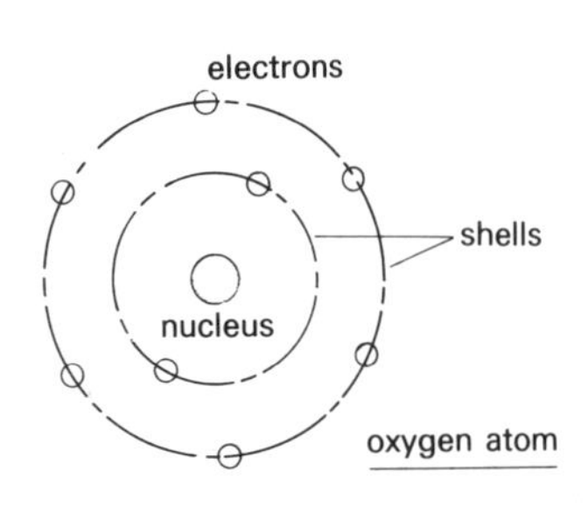

Figure 21 *Structure of the atom*

The atomic theory suggests that an atom consists of a *nucleus* which is surrounded by *electrons* orbiting in *shells*, as illustrated diagrammatically in Figure 21. The atoms of different substances have different structures. They all consist of a nucleus (but the nature of the nucleus differs), and the nucleus is surrounded by electrons rotating in their shells at high velocities. The properties of the atom depend upon the number of electrons surrounding the nucleus. It is not only the structure of an atom which varies from substance to substance, but the behaviour of the atom also differs. It is very rare for one single atom to exist alone; in most cases atoms exist in clusters, which are called *molecules*. A molecule may contain only one kind of atom as in Figure 22(*a*), or it may contain several dif-

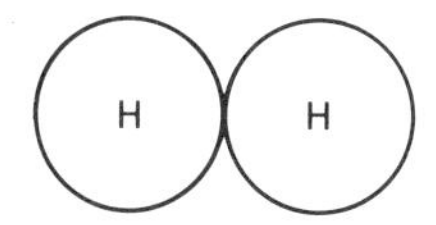

one molecule of hydrogen containing two atoms of hydrogen

(*a*)

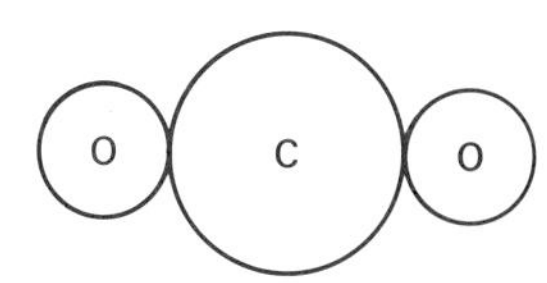

one molecule of carbon dioxide containing two atoms of oxygen and one atom of carbon

(*b*)

Figure 22 *Simple structure of molecules*

ferent kinds of atom, a very simple example being shown in Figure 22(*b*). In a very few substances, such as helium, atoms do exist alone; these lone atoms are then called molecules.

Elements, compounds and mixtures

Apart from these very rare instances, a molecule is best defined as a cluster of atoms, which may be all of the same kind, or of several different kinds. Substances which contain only one kind of atom (such as hydrogen in Figure 22(*a*)) are called *elements*. Thus a molecule is the smallest part of an element which can exist by itself. Many chemistry books contain a long list of elements in a table, known as the Periodic Table, in which (with a few exceptions) elements are arranged according to the mass of their atoms. This is a more useful classification, to engineers as well as to chemists, than classifying them as solid, liquid or gas.

There are over 100 different elements, and since an element contains only one kind of atom, there must be over 100 different kinds of atom. Atoms can be clustered together in millions of different ways, to give the large variety of substances known on earth. For example, graphite and diamond both contain only carbon atoms, but the atoms of graphite are arranged differently from the atoms of diamond. It is a sobering thought that a change in the arrangement of atoms produces materials which have vastly different properties, and upon which society puts vastly different values.

If red and blue paint are mixed together, the result is a mixture with the colouring of both. However, not all substances when mixed together retain their individual characteristics. Suppose two elements such as hydrogen and oxygen are combined together. Hydrogen burns quite easily in the presence of oxygen, and yet if two atoms of hydrogen are combined with one atom of oxygen (H_2O), the result is a substance, water, which will extinguish the majority of fires! The explanation for this apparent contradiction is found by considering how the atoms of hydrogen and oxygen are combined. Basically, the atoms have been *joined* together, so that they lose their own individual properties, and behave as part of an entirely new substance. When atoms behave in this manner, the new substance is known as a *compound*. During the joining process, or when the process is reversed to change a compound into its component elements, considerable heat energy may be given out or absorbed.

If two atoms of nitrogen and one atom of oxygen are mixed, the atoms do not combine or join together, and hence they retain their individual properties. The result, which is basically air, is known as a *mixture*. Using a similar argument, if salt and sugar are mixed together, both substances retain their original flavour and the result is a mixture.

Thus, a *compound* is composed of atoms of two or more substances which are joined together, usually producing or absorbing heat energy. A *mixture* is composed of atoms of two or more substances which do not combine. Because the atoms in a compound are joined together, the resulting properties of a compound can be quite unlike those of its component elements; on the other hand, the properties of a mixture are closely related to the properties of the components.

There is one other important difference between a mixture and a compound. In mixtures, the free unjoined atoms can be present in various proportions; in compounds, because the atoms are joined together, the composition is fixed, i.e. the elements are joined in fixed proportions. If different proportions are joined together, then a different compound is formed.

Self-assessment questions

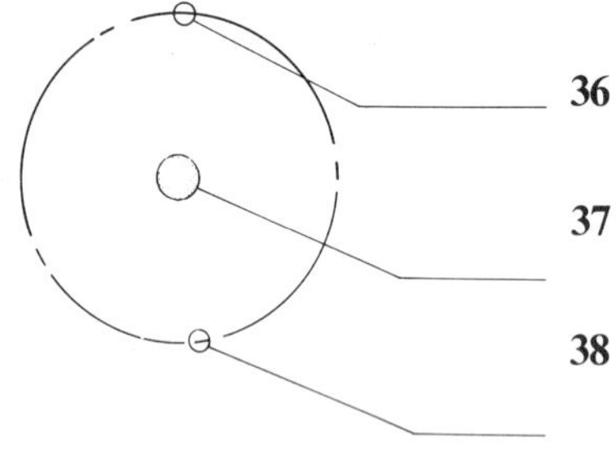

Figure 23

36 Figure 23 shows a diagram of a simple atom. Fill in the missing labels.

37 Label the following statement as either TRUE or FALSE:
'A molecule *usually* consists of at least two atoms bonded together.'

38 Statement 1: In a mixture, the components may be present in any proportions.
Statement 2: The properties of a mixture are in general derived from the properties of its component parts.

Underline the correct answer:
(*a*) Only statement 1 is true.
(*b*) Only statement 2 is true.
(*c*) Both statements are true.
(*d*) Neither statement is true.

39 Statement 1: Compounds are formed from elements by the union of atoms in fixed proportions only.
Statement 2: The properties of compounds are related to the properties of their component parts.

Underline the correct answer:
(*a*) Only statement 1 is true.
(*b*) Only statement 2 is true.
(*c*) Both statements are true.
(*d*) Neither statement is true.

40 The left-hand column of the table below contains a list of substances which are elements, compounds or mixtures. Write in the right-hand column the letter E for those substances which are elements, C for those substances which are compounds, M for those substances which are mixtures.

substance	identification E C or M
ink	
pure water	
sulphur	
salt	
gold	
carbon	
sugar	
rice pudding	
sand and gravel	

After reading the following material, the reader shall:

8 Be aware of the nature and formation of crystals.
8.1 Define a solution.
8.2 Define a suspension.
8.3 Define solubility.
8.4 Distinguish between suspensions and solutions.
8.5 List common factors influencing solubility of a solid in a liquid.
8.6 Define a saturated solution.

Suppose a small amount of granulated sugar is added to a glass of water, and the water is then stirred or shaken. The sugar gradually disappears, until after a while, the glass contains a clear liquid which looks like the original water. The sugar is mixed with the water in such a way that it cannot be seen, nor does it separate from the water. A *solution* of sugar and water has been made, by the water dissolving the sugar. The substance which is dissolved is called the *solute*, and the liquid in which it dissolves is called the *solvent*.

What is the effect of substituting sand for sugar? If the new mixture is stirred or shaken, a cloudy liquid is produced. Such a mixture is called a *suspension*, since the sand is suspended in the water. After a while, the sand gradually settles to the bottom of the glass, but a suspension can be reproduced by shaking or stirring.

The essential difference between the two experiments is that sugar dissolves in water to form a solution, whereas the sand does not dissolve. Why is it that one substance dissolves, but another does not?

There is a rather complicated answer, but the simple answer is related to the behaviour of molecules. When a solid dissolves in a solvent, the molecules of the solid are very small, and are distributed throughout the solvent in a very fine form. The molecules of both the solute and the solvent behave as though they are in the same physical state. When a solid is suspended in a liquid, the suspended particles are large gatherings of molecules, very large compared with the molecules of the liquid, and hence are not distributed in a fine form throughout the liquid. Note that both solutions and suspensions are mixtures; they are not compounds.

The conclusion to be drawn is that some substances are soluble in certain liquids, whereas others are insoluble. Water is not the only

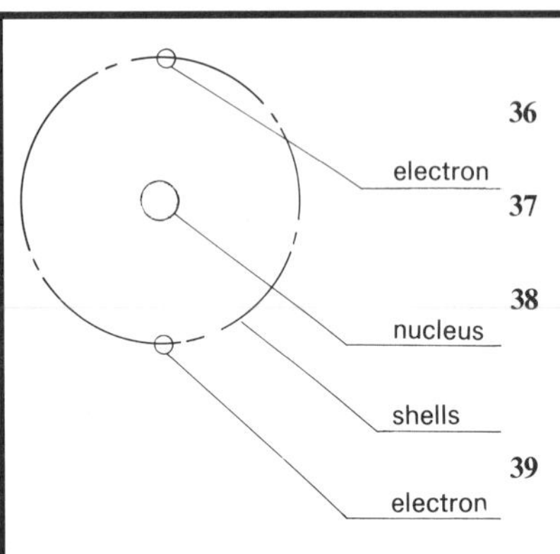

Figure 24

Solutions to self-assessment questions

36 The correct labels are shown in Figure 24, with the electron shells surrounding the nucleus.

37 The statement is *true*, since it is very rare indeed for a molecule to consist of one single atom.

38 Both statements are true, and hence the correct assertion is (*c*). The answer may be illustrated as follows: suppose sugar is added to already sweetened tea. The tea becomes sweeter, but the basic mixture of tea and sugar remains unchanged, in that the drink still tastes of tea and sugar.

39 Statement 1 is true, but statement 2 is false, and hence the correct assertion is (*a*). For example: one molecule of water consists of two atoms of hydrogen and one atom of oxygen combined together. If an extra atom of either oxygen or hydrogen is added, then the resulting compound is no longer water. The second statement is false, since the properties of a compound are very different from the properties of the component elements, e.g. the white crystalline compound called sugar consists of a black element (carbon) and two colourless gases (hydrogen and oxygen).

40 The table should be completed as follows:

substance	identification E C or M
ink	M
pure water	C
sulphur	E
salt	C
gold	E
carbon	E
sugar	C
rice pudding	M
sand and gravel	M

Ink is a mixture of a solvent and a pigment (i.e. colouring); rice pudding is a mixture of rice, milk and sugar; sand and gravel is a mixture which retains the individual properties of sand and gravel. Sugar is a compound of carbon, hydrogen and oxygen; pure water is a compound of hydrogen and oxygen; salt is a compound of sodium and chlorine.

solvent which could be used; there are many other liquids, but they do not always dissolve the same substances as water. For instance, grease is insoluble in water, but is soluble in other solvents such as petrol.

There are some important modifications which could be made to the experiment with sugar and water. These are as follows:

(i) Lump sugar can be substituted for granulated sugar. The effect is to slow down the rate of dissolving, but the sugar ultimately dissolves.

(ii) The rate of stirring or shaking can be varied. The effect of increasing agitation of the mixture is to speed up the dissolving process, since it brings more molecules of the solute into contact with more molecules of the solvent.

(iii) The temperature of the water can be increased before adding the sugar. This does not produce an acceleration in the rate of dissolving, but it does mean that more sugar can be dissolved in the same amount of water, i.e. the solubility of the sugar is increased. This does not mean that the solubility of *all* solutes in *all* solvents is increased by increasing temperature. There are solutes which behave differently from sugar; for example, the solubility of salt in water decreases slightly with increasing temperature.

(iv) Suppose the three variables discussed above are kept constant, i.e. granulated sugar is added to water at constant temperature, and the mixture is agitated mechanically at a constant rate. The final modification is to add more and more sugar to the mixture. A point is reached eventually at which no more sugar can dissolve. Such a solution is referred to as *saturated*. Thus a *solution is saturated if no more solute can be made to dissolve, with the temperature remaining constant*. The extent to which a given substance can dissolve in a particular liquid is termed its *solubility*, which is defined as *the maximum amount* (usually in grams) *of a solute that will dissolve in exactly 100 grams (0.1 kg) of solvent, at a given temperature*.

One final, and very important word concerning solutions: the discussion up to now has been confined to the solubility of solids in liquids. This is not the only kind of solution. For example, champagne is a solution of a gas in a liquid; vinegar is a solution of a liquid in a liquid; some dental fillings (e.g. mercury in cadmium) are solutions of a liquid in a solid; humid air is a solution of a liquid in a gas; and finally, brass is a solution of a solid in a solid (i.e. a solution of copper and zinc). This final example will be examined in more detail later in this section.

Self-assessment questions

41 The adjacent table lists six mixtures which are either solutions or suspensions. Use the right-hand column of the table to distinguish between solutions and suspensions. Write in the appropriate word.

mixture	solution or suspension
sea water	
wax and water	
petrol and water	
lemonade	
soda water	
chalk and water	

42 Four factors (*a*), (*b*), (*c*), (*d*) are listed below. Two of the factors affect the rate at which a solute dissolves in a solvent. One affects the amount of solute which can be dissolved. The fourth has no effect upon solubility. Circle (*a*), (*b*), (*c*) or (*d*) to identify the factor which affects the amount of solute which can be dissolved. Mark with a cross the factor which has no effect upon solubility:

(*a*) size of particles of the solute
(*b*) density of the solute
(*c*) agitation of the mixture
(*d*) temperature of the mixture

43 Define each of the following:

(i) a saturated solution
(ii) solubility

After reading the following material, the reader shall:

8.7 Identify a crystal as a regular arrangement of molecules.
8.8 Recognize that crystals have flat faces and specific angles.
8.9 Describe the process of crystallization from a solution.
8.10 Describe metals as polycrystalline substances.
8.11 Identify an alloy as two or more materials combined together.
8.12 Recognize that alloys can be solid solutions.

Having dissolved a solid in a suitable solvent, is it possible to reclaim the solid from the solution? There are two possible ways.

If a solution is saturated at a particular temperature, the solvent cannot dissolve any extra solid at that temperature. Thus, if the temperature is varied, some of the solid separates from the solution. Alternatively, if part of the solvent can be removed by some means from the saturated solution, then some of the solid could not be retained in solution. Solvent molecules can be removed by the process known as *evaporation*. During evaporation molecules of the solid steadily become surplus, and are deposited out of solution.

Hence, either by varying the temperature of the solution, or by evaporating the solvent, molecules of the solid are thrown out of solution. These molecules gather together to form crystals, the process being called *crystallization*. When liquids solidify, or when saturated solutions evaporate, there is often a tendency for the solid so formed to assume the regular shape of a crystal. Water crystallizes upon freezing into regular patterns such as snowflakes, or the needle crystals evident in an ice-lolly.

If crystals are examined under a microscope, they are found to have regular geometrical shapes, with flat sides, straight edges and no

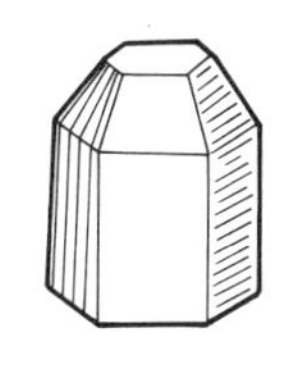

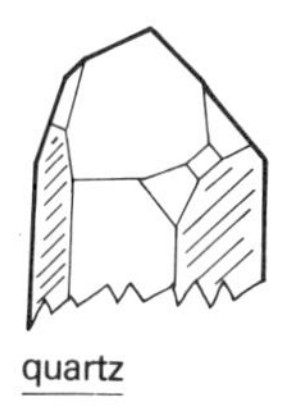

Figure 25 *Typical crystal formations*

curved surfaces. This does not mean that crystals of different materials are necessarily similar, or even that the same substance always crystallizes into the same shape. Carbon for example can crystallize as diamond or graphite, each with a different arrangement of the atoms.

There is a huge variety of shapes of crystals, a small selection being illustrated in Figure 25; what is constant is that they are regular arrangements of molecules, so that their sides are flat, their edges straight and therefore the angles between the sides are sharply defined.

All metals are crystalline in structure – indeed almost all materials, with some notable exceptions such as glass, are crystalline. Metals are often described as *polycrystalline* materials, which means that they are composed of a large number of crystals (the prefix 'poly' originates from the classical Greek language, and means 'many'). Reference was made earlier in this section to the formation of crystals as water freezes; in a similar manner, crystals are formed in metals.

All metals melt and become liquid if they are heated to a sufficiently high temperature; pure copper melts at 1 083 °C for example. If the liquid copper is allowed to cool so that it begins to solidify (i.e. it 'freezes'), the atoms of the copper group themselves together to form a crystalline structure.

Pure metals possess few mechanical properties useful to designers and engineers (copper and aluminium with high electrical conductivity are exceptions). Hence *alloys* of metals are produced, which possess enhanced properties, such as greater strength than any of the parent materials. An alloy can be described as an intimate blend of one metal, known as the 'parent' or 'base' metal, with at least one other metal or non-metal. For example, one type of brass contains two-thirds copper and one-third zinc, while steel is mainly iron with added carbon.

Earlier in this section, it was suggested that the alloy brass is a solid solution of copper and zinc. It might appear strange that a solid metal can exist in solution with another solid metal, but it may be explained in the following manner.

Copper is melted, and zinc which has a lower melting point, is added to it. The copper dissolves the zinc so that a solution of zinc in liquid copper is formed – the process is exactly analogous to the formation of a solution of sugar in water. When the liquid solution of zinc and copper is allowed to cool (remember that the temperature is over 1 000 °C), solidification begins, but the zinc does not separate out of solution, even when the temperature is reduced to room temperature. Thus a *solid solution* is formed.

Self-assessment questions

44 Complete the following statement by inserting the missing words:
A crystal is a regular shape with straight ______ and specific ______.

45 Answer TRUE or FALSE to each of the following statements:
(i) Crystallization occurs when a saturated solution evaporates.
(ii) During evaporation, crystals are formed from molecules of the solvent.
(iii) A given substance always crystallizes into the same shape.

46 Identify from this list of statements those which are applicable to alloys:
(i) The constituents of the material are a base metal and one other metal.
(ii) The constituents of the material are a parent metal and a non-metallic material.
(iii) When the two materials are combined, the strength of the resulting material is greater than that of either of the two original materials.

Solutions to self-assessment questions

41 The completed table is shown below.

mixture	solution or suspension
sea water	solution
wax and water	suspension
petrol and water	suspension
lemonade	solution
soda water	solution
chalk and water	suspension

Sea water is a solution of salt and water; lemonade is a solution of the gas carbon dioxide and flavoured water; soda water is a solution of carbon dioxide and water. Petrol finally floats on the top of water; chalk finally sinks to the bottom of water; wax ultimately floats to the surface of water. All are suspensions if agitated.

42 Factor (*d*) should be circled, and factor (*b*) should be crossed.

The size of the particles of the solute, and the speed of agitation of the mixture affect the rate at which the solute is dissolved. They do not affect the amount of solute which can be dissolved. Temperature affects the amount of solute which can be dissolved, although an increase in temperature does not always produce a greater amount of solute dissolved in the solvent. Density has no effect upon solubility.

43 (i) A solution is said to be saturated if no more solute can be made to dissolve, with the temperature remaining constant.
(ii) Solubility refers to the maximum amount of solute that will dissolve in exactly 0.1 kg of solvent, at a given temperature.

Further self-assessment questions

47 List at least three ways in which compounds can be distinguished from mixtures.

48 Give three examples of each of
(i) elements,
(ii) compounds,
(iii) mixtures.

49 List the common factors which influence the solubility of a solid in a liquid.

50 Name the constituents of two alloys which can exist as solid solutions.

Solutions to self-assessment questions

44 The statement should read:
'A crystal is a regular shape with straight *edges* (*or sides*) and specific *angles*.'

45 (i) The statement is *true*; crystallization occurs when a saturated solution evaporates.
(ii) The statement is *false*; it is the molecules of the solid which form crystals. The solvent evaporates.
(iii) The statement is *false*; certain substances crystallize into different regular shapes under different conditions.

46 *All* the statements are applicable to alloys.

Answers to self-assessment questions

47 (i) Components of a mixture may be present in any proportions; the elements of a compound are joined in a fixed proportion.
(ii) The properties of a mixture are dependent upon the properties of its component parts; the properties of a compound are vastly different from its constituent elements.
(iii) Atoms of a mixture are free; atoms of a compound are joined.
(iv) Little or no heat energy is produced or absorbed when a mixture is formed; measurable heat energy is produced or absorbed when a compound is formed.

48 It is impossible to give a complete list. In general, pure metals are elements; so are hydrogen, nitrogen etc. Many compounds have words ending in '-ate' or '-ide', e.g. calcium carbonate, sodium chloride; but other compounds such as water do not conform to this general rule. Mixtures can be recognized by the properties of their component parts – sweet tea, salt water, etc.

49 (i) Particle size of the solute.
(ii) Speed of agitation.
(iii) Temperature of solution.
(iv) Amount of solute added to solvent.

50 It is impossible to give a complete list. Some alloys which are solid solutions are brass (copper/zinc), bronze (copper/tin), white gold (silver/gold), copper and nickel.

Topic area: Energy

Section 1

After reading the following material, the reader shall:

1 Understand the concept of 'work'.
1.1 Define work in terms of force applied and distance moved.
1.2 Define the joule.
1.3 Use work diagrams to solve problems.
1.4 Draw graphs, from experimental data, of force–distance moved and relate work done with area under the graph.

The word 'energy' is used so often and in so many different ways that the very specific definition of energy used by technologists, technicians and scientists may come as something of a surprise. Energy is commonly associated with movement or activity. A drop hammer in a forge falling on to a piece of hot metal uses up large amounts of energy; heavyweight wrestlers in a ring perspire and grunt and expend much energy to impress and amuse the crowd.

The link to a precise definition of energy is contained in the word itself. The word 'energy' comes from two Greek words meaning 'in' and 'work'. To the student of physical science, energy is defined as the capacity of a body to do work. The following paragraphs explain some ideas about work; the relationship between work and energy is explored in more detail later.

Like energy, work is associated with forces and movement. Force is a concept or idea based on the direct evidence of our senses; it is possible to observe the effect of forces. Force, however, is not easily defined satisfactorily. A force is any cause (i) which may change the velocity of a moving body (ii) or which moves or tends to move the body on which it acts.

When a force is applied to a body and causes it to move, then the force is said to be doing work. Work is calculated as the product of the force acting on the body and the distance moved by the body in the direction in which the force acts.

The SI unit of force is the newton (denoted by the letter N). One newton is the force that will give a mass of one kilogram an acceleration of one metre per second every second (accelerations have units m/s^2).

In physical science, the word 'work' is used with a meaning different from that of everyday speech. If a man is holding up a heavy weight, then he is doing no work since the force exerted produces no movement. Also, it should be noted, the force exerted must be overcoming resistance for work to be carried out. This idea becomes clearer if a space craft moving through empty space is considered. Where there is no gravitational force to combat, or air resistance to overcome, the craft continues along its path with no force being exerted on it; no work is needed, since there is no resistance to motion.

If there is no motion or no resistance, then no work is done.

In SI, unit work is done when unit force moves through unit distance in the direction in which the force is acting. The unit of force is the newton, the unit of distance is the metre, so the units of work are newton metres.

Example 1
A horizontal planing machine has a stroke of 1 m, and during the cutting action the force on the cutting tool is 1 200 N. Find the work done during one cutting stroke.

Work done = force × distance moved in direction of force
= 1 200 × 1
= 1 200 Nm
= 1.2 kNm

The SI unit of work, the newton metre, is called the joule after the physicist James Prescott Joule (1818–89). Joule was largely self-taught in science. He was interested in many branches of knowledge, and was one of the first scientists to realize the importance of accurate measurement.

1 newton × 1 metre = 1 Nm = 1 joule = 1 J

Work diagrams

When a constant force acts through a distance, then the event can be plotted on a simple graph (see Figure 26). The distance is normally plotted as abscissa (horizontally) and the force as ordinate (vertically). In this case, as the force does not vary, the graph will be a horizontal straight line.

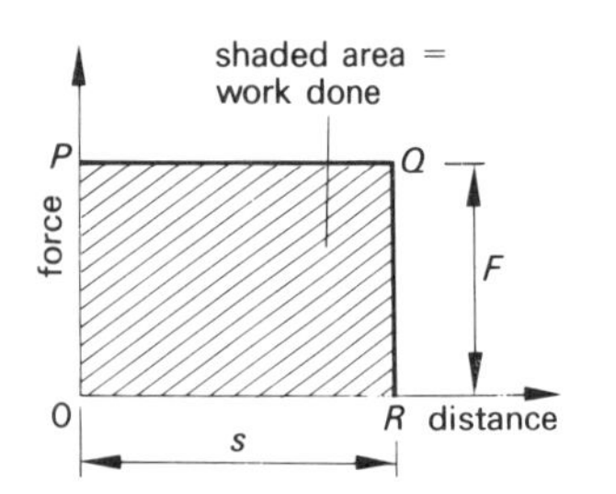

Figure 26 *Work diagram: constant force*

If the constant force be F newtons, and the distance moved be s metres, then the work done is Fs J. Clearly the shaded area in Figure 26 is also Fs.

Hence the area under a force–distance graph can be taken to represent the work done by the force.

Diagrams such as $OPQR$ are called work diagrams. For calculating the work done by a force of constant value, the diagram gives little advantage, but when the force varies, calculation of the area of work diagrams is frequently a simple way of calculating the work done.

Example 2

A spring has a stiffness so that it requires a force of 2 N to compress it 1 mm. If the spring is initially neither stretched nor compressed, find the work done in compressing it 100 mm.

Force being exerted when spring is compressed 100 mm $= 2\ (\text{N/mm}) \times 100\ (\text{mm})$

$$\text{force} = 200\ \text{N}$$

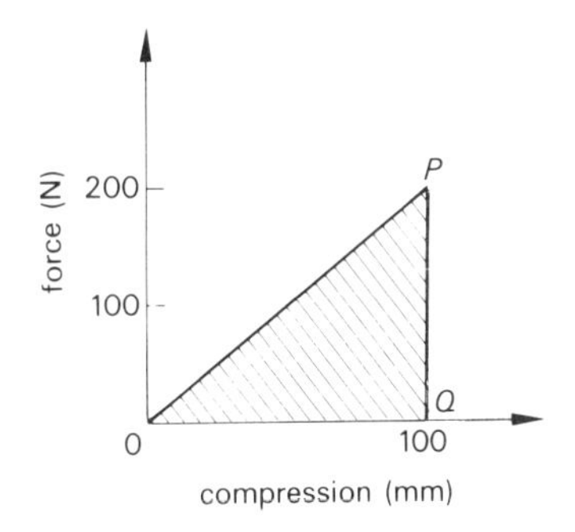

Figure 27 *Work diagram: spring force*

Figure 27 shows how the spring shortens in proportion as the compressive force is applied. The compressive force increases steadily from its initial value of zero to its final value of 200 N. The shaded area OPQ represents the work done.

$$\text{Work done} = \tfrac{1}{2} \times PQ \times OQ$$

$$= \tfrac{1}{2} \times 200\ (\text{N}) \times \frac{100}{10^3}\ (\text{m})$$

$$= 10\ \text{Nm}$$

$$\text{Work done} = 10\ \text{J}$$

Example 3

A machine tool cuts against a resistance which varies during the cutting stroke of the tool. The initial resistance is 3 000 N, and the resistance rises steadily to 9 000 N during the first 800 mm of movement. The resistance then falls uniformly to 5 000 N during the last 200 mm of the cutting stroke. Calculate, in joules, the work done during each cutting stroke.

The variation of cutting force with movement during the working stroke of the machine tool is as shown in Figure 28. (Note: area of trapezium $= \tfrac{1}{2}$ sum of parallel sides × distance between them.)

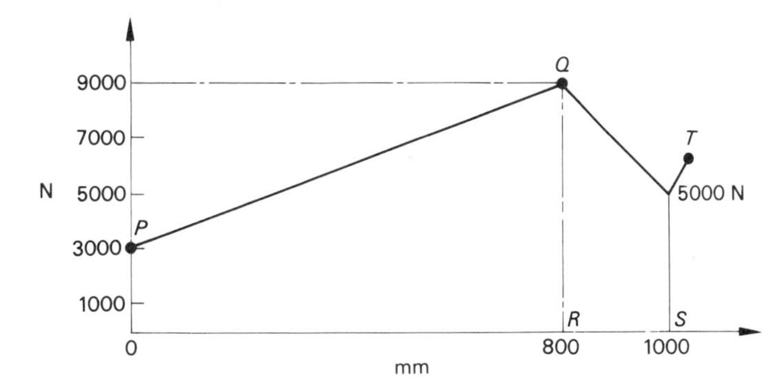

Figure 28 *Work diagram: machine tool*

$$\text{Work done per stroke} = \text{area of } OPQR + \text{area of } QRST$$
$$= \left[\tfrac{1}{2} \times (3\,000 + 9\,000) \times \frac{800}{10^3}\right] + \left[\tfrac{1}{2}(9\,000 + 5\,000) \times \frac{200}{10^3}\right]$$
$$= \left[\tfrac{1}{2} \times 12\,000 \times \frac{800}{10^3}\right] + \left[\tfrac{1}{2} \times 14\,000 \times \frac{200}{10^3}\right]$$
$$= 4\,800 + 1\,400$$
$$\text{Work done per stroke} = 6\,200 \text{ J}$$

Forces are described by a number of names. A tension is the force exerted in a chain or rope. A frictional force must be overcome whenever a body begins to move. An explosive force is felt by a person who discharges a gun.

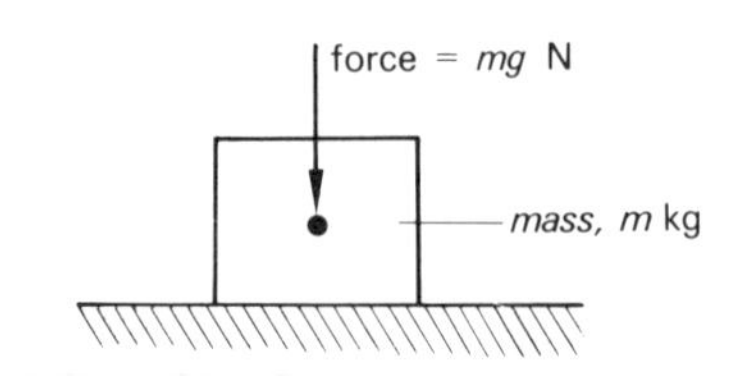

Figure 29 *Gravitational pull on a mass*

Probably the most common force is that called *weight* – the force due to gravity. A mass of m kg has a weight of mg N where g is the acceleration caused by the gravitational pull of the earth (g has a value of 9.81 m/s^2 in the United Kingdom at sea level).

When a mass m kg rests on a surface as in Figure 29, it exerts a downward force of mg N.

Example 4
A body having a mass of 40 kg is lifted through a height 20 m. Calculate the work done in kilojoules.

$$\text{Resistance to be overcome} = mg$$
$$= (40 \times 9.81) \text{ N}$$
$$\text{Work done} = \text{resisting force} \times \text{height}$$
$$= (40 \times 9.81) \times 20$$
$$= 7\,848 \text{ Nm}$$
$$= 7\,848 \text{ J}$$
$$\text{Now } 1 \text{ kJ} = 10^3 \text{ J}$$
$$\therefore \text{Work done} = 7.848 \text{ kJ}$$

Example 5
A chain hangs vertically from a drum. Each 1 m of length of the chain has a mass of 10 kg. The chain is 20 m long. The chain is wound steadily on to the drum at the top. Draw the work diagram and calculate the work done in winding the complete chain on to the drum.

$$\text{Weight of each metre of chain} = mg$$
$$= (10 \times 9.81) \text{ N}$$
$$\text{Total weight of chain} = \text{weight/metre} \times \text{length}$$
$$= (10 \times 9.81) \times 20$$
$$\text{Total weight} = 1\,962 \text{ N}$$

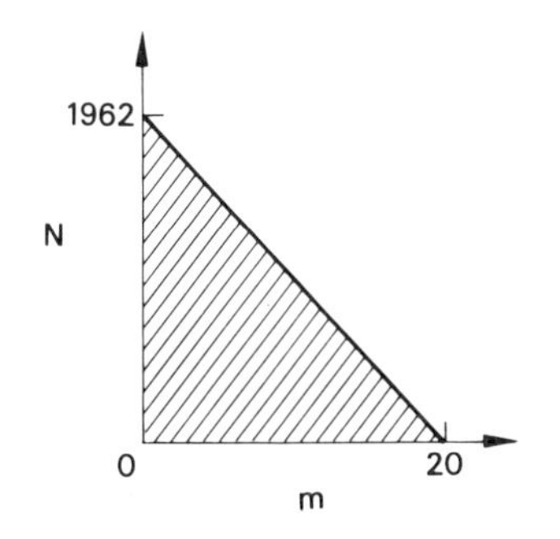

Figure 30 *Work diagram: chain problem*

Hence the initial lifting force is 1 962 N. The lifting force diminishes steadily to zero as the chain is wound on to the drum. The work diagram is shown in Figure 30.

Work done in winding up the chain is the shaded area.

$$\therefore \text{Work done} = \tfrac{1}{2} \times 1\,962 \times 20$$
$$= 19\,620 \text{ Nm}$$
$$= 19\,620 \text{ J}$$
$$\text{Work done} = 19.62 \text{ kJ}$$

The fact that the area of a force–distance diagram represents the work done is particularly useful when the force exerted varies in a non-linear manner.

Example 6

A load is moved by a tractive force measured in newtons. The force is measured at certain points in the travel of the load, and the corresponding values of the force and the distance moved are given in the table below:

force (N)	820	720	500	380	300
distance (m)	0	20	40	60	80

Draw the work diagram and calculate the work done in kilojoules.

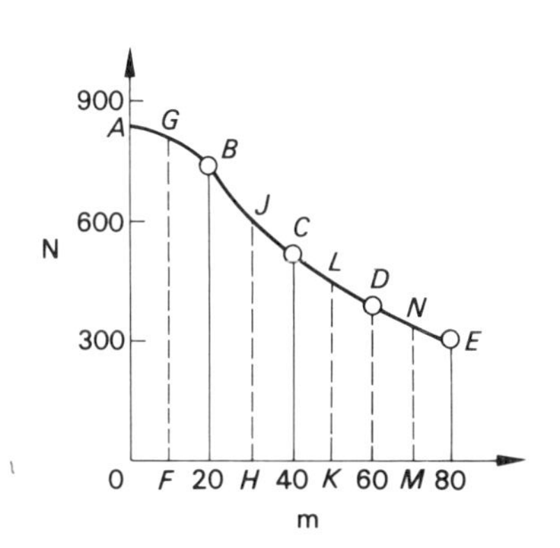

Figure 31 *Work diagram: Variable force*

The work diagram is shown in Figure 31. The points *ABCDE* have been joined by a smooth curve. A very close approximation to the average force acting can be calculated by finding the average of the 'mid-ordinates' *FG*, *HJ*, *KL*, *MN*. The base line has been divided, in this case, into four equal parts, and the mid-ordinates are the vertical lines shown dotted.

By measurement from the work diagram, $FG = 790$ N, $HJ = 610$ N, $KL = 450$ N and $MN = 330$ N.

$$\text{Average force acting} = \frac{FG + HJ + KL + MN}{4}$$
$$= \frac{770 + 610 + 460 + 330}{4}$$
$$= \frac{2\,180}{4}$$
$$= 545 \text{ N}$$
$$\therefore \text{Work done} = \text{average force} \times \text{total distance moved}$$
$$= 545 \times 80$$
$$= 43\,600 \text{ J}$$
$$\text{Work done} = 43.6 \text{ kJ}$$

Self-assessment questions

1 Complete the following statements:

(*a*) Work = force × ______

(*b*) Energy is the capacity to ______

(*c*) The force needed to support a mass of 3 kg is ______ × ______ N.

(*d*) Before work can be said to be done, a resistance must be ______

2 Define the term 'work'. When using SI units, in what units is work measured?

3 Define the unit known as the joule.

4 Underline the correct answer:
Work is best described as –

(*a*) a force

(*b*) movement of a body

(*c*) a force moved vertically

(*d*) the use of a force to move a body a certain distance

5 A force F moves through a distance L. The force is then increased to $3F$, and then moves through a further distance $2L$. Draw the work diagram to scale.

6 A tool cuts a hole in a metal casting, and the resistance to the applied force varies. In the first part of the action, the resistance rises uniformly from 4 000 N to 8 000 N and during this period the tool moves 600 mm. The resistance then falls uniformly to 4 000 N during the final 200 mm of movement. Draw the work diagram, and calculate the work done per cutting stroke in joules.

7 A helical spring is extended by a force which increases uniformly from zero to 600 N. The extension of the spring is then 200 mm.

Draw the work diagram to scale and calculate in joules the work done in extending the spring the last 80 mm.

8 Calculate the work done in kilojoules in lifting a mass of 10 kg (at a steady velocity) through a vertical height of 10 m.

9 A body moves under the influence of a variable force:

Force (N)	23	27	32	34	38	38
distance (m)	0	1.0	2.0	3.0	4.0	5.0

(*a*) Draw the work diagram.

(*b*) Calculate the average force acting: use the method of mid-ordinates and assume the points on the graph are joined by a smooth curve.

(*c*) Calculate the work done (in joules) in the total movement of 5 m.

After reading the following material, the reader shall:

2 Describes the nature and types of energy.
2.1 Name common forms of energy.
2.2 Name five forms of energy.
2.3 Identify examples of the conversion of one form of energy into another form.
2.4 Describe fuels as a source of energy.
2.5 Give two examples of conversion of heat energy to other forms of energy and vice versa.

The action of sunlight on growing things stores up potential chemical energy in starches and sugars. These provide food for animals – including man. With the energy stored in the food, animals can keep their bodies heated and carry out the activities needed to sustain life.

In explosives, chemical energy is changed into heat with extreme rapidity. Although there is more chemical energy in a kilogram of butter than in a kilogram of gunpowder, the butter does not explode because it releases its energy much more slowly.

When a mass m is raised through a vertical height h then the work done is mgh. While the mass is poised in this elevated position, the energy stored in it is called potential energy. When work is done, energy is changed from one form to another. The form of energy first recognized from this point of view was mechanical energy. A body in motion is able to displace or deform another body with which it collides. Mechanical energy is present in flowing water (which can turn a waterwheel) and in moving air (which can turn the sails of a windmill or propel a yacht).

The energy of motion, of which these are examples, is called kinetic energy. Kinetic is derived from a Greek word meaning 'moving'.

These are only a few examples of the types of energy that are met in the study of physical science. In the following paragraphs, some forms of energy frequently encountered are described.

Energy can be divided into two broad categories – kinetic and potential.

Kinetic energy

A body that is in motion possesses energy: it has the capacity to do work. By striking another body which is free to move, the moving body can exert a force and cause the second body to shift its position.

A piledriver provides a good example of the relation between kinetic energy and work. A piledriver is a machine used to produce sound foundations for structures and buildings. A large mass called a drop hammer is lifted and then allowed to fall freely on to the head of a pile (see Figure 35).

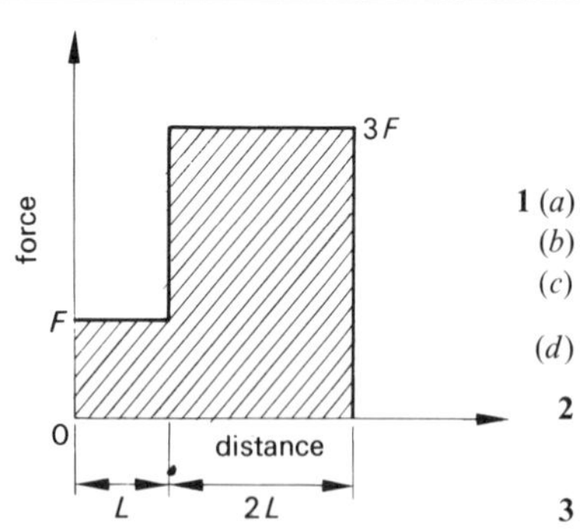

Figure 32 *Work diagram: changing force*

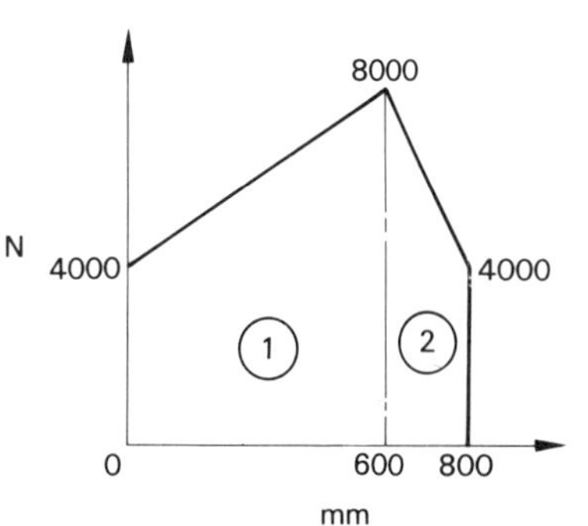

Figure 33 *Work diagram: tool*

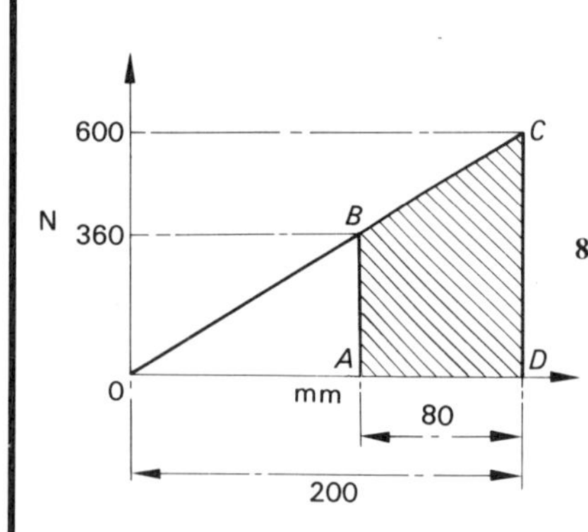

Figure 34 *Work diagram: spring extension*

Solutions to self-assessment questions

1 (*a*) Work = force × *distance moved in the direction of the force.*

(*b*) Energy is the capacity to *do work.*

(*c*) The force needed to support a mass of 3 kg is $3 \times g$ or 3×9.81 N = 29.43 N

(*d*) Before work can be said to be done, a resistance must be *overcome.*

2 Work is done when a force applied to a body causes the body to move. In SI units, work is measured in joules.

3 A joule equals one newton-metre.

4 (*d*) is the correct answer i.e. work is best described as the use of a force to move a body a certain distance.

5 See Figure 32.

6 The work diagram is shown in Figure 33.
Work done per cutting stroke = area (1) + area (2)

$$\text{i.e. } W = \left[\tfrac{1}{2} \times (4\,000 + 8\,000) \times \frac{600}{10^3}\right] + \left[\tfrac{1}{2} \times (8\,000 + 4\,000) \times \frac{200}{10^3}\right]$$

$$= \left[\tfrac{1}{2} \times 12 \times 10^3 \times \frac{600}{10^3}\right] + \left[\tfrac{1}{2} \times 12 \times 10^3 \times \frac{200}{10^3}\right]$$

$$= (\tfrac{1}{2} \times 12 \times 600) + (\tfrac{1}{2} \times 12 \times 200)$$

$$= 3\,600 + 1\,200$$

Work done = 4 800 J

7 The work diagram is shown in Figure 34.
Work done = $\frac{1}{2}$ sum of parallel sides × distance between them

$$= \tfrac{1}{2} \times (360 + 600) \times \frac{80}{10^3}$$

$$= \tfrac{1}{2} \times 960 \times \frac{80}{10^3}$$

Work done = 38.4 J

8 Force of gravity acting on mass = mg

i.e. weight of mass = 10×9.81

= 98.1 N

Work done in lifting mass = force × distance

= 98.1×10

= 981 J

Work done = 0.981 kJ

Figure 35 *Piledriver*

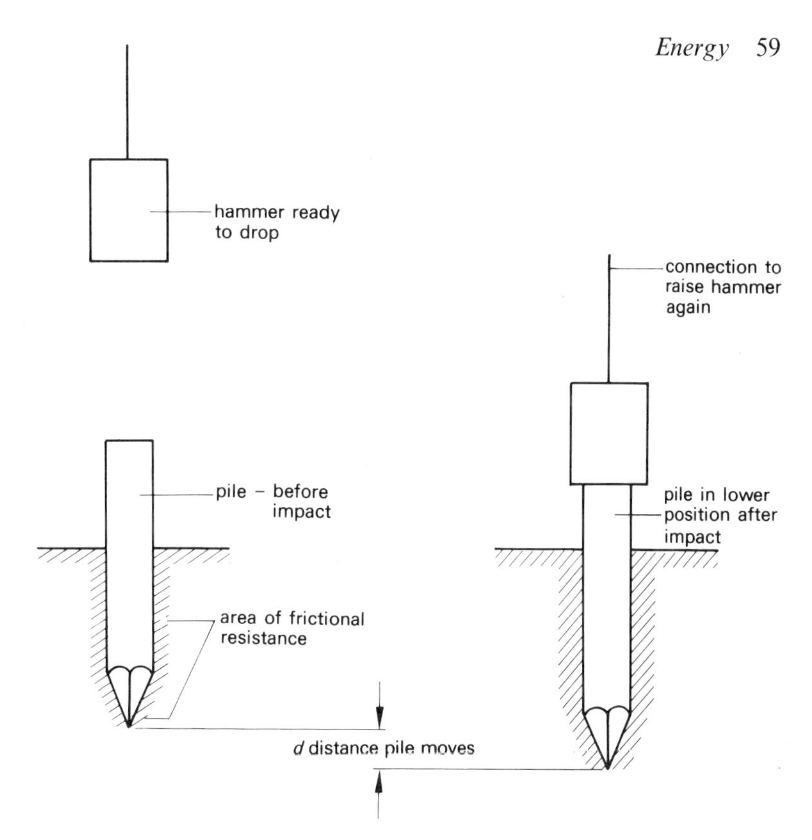

As the hammer falls and speeds up it gains kinetic energy; this is lost during the fraction of a second when the hammer hits the pile. The force exerted on the pile drives the pile into the ground against

(*a*) The work diagram is shown in Figure 36.

(*b*) The five mid-ordinates are shown as dotted lines. The values of these are 25, 30, 33, 36, 38 N

$$\text{Average force acting} = \frac{25+30+33+36+38}{5}$$

$$\underline{\text{Average force} = 32.4\ \text{N}}$$

(*c*) Work done = average force × distance moved

$$= 32.4 \times 5$$

$$\underline{\text{Work done} = 162\ \text{J}}$$

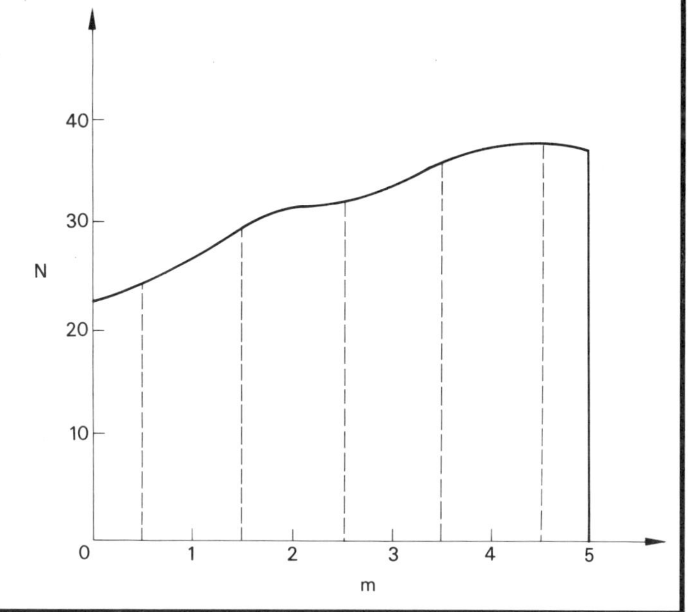

Figure 36 *Work diagram: force not constant*

frictional resistance. Work is accomplished by the energy of the falling hammer. If the distance moved by the pile is d metres and the average ground resistance overcome is F newtons, then the work done at the expense of the kinetic energy of the drop hammer is Fd joules.

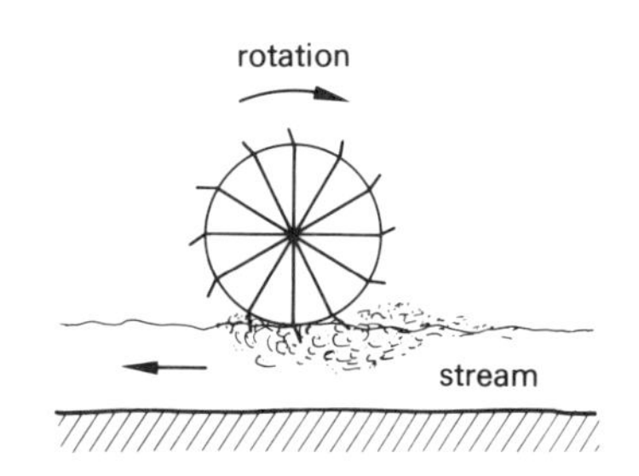

Figure 37 *Undershot waterwheel*

Men have made use of kinetic energy for a great length of time. Ancient sailing ships used the energy of the winds to move them along. Waterwheels turned by rapidly moving streams have been in existence for more than two thousand years. Windmills have a history extending for more than a thousand years, and there is now serious talk of harnessing the energy of the moving air by a large number of huge windmills which would generate electricity to feed into the national grid system.

Waves on the sea carry large amounts of kinetic energy, but unfortunately this energy so far has only been harmful, damaging ships, harbours and shore-line installations. The only use made of wave energy is to ring the warning bells of bell buoys.

Everyday experience shows that the more massive a body is and the higher its speed, the more work it will do upon striking and being slowed down by an obstacle. A body of mass m moving with a speed v possesses kinetic energy of $\frac{1}{2}mv^2$. (This formula will be derived in a more advanced unit. The reader is not expected to remember the formula for kinetic energy.) When the mass is in kilograms and the speed in metre/second, then the kinetic energy is measured in joules ($1\ \text{J} = 1\ \text{Nm}$).

Example 7

A body of mass 100 kg has a speed of 40 m/s. On striking another body, its speed falls to 20 m/s. By how much has its kinetic energy been reduced?

$$\begin{aligned}
\text{Initial kinetic energy} &= \tfrac{1}{2}mv^2 = \tfrac{1}{2}\times 100\times(40)^2\\
&= \tfrac{1}{2}\times 100\times 1\,600\\
&= 80\,000\ \text{J}\\
\text{Final kinetic energy} &= \tfrac{1}{2}mv^2 = \tfrac{1}{2}\times 100\times(20)^2\\
&= \tfrac{1}{2}\times 100\times 400\\
&= 20\,000\ \text{J}\\
\therefore\ \text{Reduction in kinetic energy} &= \text{initial kinetic energy} - \text{final kinetic energy}\\
&= 80\,000 - 20\,000\\
\text{Reduction} &= \underline{60\,000\ \text{J}}
\end{aligned}$$

Potential energy

When a body or a system has the capacity to do work by reason of its position or configuration, then the body or system possesses potential energy.

A coiled spring that is compressed or stretched possesses potential energy, for when the spring is released it can perform work. Chemical and nuclear systems can be thought of as being in possession of potential energy, for these systems are capable of doing work.

A small number of examples of potential energy are now described.

Gravitational potential energy

Grandfather clocks are usually driven by heavy weights suspended on cords. As the clock ticks on, the weights descend, and at regular intervals they need to be wound up to their highest position. Work is done in raising the weights, and energy has to be expended to carry out the work. This energy is stored in the raised weights, and as the weights descend, this stock of energy is reduced as the clock is driven. The energy stored in the weights is called gravitational potential energy, since the work done is carried out against the resistance due to the pull of gravity on the weights. (Potential energy can be compared to money kept in a bank; the money can be thought of as 'potential purchases'.)

Elastic strain energy

Grandfather clocks are driven by weights; most modern clocks are driven by springs. A spring stretched or compressed or twisted from its normal shape has potential energy stored in it. This statement is known to be true, not because anyone can *see* the energy stored in the distorted metal, but because it is known that work can be done by the spring when the strain is released. The energy available is called strain energy or elastic potential energy; it may be pictured as stored in force-fields between the molecules of the springy material.

It is often said that the energy stored in a spring drives a mechanism such as a clock. This type of expression may be good enough for ordinary, everyday conversation, but from a physical science point of view, it is worth remembering that it is forces that make things move, not energy. It is the force exerted by a wound-up spring that drives a clock, and the work done by this force uses up the elastic potential energy initially stored in the spring. Strictly speaking, it is the whole arrangement of a moving force and a checking force that stores potential energy.

Chemical energy

Oil and petrol have chemical energy that can be released and cause useful work to be performed. Explosives have chemical energy that can be released in a very short period of time. Food has chemical energy which is needed to maintain all animal life.

Chemical energy is released or absorbed when a chemical reaction occurs. A chemical reaction is one in which a number of substances combine to produce other, different, substances. For instance, when carbon and oxygen combine, large amounts of heat are released – more than enough to heat more carbon, and so keep the process going once it has been started. The burning of carbon is an example of an exothermic chemical reaction, that is, a reaction that is accompanied by the evolution of heat. This particular reaction has been studied by chemists, and their 'shorthand' way of describing the reaction is to write what they call a 'chemical equation'.

1 molecule of carbon + 1 molecule of oxygen
= 1 molecule of carbon dioxide + heat

Chemical energy can be thought of as a type of potential energy. This can be explained by considering the forces that bind the atoms together into molecules. When, during a chemical reaction, the atoms are re-arranged, then these binding forces are released, and some of the chemical potential energy is released and appears as heat.

Substances which release their potential energy as heat when they are burned are known as fuels. Common fuels – gas, oil, coal – all contain the substances carbon and hydrogen.

Nuclear energy

Nuclear or atomic energy is the only energy form known on earth which owes nothing to the sun. Radioactive atoms – such as those found naturally in uranium ore – shoot out subatomic particles with incredible speed. These particles carry enormous burdens of kinetic energy for their size. In the present state of our knowledge of atomic structures, the only possible source of the tiny particles that can be suggested is the very small massive core (nucleus) of the radioactive atom. It is possible therefore to think of the atomic core as having a great store of nuclear energy.

The discovery of radioactivity at the end of the nineteenth century revealed evidence on the atomic scale in which energy seemed to be created, but matter appeared to be destroyed at the same time. In 1905 Albert Einstein showed that mass and energy are not two different phenomena, but are interchangeable. Many examples are known today in which matter is converted directly into energy; other examples are known where energy is converted into matter.

Readers will be aware of the awesome energy released by nuclear bombs; they will also know that nuclear power stations may supply the growing energy needs of mankind in the near future.

Einstein's law states that $E = mc^2$, where E is the energy expressed in joules, m is the mass in kilograms and c is the speed of light in metre/second. The speed of light is so great (300×10^6 m/s) that a vast quantity of energy is produced by the annihilation of a very small amount of matter. The energy equivalent of one kilogram of mass is 90×10^{15} joules. This will keep 25 million one kilowatt electric fires burning for one thousand hours each. To produce the same amount of energy by chemical means would require the combustion of large amounts of fuel.

Nuclear physics is one of the most dynamic fields of modern scientific development. As a result of the investigations and experiments, mankind will be able to do much good or cause untold harm. The theories that are so far available are incomplete. It may be a long time before a grand scheme is discovered that will explain the mysteries of energy available from the nucleus.

Energy conversions

One reason for giving importance to the study of energy is that it is the topic that occurs in simply every topic in physical science. An understanding of energy is needed when considering optics, sound, electricity, mechanics and properties of matter, to name but a few. Many of the distinctions between the branches of physical science are not at all clear-cut mainly because, in any particular instance, energy is being converted into a number of different forms at the same time. Some investigators have gone so far as to say that the whole field of physical science can be thought of as the science of energy and its transformation.

There are many energy transformations that can be noted in nature – for instance the evaporation of water by solar energy, or the storage of solar energy in fossil fuels like coal and oil. In the world of technology, the idea of energy conversion is more generally applied to man-made operations in which the energy is made more usable. For instance, in the engines of motor cars, the burning of petrol converts the potential chemical energy into energy that alters the kinetic energy of the vehicle.

The law of conservation of energy states that energy cannot be created or destroyed; it can only change in form. The developments in science that have taken place this century indicate that this law is true in all cases, incorporating the idea that mass, in certain circumstances can be converted into energy.

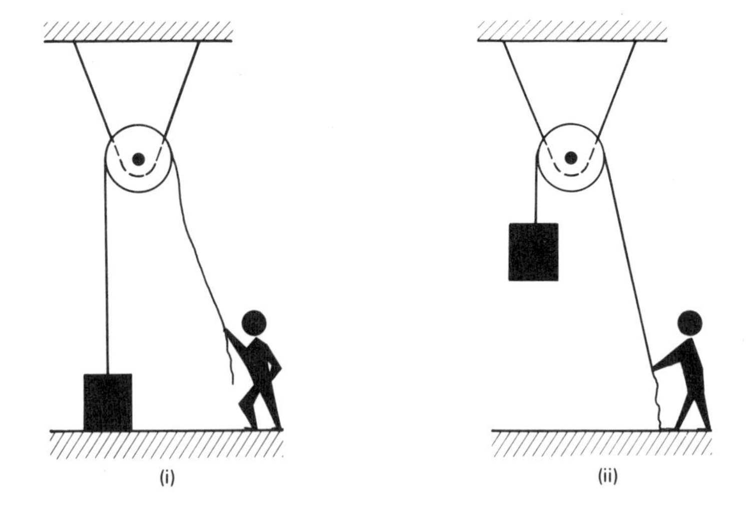

Figure 38 *Chemical energy to potential energy*

(i) *chemical energy from man to*
(ii) *gravitational potential energy*

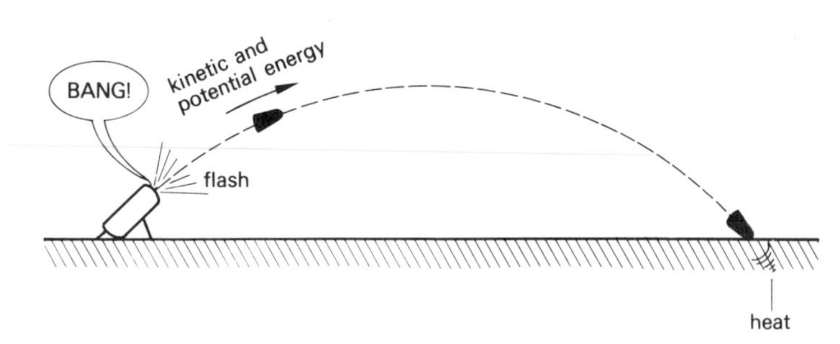

Figure 39 *Chemical energy to kinetic and potential energy*

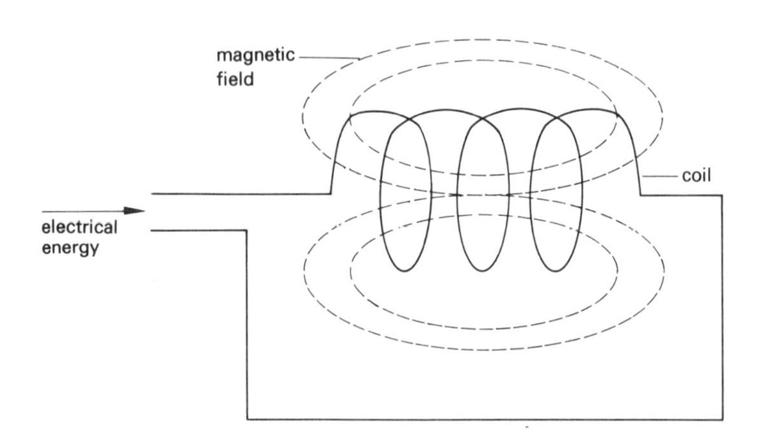

Figure 40 *Electrical energy to magnetic energy and heat energy*

Figures 38–40 show examples of energy conversion. Figure 41 shows an electric power system in which either chemical or nuclear energy is converted first into the thermal energy of hot steam, then into the mechanical energy of a turbine. This mechanical energy is converted in a generator into electrical energy. This electrical energy is then used for a wide variety of purposes.

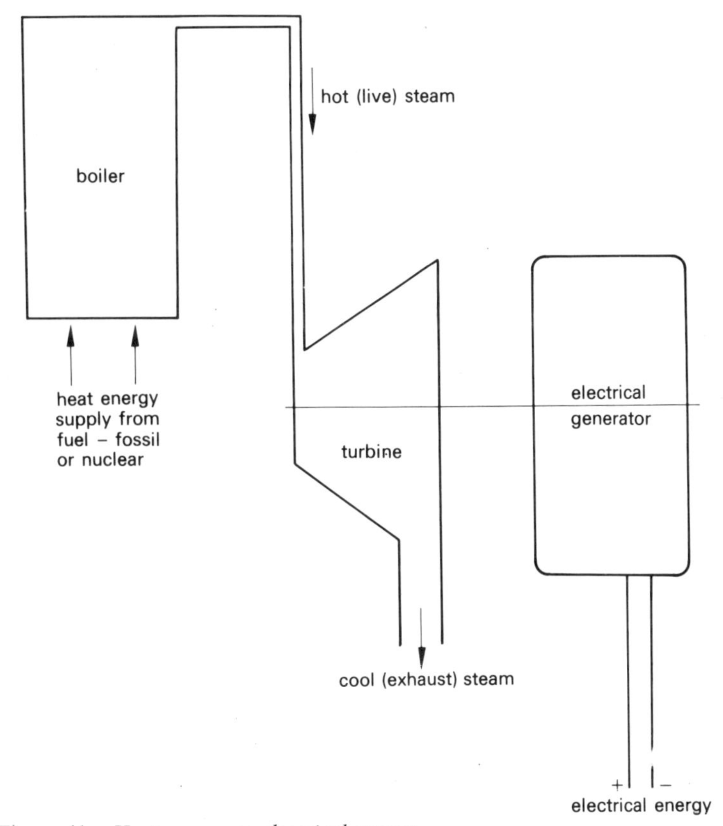

Figure 41 *Heat energy to electrical energy*

In all machines a great deal of energy is lost in the form of heat. Whenever movement is involved, energy is 'lost' in friction, which causes heating. Even in very efficient machines like electric motors, the passage of the electric current heats the wires, and this energy, although not actually 'lost', is wasted from the point of view of the person who owns the machine.

Since all forms of energy can be changed into one another, it might be thought that such interchanges can take place in any direction with equal efficiency. This is not so. The more highly organized forms of energy, such as those binding the molecules of complex chemical compounds together, are easily degraded to lower forms, but the lower forms can only be changed to higher forms with difficulty. The lowest state is heat energy, which consists of the random movement of molecules.

The ease with which energy can be degraded to heat can be understood if a pack of cards is considered. An ordered pack can be readily reduced to disorder by being shuffled, but the shuffled pack is very unlikely to be restored to its original condition by further shuffling.

Self-assessment questions

10 Name one source of energy which does not owe its origin to energy from the sun.

11 Name one element found in fossil fuels such as oil and coal.

12 List five common forms of energy.

13 Complete the statement of energy conversions on Figures 42–5.

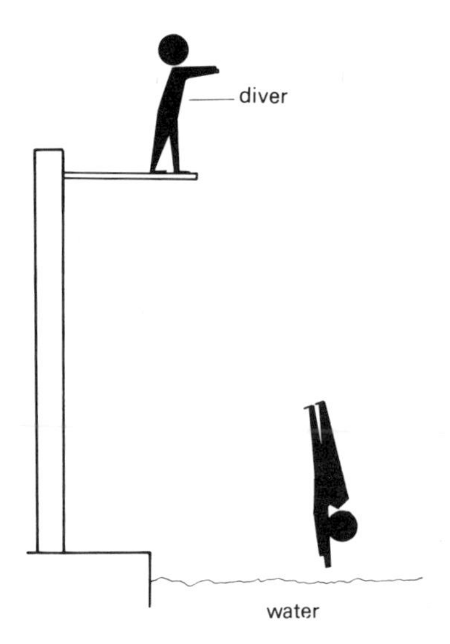

Figure 42
Gravitational potential energy converted into ________

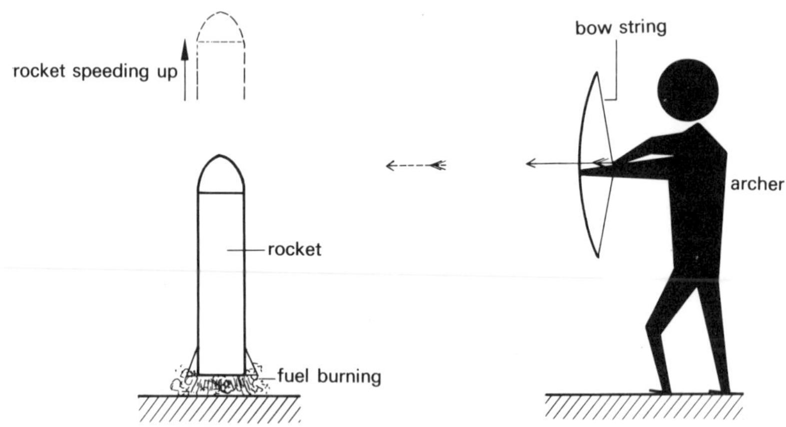

Figure 43
Chemical energy converted into heat energy and ________

Figure 44
________ *of bow string converted into kinetic energy of arrow*

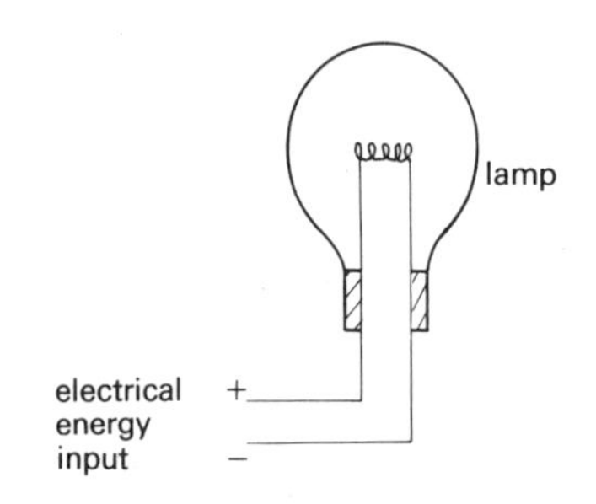

Figure 45
Electrical energy converted into ________ *and* ________

14 Fuels are a source of energy. TRUE/FALSE

15 Give two examples where heat energy is converted into other forms of energy.

16 Give two examples where energy is converted into heat energy.

After reading the following material, the reader shall:

2.6 Define the relationship between energy and work done.
2.7 Describe the relationship between energy and work done.
2.8 Define efficiency in terms of energy input and output.

Energy has been described as the capacity to do work. This is quite true, but there are cases where a system may possess an enormous store of energy that is not available to do useful work. For instance, clouds contain a vast amount of kinetic energy, but this cannot be harnessed; the waves in the sea are extremely dynamic, but their energy so far, has not benefited anyone. Because of these ideas, energy may be better defined as *that property of a system which is reduced when work is done*.

Energy must be available before work can be done. A well-appreciated example is the human body. Food is consumed, this containing heat energy. This energy is extracted and converted into mechanical energy during movements of the body.

In these modern technological times, machines are constantly being used. Machines vary considerably in their complexity. Some of the simplest machines, such as levers, have a very long history. The newest machinery used in automated factories has only been designed within recent years. Just what is a machine?

A machine is a device for overcoming a resistance at one point by the application of a force at some other point. In order to produce a force, energy has to be expended. The energy supplied to a machine is referred to as the input; the useful work performed by the machine is referred to as the output.

In practice, it is impossible to convert all the input into the desired output form. Some energy is always 'wasted' during the conversion processes. For example, the chemical energy in petrol is used to propel a motor car. After burning the fuel, some of the energy is consumed in overcoming friction in the moving parts of the car. A great deal of energy is carried away in the hot exhaust gases. Only a relatively small part of the energy available in the fuel is actually used to move the car along the road.

The difference between the input energy and the output energy represents energy which is not available for use as desired; from the practical point of view this energy is wasted. (For the motorist the considerable energy in the hot exhaust gases is wasteful and costly.) The waste energy is normally referred to as 'lost energy' or 'energy loss'.

In all machines there is movement, and wherever surfaces move relative to one another friction is bound to occur. From the point of view of the efficient working of a machine, friction is the enemy. Friction causes energy to be 'wasted'. (But it is worth remembering that without friction the world would be a very different place. Nothing moving would stop, and nothing stationary would start to move. Friction has a great impact on all aspects of physical science.)

The ratio of the energy output of a machine to the energy input is a measure of the capability of the machine to convert the energy input into the desired form. This ratio is called the mechanical efficiency of the machine –

$$\text{i.e. efficiency} = \frac{\text{output energy}}{\text{input energy}}$$

As energy is always wasted in friction and other effects, the efficiency is always less than 1.

Sometimes an alternative definition is used –

$$\text{i.e. efficiency} = \frac{\text{work got out of machine}}{\text{energy supplied to machine}}$$

Efficiency is a fraction. Some very efficient gearboxes can have an efficiency as high as 0.9 but many machines have low mechanical efficiencies.

Solutions to self-assessment questions

10 Nuclear energy.

11 Carbon; hydrogen.

12 The following list gives a good range of examples: gravitational potential energy, elastic strain energy, kinetic energy, heat energy, chemical energy, electrical energy, magnetic energy, electromagnetic energy, wave energy (including light, sound and ocean waves), nuclear energy.

13 Figure 42: Kinetic energy.
Figure 43: Kinetic energy and potential energy.
Figure 44: Elastic strain energy.
Figure 45: Heat energy and light energy.

14 True.

15 Examples which can be given include:
Heat from the sun heats the land, sea and the air. Water heats and cools more slowly than the land, and this gives rise to sea breezes near sea shores and lakeside shores. The heat energy has caused an increase in kinetic energy (the breezes).

Hot gases inside a cylinder cause a piston to exert a force. This force can cause mechanical work to be done. The heat energy has been converted into kinetic energy and potential energy.

16 Examples which can be given include:
Chemical energy is converted into heat energy during the combustion of explosives, fuels and food.

Electrical energy is converted into heat energy in an electric fire. Kinetic energy is converted into heat energy whenever friction between surfaces occurs.

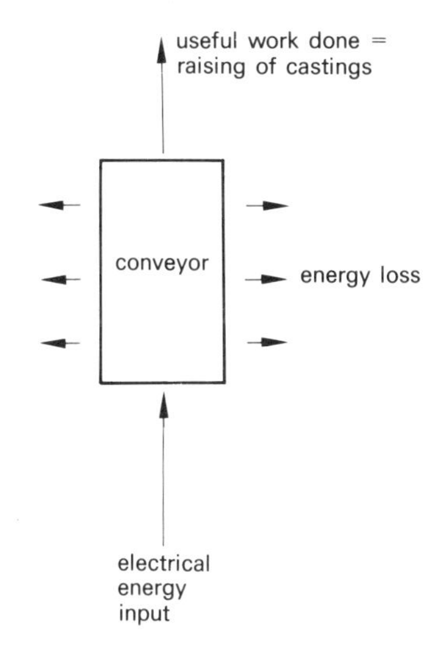

Figure 46 *Conveyor system*

It has been the practice in reference books to express efficiency as a percentage –

$$\text{i.e. (percentage) efficiency} = \frac{\text{work output}}{\text{energy input}} \times 100$$

The efficiency of the gear box would be quoted as 90%.

Example 8

A conveyor lifts 100 castings each of mass 80 kg through a height of 6 m. If the total electrical input during the operation is 9 425 J, find the efficiency of the system. Express the efficiency as a percentage.

In this case, the useful work done by the machine can be thought of as the increase in gravitational potential energy of the castings.

The force of gravity exerted on each casting is 80 g newtons, where $g = 9.81\ \text{m/s}^2$.

$$\begin{aligned}\text{The useful work done} &= \text{increase in potential energy}\\ &= (80 \times 9.81)\ \text{newtons} \times 6\ \text{metres}\\ \underline{\text{Useful work}} &= \underline{(80 \times 9.81 \times 6)\ \text{J}}\end{aligned}$$

The system is represented diagrammatically in Figure 46.

$$\begin{aligned}\text{Percentage efficiency} &= \frac{\text{useful work output}}{\text{energy input}} \times 100\\ &= \frac{80 \times 9.81 \times 6}{9\,425} \times 100\\ \underline{\text{Efficiency}} &= \underline{50\%}\end{aligned}$$

The efficiency of the system is 50%, which means that half of the energy supplied to the system is used to perform useful work. The rest of the energy is lost and eventually is degraded to heat.

Self-assessment questions

17 Define the relationship between energy and work done.

18 What is meant by the expression 'mechanical efficiency of a machine'? Explain why a machine cannot be 100% efficient.

19 When a car is driven uphill at a certain velocity it consumes fuel at a greater rate than it does when driven at the same velocity on a level road. Explain why this is so.

20 A conveyor system lifts forgings. When its useful work output in a certain time is 8 400 J, it is known that the efficiency of the system is 60%. What is the required energy input to the system?

After reading the following material, the reader shall:

3 Know the relationship between energy and power.

3.1 State that power is the measure of the rate at which work is done.

3.2 State that the unit of power is the watt.

It is important to distinguish between energy and power. Energy does not become power until it is brought under control.

Energy other than that produced by animals was rarely used before the end of the eighteenth century. In the years close to the French Revolution of 1789, James Watt and his associates were developing the steam engine. Up to that time, horses were used in large numbers to drive machinery. In particular, they were employed to pump water out of mines in the north-east and south-west of England. Watt, a keen and successful man of business, knew that he had to sell his engine in competition with the reliable horses. In order to have a measure of the competition, he decided to find out the rate at which the average horse worked. The use of 'horsepower' to rate machinery continued until recently.

Forces of nature are capable of working at enormous rates for very short intervals of time. The power of a flash of lightning can be as high as millions of horsepower; but – fortunately – the flash occupies only a tiny fraction of a second. A man can work as hard as a horse for a few seconds; normally, a man can work at about one-eighth the rate at which a horse can work.

Solutions to self-assessment questions

17 Energy can be defined as that property of a system which is reduced when work is done.

18 The mechanical efficiency of a machine is a measure of the capability of the machine to convert the energy input into the desired form. For instance, the mechanical efficiency of an electric motor gives a measure of the ability of the motor to convert the electrical input into mechanical output.

All machines are less than 100% efficient because of inevitable losses due to causes such as friction.

19 A car driven along a level road at a particular velocity is using the useful energy output from the engine to overcome frictional resistance due to the road and the air. When driven uphill at the same velocity, the energy available has to increase the potential energy of the car as well as overcome the frictional resistances. This causes an increase in fuel consumption.

20

$$\text{Efficiency} = \frac{\text{output energy}}{\text{input energy}}$$

$$\text{i.e. } 0.6 = \frac{8\,400}{\text{input energy}}$$

$$\therefore \underline{\text{Required input energy} = 14\,000\ \text{J}}$$

Power is the rate at which work is performed

$$\text{i.e. power} = \frac{\text{work done}}{\text{time taken}}$$

The unit of power in SI is defined as a rate of working of one joule per second. This unit is named the watt (symbol W) –

i.e. 1 J/s = 1 W

The name 'watt' was chosen to honour James Watt, the engineer who made such a great contribution to the development of technology.

It should be noted that the watt is the unit of all types of power.

The watt is a very small unit, and in consequence the kilowatt (kW) is frequently used –

i.e. 1 kW = 1 000 W

For even larger powers, the megawatt (MW) is used –

i.e. 1 MW = 1 000 kW = 1 000 000 W

In spite of the watt being only a small amount of power, in certain topics such as electronics, small fractions of watts are measured. These are milliwatts (mW) –

i.e. 1 mW $= \frac{1}{1\,000}$ W

(Note the very different significance of the small 'm' and the large 'M' when they precede a symbol.)

Example 9

A force of 100 N is applied to a body sliding at a uniform velocity across a horizontal surface.

Calculate

(i) The work done in moving the body 30 m,

(ii) The power in watts if the movement takes place in one minute.

(i)
$$\begin{aligned} \text{Work done on body} &= \text{force} \times \text{distance} \\ &= 100 \times 30 \\ &= 3\,000 \text{ Nm} \\ \underline{\text{Work done}} &\underline{= 3\,000 \text{ J}} \end{aligned}$$

(ii)
$$\begin{aligned} \text{Power} &= \frac{\text{work done in joules}}{\text{time taken in seconds}} \\ &= \frac{3\,000}{60} \\ \underline{\text{Power}} &\underline{= 50 \text{ W}} \end{aligned}$$

Example 10

In an engineering machine-shop a piece of metal is being planed down to the required dimensions. The constant resistance to the cutting tool is 1 000 N, and the cutting stroke is 2.5 m. The cutting stroke takes 5 seconds.

Calculate

(i) The power consumed at the tool point in watts,

(ii) The power input to the system, if the efficiency of the system is 0.6 State the power input in kilowatts.

$$\text{Work done per cutting stroke} = \text{force} \times \text{distance}$$
$$= (1\,000 \times 2.5)\ \text{Nm}$$
$$= 2\,500\ \text{J}$$
$$\therefore \text{Power consumed at tool point} = \frac{\text{work done}}{\text{time taken}}$$
$$= \frac{2\,500\ \text{J}}{5\ \text{s}}$$
$$\underline{\text{Power consumed} = 500\ \text{W}}$$
$$\text{Efficiency} = \frac{\text{output}}{\text{input}}$$
$$\text{i.e. } 0.6 = \frac{500\ \text{W}}{\text{input}}$$
$$\therefore \text{Power input} = \frac{500}{0.6}$$
$$= 833\ \text{W}$$
$$\text{But } 1\ \text{kW} = 1\,000\ \text{W}$$
$$\therefore \text{Power input} = \frac{833}{1\,000}$$
$$\underline{\text{Power input} = 0.833\ \text{kW}}$$

Self-assessment questions

21 Define

(*a*) work,

(*b*) energy,

(*c*) power.

State the units in which power may be measured.

22 A mass of 500 kg is lifted vertically at a rate of 2 m/s. Calculate the power developed in kilowatts.

23 A pumping engine is to raise 100 kg of water per minute through a height of 100 m. If the efficiency of the system is 50%, calculate the power input in kilowatts. Take g as 9.81 m/s^2.

After reading the following material, the reader shall:

4 Understands the concepts 'heat' and 'temperature'.
4.1 Define heat.
4.2 Define temperature.
4.3 Describe absolute zero of temperature.
4.4 Differentiate between heat and temperature.
4.5 Define the fixed points on the Celsius scale.

In the late eighteenth century the leading natural philosophers (people who would nowadays probably be called scientists) were completely puzzled by the problems associated with 'heat'. This was a period when the western world was in a ferment; all kinds of ideas – political, economic, scientific – were discussed at length. But the men following the intellectual giants such as Newton and Leibnitz found great difficulty in explaining a 'something' that could be produced by combustion (of wood or coal), by friction (found whenever one surface slides over another), by the rays from the sun, from working metal into shape (e.g. by hammering) or by rapidly reducing the volume of a gas in a cylinder. Associated with the question 'what is heat?' was another problem – how was heat transferred from one body to another?

Observations over a period of time show that some of the more common effects that heat has on bodies are:

(i) To make them glow when heated strongly enough. Iron for instance, first glows a dull red. At about 900 °C it becomes bright red-hot, at about 1 200 °C it is yellow-hot and above 1 500 °C the iron glows white-hot. (Highly experienced craftsmen can tell by the colour of a metal what working processes are suitable at a particular instant.)

(ii) To cause expansion. The amount by which a body changes size depends upon the original size, the change in temperature that occurs, and the material of which the body is made.

(iii) To make them change their state, e.g. to turn ice into water, or to turn a liquid into a gas. (Many recent books call these 'changes of phase'.)

(iv) To cause changes in the properties of a material. For instance, the resistance offered to the flow of electricity in a given conductor varies with the temperature.

(v) To bring about chemical changes. This particular effect of heat has been of great importance in the development of those polymers called 'thermosetting'.

The idea that heat is a kind of motion of a body's particles has a long history. The ancient Greeks noted the extreme mobility of flames, and associated heat with movement. The growing body of scientists early in the nineteenth century worked hard on the problems of heat, but it was not until about 1850 that the work of the German scientist Clausius finally showed conclusively that heat is a form of energy.

A hot body (such as a mass of highly compressed gas in a diesel engine cylinder) is a store of heat energy, and some of this energy can be converted into mechanical work when the mass is cooled under controlled circumstances.

The higher the temperature of a body, the more heat it possesses. But it cannot be said that one body possesses more heat than another body just because its temperature is higher. A cup of hot tea is plainly at a higher temperature than a bucket of cool water; yet the larger quantity of cool water will certainly melt more ice than the small volume of tea. Despite its higher temperature, the tea possesses less thermal energy than the cool water.

Solutions to self-assessment questions

21 (*a*) Work is done when a force applied to a body causes it to move.

(*b*) Energy is that property of a system which is reduced when work is done.

(*c*) Power is the rate at which work is preformed. Power may be measured in mW, W, kW, MW.

22

$$\text{Force needed to lift mass} = mg$$
$$= 500 \times 9.81 \text{ N}$$
$$\text{Work done on mass during 2 m lift} = (500 \times 9.81) \times 2$$
$$= 9\,810 \text{ Nm}$$
$$= 9\,810 \text{ J}$$

But this is the work done per second

$$\therefore \text{Power developed} = 9\,810 \text{ W}$$
$$\underline{\text{Power developed} = 9.81 \text{ kW}}$$

23 The potential energy of the water is increased by being raised. Increase in potential energy of 1 000 kg of water raised by

$$100 \text{ m} = (1\,000 \times 9.81) \text{ N} \times 100 \text{ m}$$
$$= 981 \times 10^3 \text{ Nm}$$
$$= 981 \times 10^3 \text{ J}$$

This is the energy which is usefully employed i.e. it is the output of the system.

$$\text{Now, efficiency} = \frac{\text{energy output}}{\text{energy input}}$$
$$\text{i.e. } 0.5 = \frac{981 \times 10^3 \text{ W}}{\text{input}}$$
$$\therefore \text{Energy input} = \frac{981 \times 10^3}{0.5}$$
$$= 1\,962 \times 10^3 \text{ J}$$

This energy input takes place in one minute (i.e. the time to raise the water 100 m).

$$\text{Energy input per second} = \frac{1\,962 \times 10^3}{60}$$
$$= 32.7 \times 10^3 \text{ W}$$
$$\underline{\therefore \text{Power input} = 32.7 \text{ kW}}$$

Force is a physical quantity that can be readily recognized by the senses. The effects of forces can be observed – they cause bodies to move, stretch materials etc. Temperature, too, is a physical quantity that appeals to the senses. Hot material is uncomfortable to the touch, and radiant heat from a glowing metal ingot must be guarded against. Essential to the idea of temperature is the fact that heat flows from a hotter body to a cooler body.

A review of the ideas put forward can provide a working definition of 'heat': *Heat is energy which is in the process of transfer between a body or system and its surroundings. The transfer is brought about by temperature differences.* The word system is included, for heat is sometimes transferred from a number of parts e.g. the parts of a petrol engine.

Although temperature differences are readily detected by the senses, the sense of touch is a most unreliable guide to the temperature of a body. To measure temperatures an instrument is needed – a thermometer. In the past temperatures have been measured on a variety of scales, but in those parts of the world where SI is employed, the unit of temperature is now the kelvin (K). It should be noted that a temperature change of one kelvin is exactly the same as a temperature change of one degree celsius.

To *measure* temperature changes some reference point has to be chosen, and the temperature of melting ice has frequently been used.

The temperature of a body can be thought of as the degree of hotness relative to some chosen point. The temperature does not depend on the size or physical nature of the body.

Some of the difficulties associated with the ideas of heat and temperature are eased when the molecules making up a body are considered. The molecules in a gas are widely spread and can move freely. In a liquid the molecules are tighter packed, but still have much freedom of motion. In solids the molecules are more completely fixed; they vibrate continually about an average position, but are packed close together in comparison with those in gases and liquids.

Suppose the temperature of a body were reduced to such a point that all the molecular motion ceased. At this point the kinetic energy of the molecules would be zero, i.e. the body would possess no heat energy. This is the lowest temperature that can even be imagined; it is called *absolute zero* and has the value 0 K (note it is not written 0 °K). Absolute zero is 273.15 K below the temperature of melting ice.

In experimental work scientists have produced temperatures very close to absolute zero. Some odd effects have been noted when materials are at extremely low temperatures. For instance, the resistance to current flow almost disappears in some materials, and a current circulates almost indefinitely without aid from a battery or other power source. Some fluids (such as liquid helium) cannot be contained in an open vessel; the super-cold fluid creeps up the sides and escapes over the top!

Although there is an absolute minimum temperature there does not seem to be a maximum temperature.

On earth, air temperatures of 75 °C (348 K) have been recorded in deserts. The melting temperature of tungsten is about 3 643 K (3 370 °C); this is the highest melting point of all known metals. The sun, one of the cooler stars, has a surface temperature of 6 000 K, whilst its temperature at the centre is estimated to be 10 000 000 K. This enormous temperature is reached on earth only when fusion takes place as in a hydrogen bomb.

Heat and temperature, then, mean different things, but the two concepts are closely inter-related. To gather together the points from the previous descriptions, temperature is a term used to indicate the hotness or coldness of a body. The greater the heat in a body, the higher its temperature is said to be. Heat will only flow between bodies when there is a temperature difference, and heat cannot flow from a body at a lower temperature to one at a higher temperature.

An instrument used for the measurement of temperature is called a thermometer; there are many different types, used in widely varying situations. Probably the most familiar type is the mercury-in-glass thermometer. This instrument works on the principle that mercury expands very steadily as it is heated.

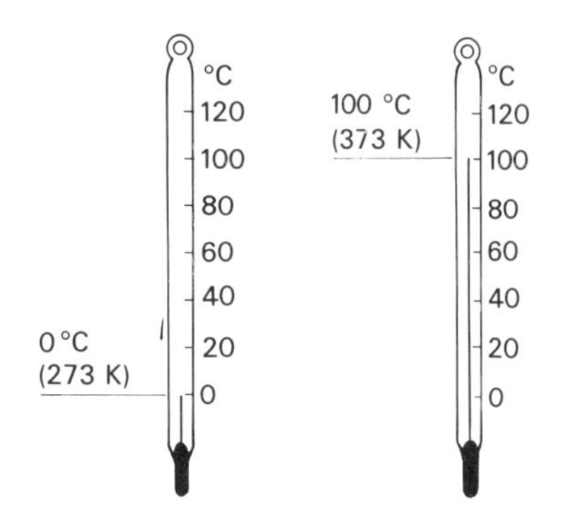

Figure 47 *Temperature scales*

The common thermometer consists of a glass tube of small uniform bore, with a bulb of mercury at the lower end. If this bulb is immersed in pure ice when it is melting, the mercury column will stay at one height. On the Celsius scale this point is labelled 0 °C (see Figure 47). The instrument is then placed in pure boiling water, and a mark made at the new height of the mercury column. This mark is labelled 100 °C (see Figure 47). The distance between the two marks is then evenly divided, normally into intervals that are easily read.

Mercury is a good material to use since it does not stick to the glass and cause inaccurate readings. Although it is a metal, it remains a liquid over a wide range of temperature. Mercury-in-glass thermometers cannot, however, be used in very cold circumstances as mercury freezes at about 235 K (−38 °C).

After reading the following material, the reader shall:

5 Know the meaning of specific heat capacity.
5.1 Define specific heat capacity.
5.2 Solve problems associated with mass, specific heat capacity, and temperature change.

Designers frequently face problems that involve calculations of heat quantity. Consider the following. An experimental electric furnace is rated at 4 kW, and can be assumed to have an efficiency of 0.75. The furnace is to heat a 1 kg mass of aluminium to a temperature of 1 000 °C. Assume that the initial temperature of the aluminium is 20 °C. How many minutes does the metal take to reach the desired temperature?

Clearly the furnace designer must take into account the particular material which is to be heated. Different substances require different amounts of heat to raise the temperature of unit mass by one degree. *The quantity of heat needed to raise the temperature of 1 kg of a substance by 1 K (i.e. 1 °C) is called the specific heat capacity of that substance.* The specific heat capacity of any material has to be found by experiment; it cannot be calculated by theoretical considerations.

The units of specific heat capacity are joules per kilogram per kelvin. This is written, in SI, J/kgK. As one kelvin is equal to one degree celcius, it makes no difference to calculations if the units of specific heat capacity are given as J/kg°C. Many books give values in J/kg°C, and the values in the table below are listed in both SI and celsius units.

substance	J/kg K	J/kg °C
lead	130	130
mercury	140	140
copper	390	390
iron	440	440
aluminium	900	900
magnesium	1 050	1 050
turpentine	1 800	1 800
paraffin oil	2 100	2 100
water	4 200	4 200

Experimental results give the following values of specific heat capacity for a few materials. In fact, the specific heat capacity does not remain constant, but varies somewhat with temperature. These variations are relatively slight, and need not concern the reader at this stage.

The very high figure for water will be given consideration later.

The specific heat capacity is the heat required to heat unit mass through one degree. It follows that if the specific heat capacity of a substance is c J/kg°C, then the heat needed to raise the temperature of m kilograms of the material by t degrees is mct joules.

To return to the problem about the furnace. The value of the specific heat capacity of aluminium can now be used.

Useful energy available to heat the aluminium = 0.75 of 4 kW
= 3 kW
∴ Useful energy = 3 000 J/s

Consider what happens in one second in the furnace.
Heat given to metal $= mct$ where t is temperature rise per second.

$$\text{i.e. } 3\,000 = 1 \times 900 \times t$$

$$\text{i.e. } t = \tfrac{3\,000}{900} = \tfrac{10}{3} \text{ degrees/second}$$

$$\text{Temperature rise/minute} = \tfrac{10}{3} \times 60 = 200\ ^\circ\text{C}$$

$$\text{Temperature rise needed} = 1\,000 - 20 = 980\ ^\circ\text{C}$$

Since the temperature rises by 200 °C per minute, the time taken to rise by 980 °C is $\frac{980}{200} = 4.9$ minutes.

Simpler problems are set for students; this problem has been detailed in order that the reader can see that physical science has many extensions into the real world of science and technology.

Example 11
Calculate the amount of heat needed to raise the temperature of 2 kg of copper by 20 °C.

$$\text{Heat needed} = mct$$

$$\text{From table, } c = 390 \text{ J/kg}^\circ\text{C}$$

$$\therefore \text{Heat needed} = 2 \times 390 \times 20$$

$$\text{Heat needed} = 15\,600 \text{ J}$$

Example 12
How many kilograms of water can be raised from 15 °C to 35 °C by the absorption of 840 kJ?
The value of c, from the table, is 4 200 J/kgK

$$\text{Increase in temperature} = 35 - 15 = 20\ ^\circ\text{C} = 20 \text{ K}$$

$$\text{Heat absorbed} = 840 \text{ kJ} = 840\,000 \text{ J}$$

$$\text{Since heat absorbed} = mct$$

$$840\,000 = m \times 4\,200 \times 20$$

$$\text{i.e. } m = 10 \text{ kg}$$

It will be noted from the table that the specific heat capacity of water is by far the highest listed. There are very few substances that require so much heat to bring about a particular temperature change.

This has many practical implications; for example the water cooling system of motor cars would have to be much more bulky if the specific heat capacity were less.

Water has properties very different from most substances. These peculiarities have a great influence on life on this planet; details of the properties of water will be given in following parts of this topic area.

After reading the following material, the reader shall:

6 Know the meaning of the terms 'sensible heat' and 'latent heat'.
6.1 Define sensible heat.
6.2 Define specific latent heat of fusion.
6.3 Define specific latent heat of vaporization.
6.4 Define specific latent heat.
6.5 Differentiate between sensible heat and latent heat.
6.6 Solve simple problems related to specific latent heat.

Changes of state (or phase)

All gases can be changed into liquids, and all liquids changed into solid form if the temperature is reduced sufficiently. Many substances can exist as a solid, as a liquid or as a gas. These are considered to be different *states* of matter, rather than different kinds. The transition from one state to another is called a *change of state*. (Some books call this a 'change of phase'; in this book the expression 'change of state' is retained.)

Consider what happens when ice below 0 °C (273 K) is heated at a constant rate. The ice warms up, then turns into water, which in turn warms up, and turns into steam. This steam can be heated up if it is contained in a vessel and the constant heat supply maintained.

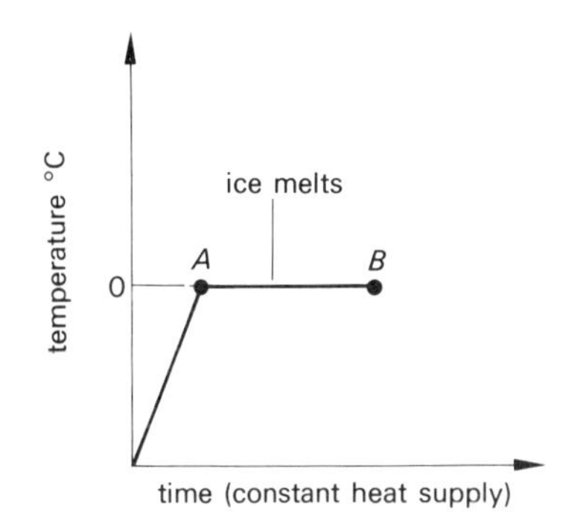

Figure 48 *Temperature–time: ice to water*

Consider these events in turn. Starting with a dish of chipped ice at a temperature below freezing point, the supply of heat causes a rise in temperature. This rise continues steadily until the temperature 0 °C is reached. Then, although the heat supply is kept steady, the temperature rise ceases as the ice begins to melt. If the ice–water mixture is kept thoroughly mixed, no temperature change occurs until all the ice has disappeared. This is represented in Figure 48.

When a solid (such as ice) is heated the molecules vibrate more vigorously, but their average position remains unchanged. During the change of state from solid to liquid (i.e. during melting) the temperature remains constant. The heat energy absorbed during this period causes the molecular vibrations to become so furious that the molecules break away from their fixed positions, and move about much more freely. When a substance is solid, the energy of the molecules is mostly vibrational. When the solid melts, the molecules of the liquid are moving rapidly, and their energy is mostly kinetic.

Maintaining the constant supply of heat next causes the water to rise in temperature until a temperature of 100 °C (373 K) is reached. Bubbles of vapour form all through the water at 100 °C, and the liquid is described as boiling. The temperature remains constant

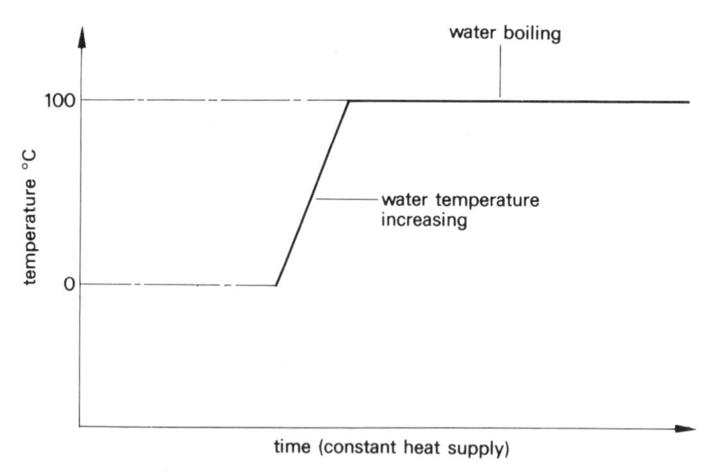

Figure 49 *Temperature–time: water to steam*

until all the water has been turned into vapour. This is shown in Figure 49.

On heating a liquid, the kinetic energy of the molecules is increased. Finally the liquid turns into a vapour when the energy of the molecules is so great that they break all bonds between them and rush about at great speed in a completely random fashion.

If the water vapour – steam – is heated in a vessel its temperature rises indefinitely. There is no subsequent change of state. Steam heated in this manner is called superheated steam; it is the type of steam generated in power stations.

The series of processes producing superheated steam from the original ice is shown in Figure 50.

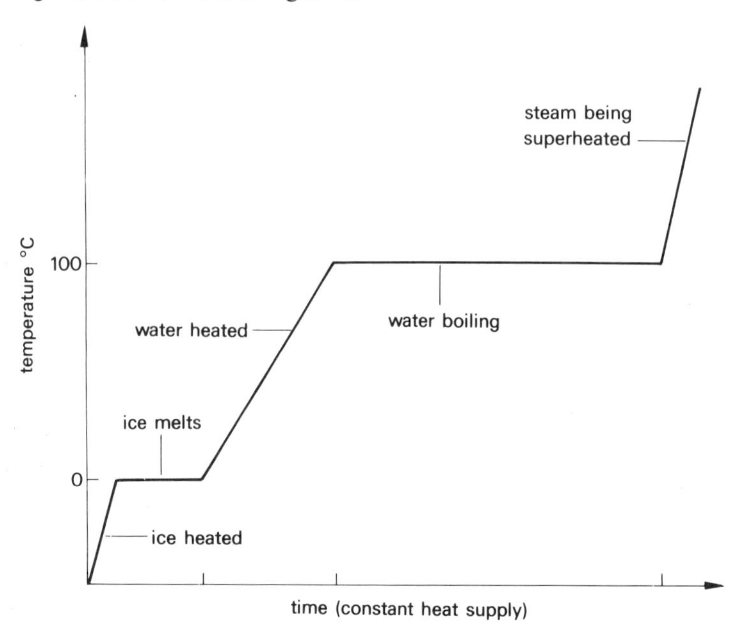

Figure 50 *Temperature–time: conversion of ice to water and then to steam*

If the changes of state are reversed, then heat is given out. The condensation of steam to water liberates large amounts of heat, and freezing of water also releases heat.

Changes of state for a number of other substances follow similar patterns to that described for ice–water–steam. Most substances have definite temperatures at which they melt and boil.

Summary

(i) When a substance is in the process of changing its state (e.g. from solid to liquid) the heat energy input does not cause any change in temperature. Similarly, when a substance is in the process of changing from gas to liquid, the heat energy output does not cause any temperature change. The heat energy – referred to as 'latent heat' – has, in these cases been used to bring about a new arrangement of the molecules.

(ii) When a substance absorbs or rejects heat whilst remaining in one particular state (e.g. stays liquid), then the thermal energy interchange causes a change in temperature.

Sensible heat is that heat which causes a change in temperature. The word *sensible* is used since the rise in temperature can be noted by the senses (e.g. by touch or by observing a change in the reading of a thermometer).

On Figure 48 there is no change in temperature during part *AB*; there the solid is turning into a liquid, or if heat is being rejected, a liquid is turning into a solid. The heat absorbed or given out during such a change is called the *latent heat of fusion.* This quantity can only be determined by experimental means. For example, if 1 kg of ice at 0 °C is melted into water at 0 °C, the latent heat absorbed has been found to be 335 kJ. The *specific* latent heat of fusion of ice is thus 335 kJ/kg. (Note: The word 'specific' refers to 1 kg of a substance.)

Example 13
A particular aluminium alloy changes from solid to liquid at 660 °C; the specific latent heat of fusion is 410 kJ/kg. The specific heat capacity of aluminium is 0.9 kJ/kgK.

How much energy, in kJ, is needed to melt a casting of mass 10 kg, if the casting has a temperature of 10 °C when it is fed into the furnace?

Energy needed = energy to raise the metal up to the melting temperature + energy to convert the solid metal into liquid.

Energy to raise the metal up to 660 °C = mass × specific heat capacity × temperature rise

= 10 kg × 0.9 kJ/kgK × (660 − 10) K

= 5 850 kJ

Energy to melt 10 kg of metal = mass × specific latent heat of fusion

= 10 kg × 410 kJ/kg

= 4 100 kJ

∴ Energy needed = (5 850 + 4 100) kJ

Energy needed = 9 950 kJ

During a change of state from a liquid to a vapour, the heat energy performs two tasks: (i) it overcomes the forces which keep the molecules fairly close together; (ii) it exerts a force on the atmosphere surrounding the new vapour, and pushes back the atmosphere to allow the new vapour to take its place. To appreciate the considerable change in volume which occurs on the evaporation of water, it may be noted that the volume of steam produced (at normal temperature and pressure) is about 1 700 times as great as the volume of water from which it was formed.

The specific latent heat of vaporization of a substance is the amount of heat needed to change 1 kg of a substance from liquid to vapour, the temperature remaining constant.

For example at standard atmospheric pressure, 2 250 kJ are needed to turn 1 kg of water at 373 K into steam at the same temperature and pressure. This is a very large amount of energy, and it gives an idea of why a scald caused by steam is so damaging. When a person is accidentally brought into contact with steam, this condenses and gives out very large amounts of energy which heat the person, causing scalding.

Example 14

(i) Find the amount of energy in joules needed to raise the temperature of 5 kg of water from 50 °C to 100 °C, the pressure being standard atmospheric.

(ii) Find the additional energy needed to convert the water at 100 °C into steam at 100 °C.

From previous table, specific heat capacity of water is 4 200 J/kgK.

(i) Heat needed to raise temperature of water from 50 °C to 100 °C = mass × specific heat capacity × temperature rise

= 5 kg × 4 200 J/kgK × (100 − 50) K

= 1 050 000 J

Energy needed = 1.05×10^6 J

(ii) As given previously, specific latent heat of evaporation of water is 2 250 kJ.

Heat energy needed to convert 5 kg of water into steam = mass × specific latent heat of evaporation

$= 5 \times 2\,250 \times 10^3$ J

$= 11\,250 \times 10^3$ J

Energy needed $= 11.25 \times 10^6$ J

Self-assessment questions

24 Write explanatory notes on the following:

(*a*) Large quantities of water placed in an unheated greenhouse help to prevent frost from damaging the plants.

(*b*) A steam burn is usually much worse than a burn caused by hot water.

(*c*) A sponge pudding is much less likely to burn the mouth than a jam pudding at the same temperature.

25 Define 'heat'.

26 Define 'temperature'.

27 Describe absolute zero of temperature.

28 What is meant by a change of state? Are these changes chemical or physical?

29 Select the most appropriate answer.

Heat is most closely related to:

(*a*) temperature

(*b*) energy

(*c*) chemical changes.

30 Complete the following statements

(i) The amount of heat needed to change 1 kg of a substance from liquid to vapour at constant temperature is called ______

(ii) The amount of heat needed to change 1 kg of a substance from solid to liquid is called ______

(iii) Sensible heat is that heat which ______

31 Select the most appropriate answer.

In order for a liquid to be converted into a vapour,

(*a*) heat must be absorbed.

(*b*) heat must be given out.

(*c*) the temperature must rise.

(*d*) the temperature must fall.

32 How much heat in megajoules is needed to heat 10 kg of water through a temperature rise of 50 °C? Take the specific heat capacity of water to be 4 200 J/kgK.

33 Calculate the heat energy required in MJ to melt a complete cupola charge of cast iron given the following details:
Mass of cupola charge 1 000 kg.
Initial temperature of cast iron 15 °C.
Melting temperature of cast iron 1 200 °C.
Specific heat capacity of the iron 0.5 kJ/kgK.
Specific latent heat of fusion of the iron 100 kJ/kg.

After reading the following material, the reader shall:

7 Describe the three methods of heat transfer.
7.1 Define
(*a*) conduction
(*b*) convection
(*c*) radiation.
7.2 Describe conduction.
7.3 Describe convection.
7.4 Describe radiation.

To a visitor from Space, our houses and factories would appear to be 'controlled environment boxes'. In these places people shield themselves from the changes in the air around them, and try to create comfortable places in which they can carry on with their activities.

When atmospheric conditions bring rain or snow, the walls and roofs keep the interior relatively dry. When air currents started by the heat from the sun cause winds, then the walls of the boxes create islands of still air in which work and living continue without disturbance. When it is very cold outside, the heat can be turned on to maintain a tolerable temperature; and if in summer the air outside becomes very hot, air-conditioning units can be used to reduce the temperature and humidity inside the box.

We are all familiar with the fact that heat energy can be transferred from one place to another. When a pan is placed on a lighted gas cooker, heat energy from the burning fuel passes through the base of the pan to the liquid inside. Air heated by a gas fire sets up warm currents of air. Sunbathers enjoying the warmth from the sun's rays know that heat is being passed to them. These three examples illustrate the three methods by which heat energy can be transferred from one situation to another. Householders, factory owners, heating engineers are among the many people who need an understanding of the methods of heat transfer. By using heat energy with understanding, it is possible to make efficient and economical use of vital, diminishing fuel resources.

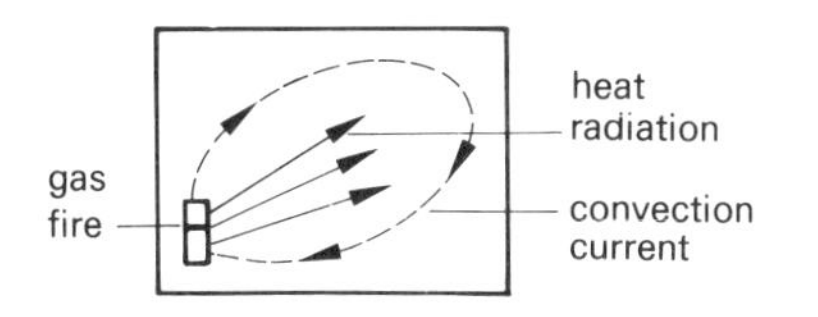

Figure 51 *Transmission of heat energy*

Each kind of energy has its own particular methods of transmission. Thermal energy is transmitted by conduction, by convection and by radiation.

Figure 51 shows diagrammatically a gas fire burning in a room. Heat is produced by the combustion of the gas, and parts of the fire glow with heat. Most of this heat radiates away from the fire as infra-red rays. The rays transfer heat on to whatever objects they happen to strike. Also, the air near the gas flames becomes heated, expands and rises towards the ceiling. Cooler, denser air flows in to take its place; this air is in turn heated, and a convection current is set up. Some of the heat released during the burning of the gas causes the body of the fire to warm up. Although the frame of the fire is normally made of a ceramic material, heat is conducted to all parts of the gas fire. Frequently the frame of the fire is too hot to touch with comfort.

The heat from the burning fuel has been transmitted by radiation, convection and conduction.

Conduction

Conduction is the flow of heat through a body, heat being transferred from the parts at higher temperatures to the parts at lower temperatures.

This method of heat transfer is normally associated with solids. Molecules in solids are packed tighter than those in liquids or gases. Each molecule in a solid vibrates about a fixed position. When heat energy is supplied to one part of the body, then the neighbouring molecules vibrate more vigorously. These larger vibrations transfer to near-by molecules, and so heat is transferred from point to point.

It should be noted that metals are normally good conductors of both heat and electricity. This is related to the fact that any metal in the solid state contains large numbers of 'free electrons'. An atom consists of a nucleus and a number of extremely tiny particles called 'electrons', the electrons whirling about the central nucleus in orbits which are normally called 'shells'. Metals obtain their 'metal character' because they have one or more 'loose' electrons in their outer shell. Heat is conducted easily in metals because when heat energy is supplied, the loose, speeded-up electrons can readily move from one atom to strike a neighbouring atom. This motion (heat) is passed from one part of the body to another.

The ease with which metals conduct heat is illustrated in the following example. If, on a cold day, a fitter picks up a screwdriver with his hand touching both the wooden handle and the metal blade, then the metal feels much colder. The metal, a much better heat conductor than the wood, conducts the heat away from his hand much more quickly.

Generally speaking, metals are good conductors of heat. The fairly common metals, copper and aluminium are excellent conductors, and because of this are used extensively where heat transfer is required. Liquids (excluding mercury) are normally poor conductors, and gases are extremely poor conductors of heat. (Copper for instance, conducts heat 16×10^3 times more readily than air.) Very

Solutions to self-assessment questions

24 (*a*) The latent heat of fusion of water is high, and the large quantity of water absorbs much heat before the temperature in the greenhouse begins to fall.

(*b*) When steam condenses on a person, it gives out its latent heat of evaporation. When hot water cools on a person, it gives out sensible heat. The latent heat of evaporation is much higher, and in consequence a steam burn is likely to be much more serious

(*c*) Jam contains a high proportion of water. Since the specific heat capacity of water is higher than the specific heat capacity of the sponge pudding material, the sponge pudding is less likely to burn the mouth.

25, 26, 27 The reader is referred to the details in the text.

28 Changes of state are changes between solid and liquid, and between liquid and vapour.

The changes are physical, since the substance does not change its chemical composition.

29 (*b*).

30 (i) The amount of heat needed to change 1 kg of a substance from liquid to vapour at constant temperature is called *the specific latent heat of vaporization.*

(ii) The amount of heat needed to change 1 kg of a substance from solid to liquid is called *the specific latent heat of fusion.*

(iii) Sensible heat is that heat which *causes a change in temperature.*

31 (*a*).

32 Heat needed = mass × specific heat capacity × temperature rise

$= 10\text{ kg} \times 4\,200\text{ J/kgK} \times 50\text{ K}$

$= 2\,100\,000\text{ J}$

Heat needed = 2.1 MJ

33 Energy needed = energy to raise metal up to the melting temperature + energy to melt the cast iron into a liquid

Energy to raise the metal up to the melting temperature

= mass × specific heat capacity × temperature rise

$= 1\,000\text{ kg} \times 0.5\text{ kJ/kgK} \times (1\,200 - 15)\text{ K}$

$= 1\,000 \times 0.5 \times 1\,185$

$= 592.5 \times 10^3\text{ kJ}$

= 592.5 MJ

Energy to melt 1 000 kg of cast iron

= mass × specific latent heat of fusion

$= 1\,000\text{ kg} \times 100\text{ kJ/kg}$

$= 100 \times 10^3\text{ kJ}$

= 100 MJ

∴ Energy needed = (592.5 + 100) MJ

Energy needed = 692.5 MJ

poor conductors are called insulators. Many porous substances (e.g. asbestos, firebrick, cellular blankets) depend primarily for their insulating properties on the air which is trapped in the spaces.

Water and ice are both very poor conductors of heat. Because ice is less dense than water, any ice that forms during winter floats on the top of the water (e.g. on a pond or lake). The water thus freezes from the surface downwards. Since the ice crust has low conductivity, the thickness of the ice does not increase quickly. It is unusual for lakes in this country to freeze solid, and in consequence frogs and fishes can survive below the ice at such uncomfortable temperatures as 4 °C! The insulating properties of the ice are the main reason why they can survive a cold spell.

Convection

Gases and liquids are collectively referred to as fluids; they have the common property that they can flow. When fluids are heated (or cooled) they move in a continuous current; these currents are called convection currents.

For instance, convection currents over the surface of the earth carry moist warm air upwards. At the higher altitudes, where the temperature is lower, the air cools. If the air cools below its dew point then clouds form from the water droplets. Another instance is a saucepan of water heated from below; convection currents carry the heated water towards the surface whilst cooler water sinks to be heated up in its turn.

The transmission of heat energy by convection is much more rapid than conduction. The energy is transferred just as quickly as the current moves. (In conduction the energy transmission is caused by the vibration of molecules.)

Convection is the transmission of heat by the movement of a fluid, this movement being caused by temperature differences within the fluid.

Two further examples of convection currents from domestic situations will summarize the idea of convection.

In a central heating system, the 'radiators' are stood on the floors of the areas requiring heating. The heat emitted by the 'radiators' generates a convection current by heating the air and causing it to rise. 'Radiators' are not placed near the ceiling because the convection currents would not then heat the main body of the room; the warmed air would remain above the space needing heating.

In a domestic refrigerator, the cooling unit is situated near the top of the cabinet. The air immediately surrounding the cold pipes becomes heavier as it contracts, and sinks towards the floor of the refrigerator. Warmer, less dense air is pushed upwards, and in turn is cooled. A cold convection current is thus started, and the whole volume of the refrigerator is kept at a low temperature.

Radiation

So far, descriptions have outlined how heat can be transmitted by conduction and convection. Both of these methods of heat transfer use a medium – a solid or a fluid whereby the heat can be transmitted. Now the heat energy which reaches the earth from the sun cannot have arrived by either conduction or convection. The space between the sun and the earth is a virtually perfect vacuum; there are almost no molecules to collide and conduct the molecular motion called heat. Also, if space is to all intents and purposes empty, there can be no convection currents. Heat energy from the sun (and similarly most of the heat from a fire) must therefore be transmitted by some method other than conduction or convection. This mode of heat transfer is called radiation.

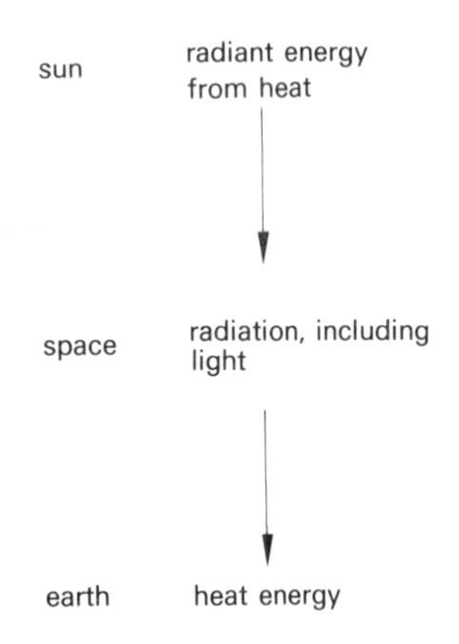

Figure 52 *Heat transmission from the sun to the earth*

A great number of experiments have been carried out on radiation; it is now generally believed that radiant heat energy is transmitted as electromagnetic waves. Radiant heat waves are an example of the transmission of electrical energy, the waves travelling at a velocity equal to 300×10^6 m/s.

On the sun, the huge amounts of heat energy released during nuclear changes are changed into radiant energy (of which light is one form). This energy takes about 500 seconds to cross the 150×10^3 kilometres between the earth and the sun. The process is shown in Figure 52.

From one point of view, radiation is by far the most important method of heat transmission, for the energy from the sun is all-important to life on earth. This tremendous flow of energy warms the earth and produces, through photosynthesis, all the food, fuel and oxygen upon which life depends. If all the energy from the sun reaching the earth in just fifteen minutes could be captured and used, there would be no need for other fuel resources for a whole year!

Man has tried for many centuries to make direct use of the sun's radiant energy, but with only little success. For more than a hundred years, efforts have been made to use solar energy to produce drinking water from sea water. In sunny Israel about 1.5% of the country's energy consumption is via solar panels which absorb the radiant heat energy from the sun. One ingenious invention at Tel Aviv University is a solar-powered car; its special batteries are partially powered from solar panels on the roof of the vehicle.

Modern man continues to strive to fulfil the dreams of his ancestors; perhaps necessity will force us to make better use of the radiant heat energy available.

Radiation is the transfer of heat energy from a hot body to a cooler body by means of electromagnetic waves, the space between the bodies being little affected by the heat transfer.

After reading the following material, the reader shall:

7.5 Give an example of heat transfer by each of these processes (conduction, convection and radiation).

7.6 Give examples of where insulators are used to restrict the flow of heat.

7.7 Discuss the use of insulation to conserve fuel in a heating installation.

Furnaces in factories make use of the ideas of conduction and convection. A draught of air is blown through the flames of the burning fuel. The hot gases produced by combustion circulate round the metal pipes that contain the liquid which is to be heated, and heat is conducted through the metal to the liquid inside.

Then the gases – still hot – pass through a chimney into the atmosphere outside. The flow of the exhaust gases is maintained upward from the chimney by convection currents being set up. The hot exhaust gases are less dense than the air at the top of the chimney, and thus continue to rise.

The railway authorities have attempted to use heat radiation to detect the axles of railway wagons which are overheating through lack of lubrication. This work is still proceeding, and could have considerable influence on the ways in which rolling stock is maintained.

Insulation

There are numerous materials through which heat passes very slowly. In fact, there are many more poor conductors than good conductors. Poor conductors, called insulators, include rubber and asbestos. Wool and asbestos are insulators mostly because they include air. Air is a poor conductor because, like all gases, its molecules are widely spaced, and so vibrations are not easily passed on.

The most effective insulating materials are those of a 'fluffy' type like glass wool. The glass itself is a poor conductor, and the air trapped between the fibres is even less able to transmit heat. The air is trapped in 'dead air spaces' and cannot circulate; if it could, convection currents would convey the heat much more rapidly from point to point.

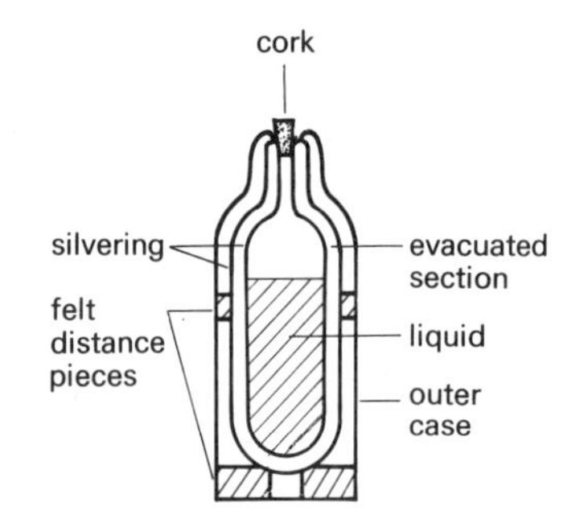

Figure 53 *Thermos flask*

The thermos flask is a familiar object. It is designed to store either cold or hot liquids, by restricting heat transmission as much as possible.

The flask (see Figure 53) consists of a double-walled bottle with a vacuum space between the walls. There is little transfer of heat by conduction because the evacuated space has few molecules to transmit the heat. Clearly convection currents cannot be very strong in a space from which nearly all the air has been extracted. Radiation is reduced to a minimum by silvering the inside surfaces as indicated in the diagram.

This type of flask was first designed in 1892 by James Dewar; it was originally used for storing liquid oxygen at very low temperatures.

Heat insulation is used in many situations; one primary concern is the saving of energy and thus cost by householders and industrialists. Many modern houses have double-glazed windows. These consist of two sheets of glass, with a stagnant volume of air between them. The heat loss from the windows is much less than it would be with a single sheet and in consequence it is less costly to keep the rooms warm. In a great number of factories steam is used in manufacturing processes. The steam pipes are lagged with insulating material (such as asbestos) so that the steam stays in the condition required, and also so that the cost of producing steam is kept down.

Self-assessment questions

34 Two identical houses, side by side, are observed during a cold spell in the winter. House A has snow on the roof. House B has no snow on the roof. In which house are the occupants more likely to be comfortable and also pay lower fuel bills?

35 Give an example of heat transfer by conduction.

36 Give an example of heat transfer by convection.

37 Give an example of heat transfer by radiation.

38 Give two examples where insulation is used to restrict the flow of heat.

39 Define conduction.

40 Explain the following:

(*a*) A stopper can sometimes be freed by heating the neck of the bottle.

(*b*) The barrel of a bicycle pump becomes hot when the pump is used to inflate a tyre.

41 Tick the appropriate column.

materials	good conductor	poor conductor
gold zinc cotton iron glass wool brick oxygen		

42 Explain why a tiled floor feels much colder to the touch than a carpet in the same room.

43 Define convection.

44 The transfer of heat by conduction is a much slower process than heat transfer by convection.
TRUE/FALSE

45 Sketch a diagram of a refrigerator and draw on it the convection current inside the cabinet.

46 Explain why immersion heaters are placed low down in hot water tanks.

47 Define radiation.

48 'Radiators' in a central heating system heat near-by air by two methods of heat transfer. What are they?

49 Heat energy from the sun reaches the earth by means of electromagnetic waves.
TRUE/FALSE

After reading the following material, the reader shall:

8 State and describe the effects of heat.
8.1 State the effects of heat on the physical dimensions of solids, liquids and gases.
8.2 Give examples of
(*a*) practical applications
and (*b*) design implications of thermal movement.

Most substances expand when heated. This is true whether the substance is in the gaseous, liquid or solid state.

Solutions to self-assessment questions

34 House A is well insulated: the snow on the roof melts only slowly; heat is not easily conducted through the roof to the snow. House B is poorly insulated: heat leakage melts the snow readily. In an un-insulated house, there are likely to be places that feel uncomfortably cold.

Thus in the well-insulated house A the occupants are likely to be more comfortable, and pay smaller bills.

35, 36, 37, 38 Answers to these questions will be found in the text. Readers will be able to give many other instances.

39 Conduction is the flow of heat through a body, heat being transferred from the parts at higher temperatures to the parts at lower temperatures.

40 (*a*) On heating the neck of the bottle, some heat penetrates to the inner wall of the bottle, in spite of the glass being a poor conductor. The inside of the bottle expands slightly, and the stopper is released.

(*b*) During the stroke compressing the air inside the pump, the air molecules are 'squeezed' together, and the extra energy they now possess increases their temperature. The air, now at a higher temperature than the atmosphere outside, conducts heat through the walls of the pump barrel.

41

material	good conductor	poor conductor
gold	√	
zinc	√	
cotton		√
iron	√	
glass wool		√
brick		√
oxygen		√

42 The tiled floor and the carpet are at the same temperature. However, the carpet conducts heat much more slowly than the tiles, and in consequence the tiles feel much colder when touched.

43 Convection is the transmission of heat by the movement of a fluid, this movement being caused by temperature differences in the fluid.

44 True.

45 Figure 54 shows that the convection current draws warm air up to the cold area at the top of the cabinet.

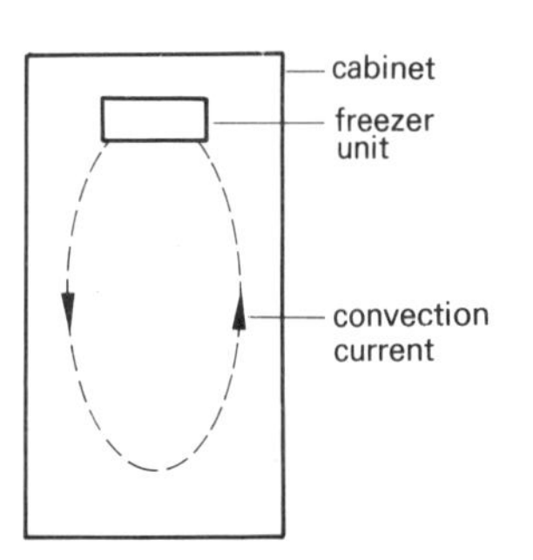

Figure 54 *Convection current in a refrigerator cabinet*

46 The water close to the immersion heater is warmed by conduction. Becoming less dense, the warmed water rises and is replaced by colder water from above. If the heater were placed at the top of the tank, the water would be heated only slowly by conduction.

47 Radiation is the transfer of heat energy from a hot body to a cooler body by means of electromagnetic waves, the space between the bodies being little affected by the heat transfer.

48 Radiation and convection.

49 True.

Gases inside the cylinders of motor car engines expand rapidly when heated by the combustion of fuels. The expansion of the liquid metal, mercury, enables fairly simple but accurate thermometers to be made. However it is when dealing with solids that engineers and designers face the majority of their problems.

The molecules of a solid are packed fairly closely together. The molecules have mutual attraction for each other, and this helps maintain the molecular pattern. When thermal energy is supplied, the molecules absorb the energy and vibrate at an increasing rate. As a result of this increased energy, the molecules are pushed a little farther apart and thus occupy a slightly greater volume. Expansion is caused by an increase in the spaces between the molecules.

Designers of motor car engines have to ensure that they allow for the different expansions of the cylinders and the pistons. Experience and calculations help them to ensure that the lubrication does not break down, and that the pistons do not 'seize-up' inside the cylinders.

Great forces are exerted when solids expand or contract. In years gone by, railway engineers laid the lines with a small gap between each length so that the lines could safely expand when the weather became hot. Different methods of allowing for line expansion have now been developed.

The forces exerted when materials change their dimensions can be put to good use. For example, steel tyres are shrunk on to the wheels of railway carriages. The tyre is machined so that its internal diameter is slightly less than the external diameter of the wheel. The tyre is then heated so that it will just fit round the wheel. When the tyre cools, it grips on to the wheel extremely hard. Generally no other means of securing the tyre is needed.

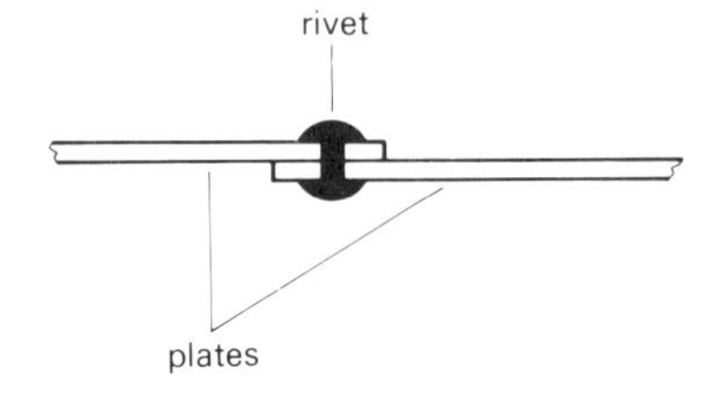

Figure 55 *Riveted plates. When the rivet cools the plates are pressed together*

Riveting is a method of joining either metals or non-metals together by a permanent form of clamping. An important application of riveting is the hot riveting of steel plates with steel rivets. This is still used in some shipbuilding and engineering activities. The rivets are heated to a red heat (judged by the craftsman) and then inserted through the holes in the plates. The end of the rivet is then hammered into a mushroom shape (see Figure 55). As the hot rivet cools, it contracts, making the joint between the plates more secure.

Measuring instruments are affected by the effects of heat on the solid parts of which they are made. Most instruments are calibrated to be accurate at a temperature of 20 °C (293 K). At all other temperatures the instruments yield slightly inaccurate readings.

It has been pointed out that water – one of the most important substances on earth – has unusual properties. It has high values for its specific heat capacity, and for both its specific latent heats. It has, however, a much more unusual property: over a small range of temperature, it actually expands as it cools. When it cools between 4 °C and 0 °C, water expands; the ice resulting occupies a greater volume than did the water at 4 °C. If the water is contained in a pipe with the ends closed (as by a tap or valve) the great forces exerted by the expanding solid can crack the pipe. Only when the temperature rises and the ice melts can the damage done to the pipes be assessed.

Water has this unusual property partly because the ice-crystal has an uncommonly open framework, riddled by a large number of channels. At low temperatures the molecular groups in water behave in ways different from most materials, and this effect, combined with the open crystal lattice, helps to explain why ice is less dense than water at 4 °C. If there is a possibility of water freezing in pipes, then engineers have to make careful allowances in their designs.

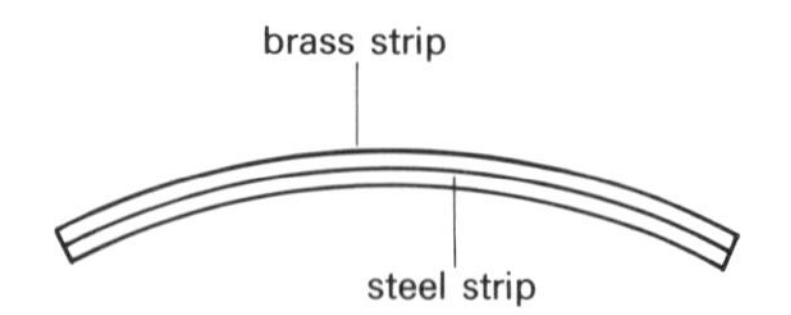

Figure 56 *Bending of a bi metal strip after heating*

Different metals expand by different amounts when subjected to the same rise in temperature. If two metal strips (say steel and brass) are riveted together to form a compound bar, then, when heat is applied, the bar bends (see Figure 56). In this case the bar takes up the shape shown because brass expands more readily than steel.

These compound bars are called 'bi-metallic strips', and they are used in many thermostats (devices for keeping a temperature steady).

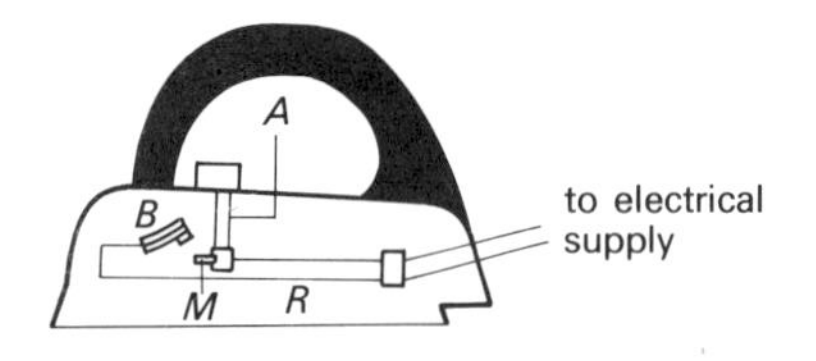

Figure 57 *Bi-metal strip in an electric iron*

Figure 57 shows the temperature control in an electric iron. The desired temperature is set by the adjusting screw A; this has been calibrated from experiments carred out beforehand. When the current is switched on, electricity flows through the resistor R and the working parts of the iron heat up. The bi-metallic strip B is in the circuit. When it is heated to the desired temperature, it curves sufficiently to open the contacts at M so the current ceases to flow. When the strip cools enough for the contact to be re-established, the current flows again. Thus the temperature of the iron is maintained close to the desired temperature.

Simple thermostats working on the same principles are used in refrigerators, central heating systems, gas cookers and hot water installations.

Thermostats for industrial and laboratory use are normally much more elaborate devices.

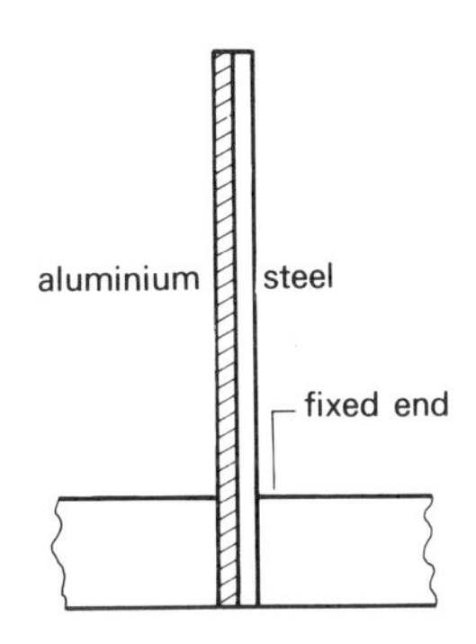

Figure 58 *Bi-metal strip*

Self-assessment questions

50 Aluminium expands more than steel when heated. When the bi-metallic strip shown in Figure 58 is heated, does it bend to the left?

51 (*a*) When a block of metal is heated the density decreases because metals expand when heated. TRUE/FALSE

(*b*) When a solid is heated it increases in volume because the molecules of a solid expand when heated. TRUE/FALSE

52 Draw a labelled diagram of a device which rings an alarm bell if the temperature in a greenhouse falls below a certain value.

53 Explain how – if at all – a rise in temperature of a flat circular steel disc affects

(i) its volume
(ii) its mass
(iii) its density
(iv) its outside diameter.

54 Give one practical application involving thermal movement of

(i) a liquid
(ii) a metal.

Try to quote applications which have not been mentioned in the text.

Solutions to self-assessment questions

50 No. The bimetal strip bends so that the top moves to the right. The extra expansion of the aluminium allows it to 'ride' over the steel.

51 (*a*) True.

The volume increases; the mass remains the same. Since density is $\frac{\text{mass}}{\text{volume}}$, the density decreases.

(*b*) False.
Expansion takes place because the spaces between the molecules increase in size.

52 The device would be similar to that shown in the electric iron.

53 (i) The volume increases when the disc is heated.
(ii) The mass remains the same.
(iii) The density decreases (see the answer to Question 51(*a*)).
(iv) The outside diameter increases.

54 Among the many possible answers that the reader may suggest, it is likely that some will be similar to the few suggestions below.

During a soldering operation the heat applied to the solid solder turns it into a liquid. The liquid is directed into those places where metal is to be joined.

Liquid iron is poured into a mould, where it sets into a casting of the shape required. As it solidifies the iron shrinks away from the sides of the mould, allowing the casting to be easily removed.

Metal spectacle frames are gently heated by warm air so that the openings increase slightly in size. The lenses are placed in position. They are firmly but gently retained by the metal frame gripping them as the metal cools and contracts.

Topic area: Waves

After reading the following material, the reader shall:

1 Describe waves and their behaviour.
1.1 List simple examples of wave motion.
1.2 Explain, using a simple diagram, the meaning of
(*a*) wavelength and
(*b*) frequency.
1.3 State the unit of frequency as the hertz.
1.4 Solve simple problems using velocity = frequency × wavelength.
1.5 Describe how waves are reflected.
1.6 Describe how waves are refracted.

Forces are what make objects move or cause work to be done. Forces are frequently derived from energy changes. For example, some of the kinetic energy of a tennis racquet is transmitted to the ball and makes it move with high velocity. Electrical energy can make an electric motor turn and perform mechanical work. If a pool of water is struck by a falling raindrop, the energy changes cause waves to spread in all directions from the point of impact.

Waves in water are obvious because they appeal to the sense of sight, and can be easily recognized as some sort of vibration. Waves which carry sound to the ear can be detected, but their form of vibration is nothing like so obvious as the ripples on a pool.

Sound waves are not alone in affecting the senses. Heat and light waves are both forms of electromagnetic waves. The nature of these has still to be fully explained, but there is no doubt that heat and light waves are at least partly vibrational in nature.

Electromagnetic waves are travelling magnetic and electrical disturbances. A broadcasting station causes powerful disturbances of this type to travel in all directions from the transmitter. These disturbances – called radio waves – can be detected in a receiving aerial wire. In the receiving wire the disturbances cause an electrical current to oscillate. Energy conversion devices change these oscillations into sound, and a radio is working.

Ultra-violet waves are invisible electromagnetic waves. They affect photographic film, they cause sun burn, and they produce vitamin D in the body. X-rays and gamma rays are also electromagnetic waves; these waves have been put to excellent use in the medical field.

Earthquakes occur at the rate of about four hundred a week. Most of these are, fortunately, only minor 'quakes, but on average one large earthquake happens each week. It is not surprising that sometimes a disaster happens. The earthquakes cause waves of different types to be transmitted through the earth. Victims of major earthquakes frequently state that the vibration of the earth is a most terrifying sensation.

Waves in water can travel at very different speeds. This can be readily observed both in the sea and in pools of water. Sound waves, however, travel through a particular medium at a velocity which depends upon the condition of the medium. Sound passes through air at sea level and standard atmospheric conditions, for instance, at about 335 m/s. Electromagnetic waves (light waves, radio waves, X-rays, infra-red waves, ultra-violet waves, gamma rays etc.) all travel at the same speed. This is about 300 000 000 m/s. (Electromagnetic waves have a velocity almost a million times that of sound in air.)

All waves exert pressure on the surface on which they impinge. In normal circumstances, the pressure created on the ear by a sound wave is very small and not easy to detect. The maximum pressure exerted on the surface of the earth by the electromagnetic radiation from the sun is about 23 N on each square kilometre.

There are really two broad categories of wave: (i) electromagnetic and (ii) dynamical.

Electromagnetic waves (e.g. light waves, radio waves) do not need a physical medium. They can pass through a vacuum; it is well known that radio signals can be detected from space. Dynamical waves (e.g. sound waves, water waves, waves in a rope) always have a material medium.

For all waves there is a simple mathematical relationship between the velocity of a wave, the frequency of the waves, and the distance between waves.

Suppose that a succession of waves is moving across a container. A cross-section would look like Figure 59.

The wavelength, λ (lambda) is the shortest distance between any two particles which are in the same phase of their oscillation. It is common to measure the wavelength between any two crests, or any two troughs, but the distance XX′ is also equal to the wavelength. The

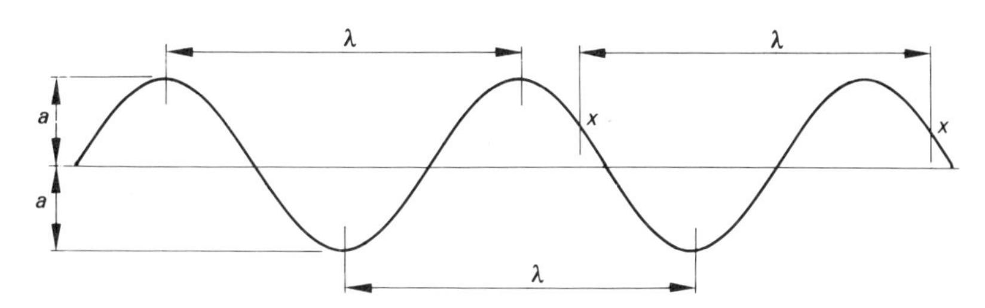

Figure 59 *Diagram showing wavelength and amplitude of a wave*

wavelength has units of length (i.e. metres). The maximum displacement of any particle from its mean position is called the amplitude. This is shown by the letter a on the diagram.

If f waves pass any point in one second, then the frequency is f waves per second. The unit of frequency equal to one complete cycle per second is called the hertz (Hz). The unit is named after the German physicist Heinrich Hertz (1857–94). He spent much of his short life working on problems of wave motion. Now if the wavelength is λ m and the frequency of the waves is f per second, then the wave must have a velocity of $f\lambda$ m/s.

As indicated on the diagram, the velocity of the wave is the speed with which its outline is travelling in the direction of the wave.

If the velocity of the wave is v, then it follows that $v = f\lambda$.

Example 1

The waves on a pool travel a distance of 0.4 m in 2 seconds. The wavelength λ is observed to be 0.04 m. Calculate

(i) The speed in m/s at which the wave front is travelling.

(ii) The frequency in hertz (Hz).

(i)
$$\text{Velocity of wave front} = \frac{\text{distance travelled}}{\text{time taken}} = \frac{0.4}{2}$$
$$\underline{\text{Velocity} = 0.2 \text{ m/s}}$$

(ii)
$$v = f\lambda$$
$$\text{i.e. } 0.2 = f \times 0.04$$
$$\therefore f = \frac{0.2}{0.04}$$
$$\underline{\text{Frequency} = 5 \text{ Hz}}$$

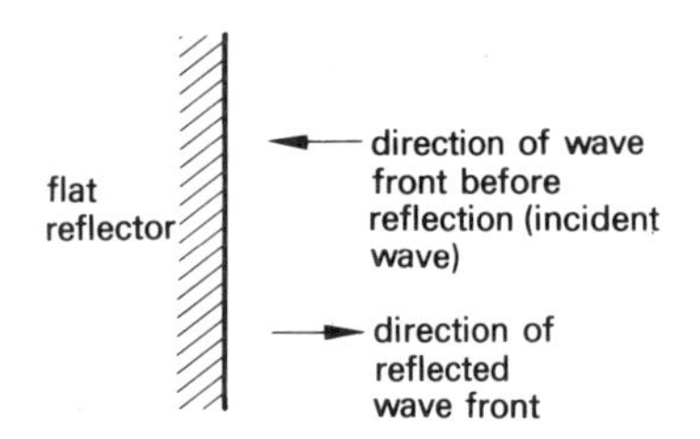

Figure 60 *Reflection at a flat surface*

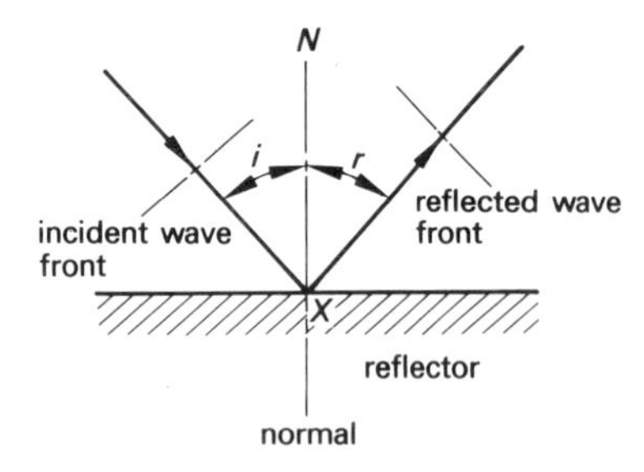

Figure 61 *Reflection of an oblique incident wave*

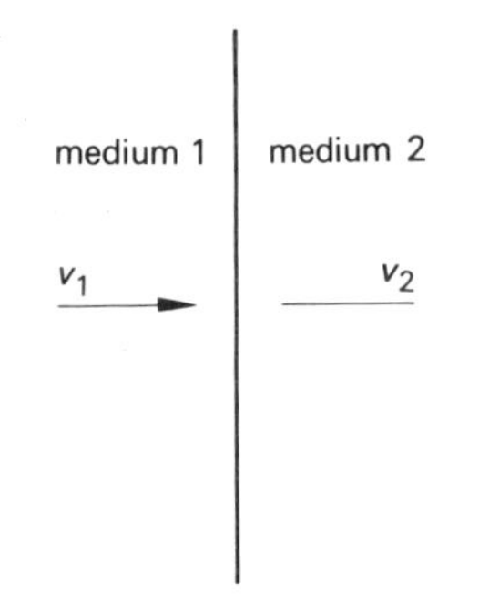

Figure 62 *The velocity of wave is altered on entering medium 2. The new velocity depends on the properties of medium 2*

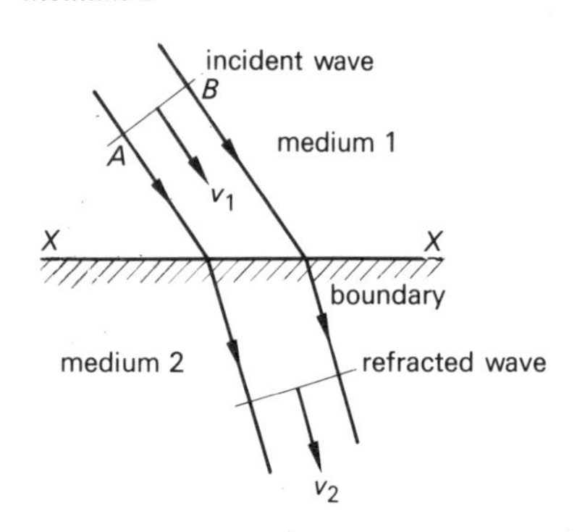

Figure 63 *Refraction at a boundary*

There are certain ways in which all waves are alike. All waves can be reflected. All waves are refracted, i.e. their paths alter when the waves pass a boundary. Other features which occur in all waves will be considered in other units.

Reflection

It is well known that mirrors reflect light. Light is one form of wave motion and it is therefore not surprising that other waves can be reflected.

If a wave front strikes a flat reflector at right angles (called 'normal' to the reflector), then the incident waves and the reflected waves move as indicated in Figure 60.

If a wave front strikes a reflecting surface obliquely, then the wave is reflected as shown in Figure 61. Observations show that the angle of incidence i is equal to the angle of reflection r. Note that line X′N is called the normal to the reflecting surface.

Figures 60 and 61 indicate what happens when waves are reflected by smooth, regular surfaces. When wave fronts strike rough surfaces, they are reflected in a 'mixed-up' fashion, and clear simple reflections are not produced.

Refraction

If a ray of light passes through air until it reaches a window and some of the light passes through the glass it is said that the light passes through medium 1 (the air), reaches a boundary (where the glass and air meet), and enters medium 2 (the glass). Figure 62 shows what happens when a wave strikes normally on the boundary, and some of the wave carries on into medium 2. The frequency of the waves is not altered when they cross the boundary. Since $v = f\lambda$, and v_1 and v_2 are different, it follows that the wavelength of the wave changes as it enters medium 2.

When a wave strikes obliquely on the boundary between two media, then that portion of the wave motion which passes into medium 2 travels in a different direction from that of the incident wave. Figure 63 shows how an incident wave *AB* strikes a boundary *XX* and travels in a new direction in medium 2. This change of direction is called *refraction*.

After reading the following material, the reader shall:

2 Know that sound is a form of wave motion.
2.1 State that sound is produced as a result of vibration.
2.2 Describe sound as a pressure wave.
2.3 State that sound has a finite velocity, the value of which depends on the medium.
2.4 Describe the reflection of sound.
2.5 Describe the refraction of sound.

The word *sound* is the name given to the sensation perceived by the ears. Sound is a form of energy in the form of mechanical vibrations. These vibrations in the air affect the ear drum; messages are passed to the brain; and there the sound is interpreted.

Sound has played a very important part in the development of mankind. For the vast majority of the time that man has existed, speech has been his main means of communication. (Writing has a short history of a few thousand years.) Because early man developed speech, he was able to pass on ideas and information in a manner superior to all other animals. The interpretation of sounds was one of the giant strides in the evolution of civilization.

Sound vibrations usually originate in the movements of parts of a solid body. The plucking of a guitar string causes vibrations, and these can readily be detected by touching the string after energy has been imparted to it. When a heavy bell is struck by its clapper, some of the kinetic energy is changed into sound energy; once again the vibrations can be detected by touch.

Sound may be defined as pressure changes in an elastic medium. These pressure changes can travel through gases, liquids, and solids.

In modern times people are frequently concerned with an excess of sound; the study of noise control has become increasingly important. Sound vibrations transmitted through solid parts of buildings (especially through elastic materials like steel) can make life very unpleasant. The ways in which sound vibrations can be 'damped' still pose many problems for designers.

Air is by far the most common medium through which sound is conducted; because air is readily set in motion, hearing is possible.

For people with normal hearing a vibration with a frequency of less than 16 Hz (i.e. 16 vibrations per second) cannot be perceived. At the other end of the scale, a sound of frequency greater than 20 000 Hz is also inaudible. Some animals, such as dogs, can hear sounds at frequencies higher than man can detect.

Vibrations having frequencies greater than 20 000 Hz are included in the science of ultrasonics. Ultrasonic vibrations carry high energy and in common with all sound waves travel in straight lines. So far, they have had some applications in medicine and in industry. In foundries, for instance, ultrasonic waves are used to check whether a casting has hidden flaws. Research will doubtless find many more applications for ultrasonic waves in the future.

Supersonics is the study of the movement of bodies at speeds greater than the speed of sound in air. Military aircraft have flown at supersonic speeds for a long time, but the development of supersonic civilian aircraft has had a chequered career. A great deal of development is needed before these aircraft can be accepted as safe, economical, and emitting acceptable noise levels.

Sound is a dynamical wave; it must have a medium to transmit it. This can be demonstrated by placing an electrical buzzer in a chamber from which the air is progressively removed. As the chamber is exhausted, the sound perceived by the ear becomes less and less, even though the buzzer can be seen to be busily working.

The transmission of sound through a material is partly due to the springiness (i.e. the elasticity) of the substance through which it travels. Sound waves are back-and-forth waves; these are called longitudinal waves.

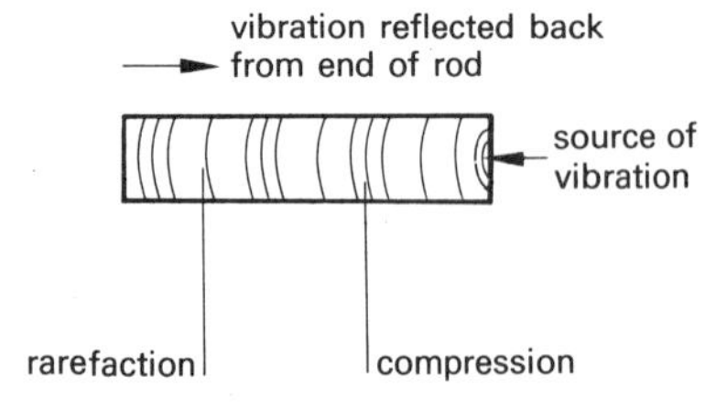

Figure 64 *Transmission of sound through a medium*

If a source of vibration is pressed against a metal rod (see Figure 64), the end of the rod is made to vibrate. This causes a series of compressions to be pulsed along the rod. These compressions travel as waves do, and the distance between areas of compression is the wavelength. Equally spaced between the compressions are the rarefactions; these are areas where the rod is stretched slightly by the vibrational energy. Similar happenings take place when air is set in wave motion. The air between the sound source and the listener does not move bodily forward; the particles of air move slightly from side to side about their average position.

Like all other waves, sound waves exert a pressure on the surfaces on which they fall. In normal circumstances, pressure changes due to sound are very small. A sound loud enough to cause pain in the ear causes a pressure change about one million times as great as the pressure change brought about by a sound that is just audible.

It always takes time for sound to travel from one point to another. Compared with many other types of waves, sound waves travel fairly slowly. In air (at standard atmospheric conditions), sound travels at about 335 m/s, in lead at 1 220 m/s, in glass at 4 880 m/s and in aluminium at 5 190 m/s. The speed of sound depends on the mechanical properties of the medium through which it is transmitted.

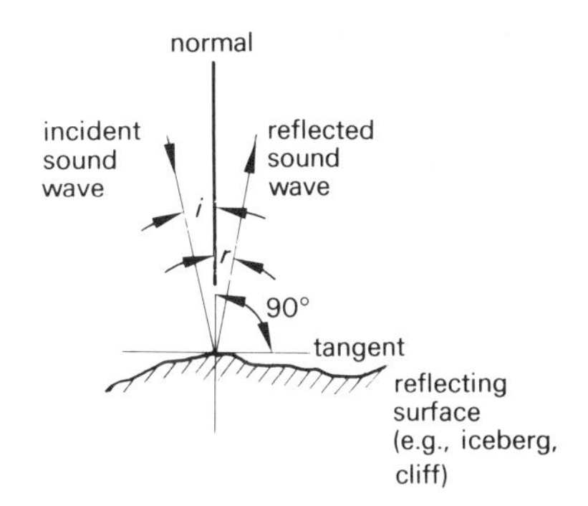

Figure 65 *Reflection of sound waves*

Like all other waves, sound waves can be reflected and refracted. When airborne sound meets a solid material with a hard and non-porous surface, much of the sound energy is reflected. As shown in Figure 65 the incident and reflected rays make equal angles with the normal to the reflecting surface. If the angle $(i+r)$ is not too large, then the reflected sounds – called echoes – can be heard.

Echoes have been put to practical use in a number of cases. Ships can detect dangers such as icebergs, particularly when it is dark or foggy.

The oldest method of sounding – ascertaining the depth of water below a ship – is by means of a 'lead-line'. This is a rope with a weight attached to its end. However, sounding in deep oceans needs very different methods. (In some parts of the Pacific Ocean the depth exceeds 11 000 metres.)

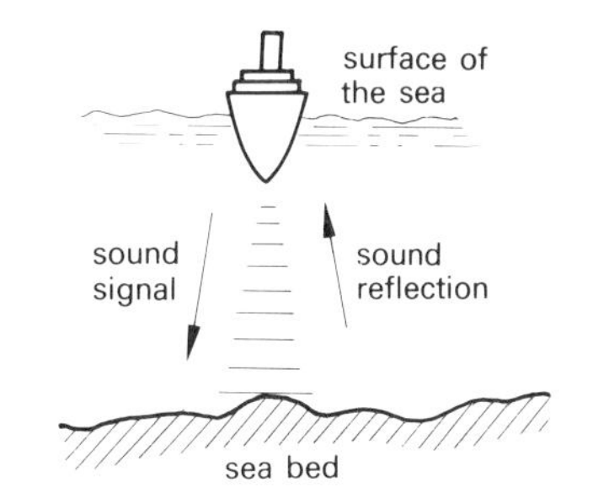

Figure 66 *Charting the sea bed by echo-sounding*

Nowadays a technique called echo-sounding is used to map the ocean depths. A sound signal is sent out from an apparatus in the hull of the ship. The echo from the sea floor is received by a hydrophone, a microphone adapted for marine use. The depth of the water can be calculated from the time taken for the sound to do the double journey, the speed of sound in sea water having been previously determined. Echo-sounding is also used to locate submerged objects such as wrecks and shoals of fish.

The refraction of sound has nothing like the practical importance of reflection. Waves are refracted (or bent) when they move from one medium into another. Sound waves travel more rapidly through warm air than through cold air. If the air in one place is warmer than in another, then the 'medium' has different properties and the sound waves will no longer travel in a straight line.

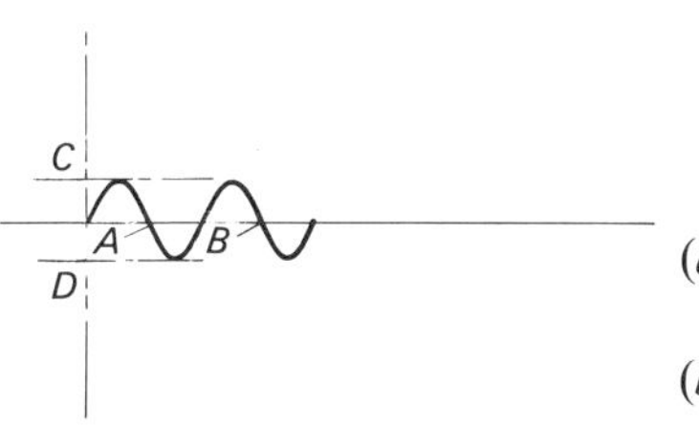

Figure 67 *Wave motion*

Self-assessment questions

1 List six examples of wave motion.

2 In Figure 67

(*a*) The distance AB is the wavelength of the wave. TRUE/FALSE

(*b*) Wavelength is normally indicated by λ. TRUE/FALSE

(*c*) Frequency is measured in hertz. TRUE/FALSE

(*d*) The amplitude of the wave is CD. TRUE/FALSE

3 Complete the statement:
Wave velocity = ______ × ______ .

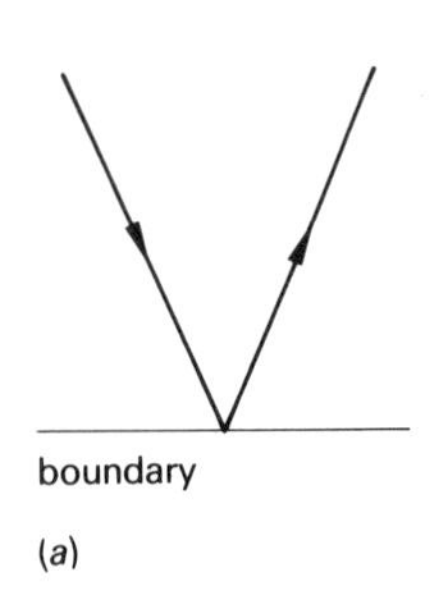

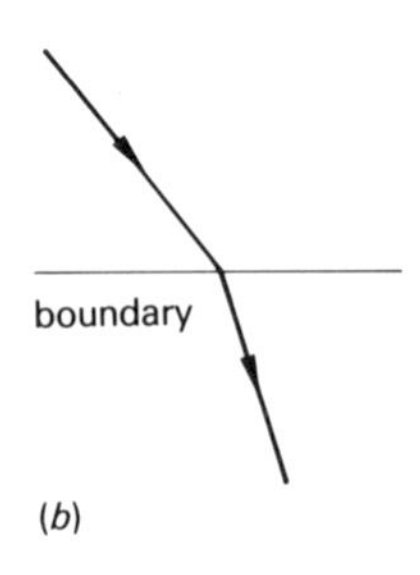

Figure 68 *Wave deviation diagrams*

4 Calculate the wavelengths in air and in water of sound having a frequency of 100 Hz. Speed of sound in air is 335 m/s. Speed of sound in water is 1 360 m/s.

5 Calculate the frequency of a radio station transmitting on a wavelength of 150 metres. Speed of radio waves is 300×10^6 m/s.

6 Study Figure 68. Which diagram represents reflection and which represents refraction of a wave?

7 Figure 69 shows a straight wave front approaching a solid barrier. Make a sketch showing how the wave front moves after reflection.

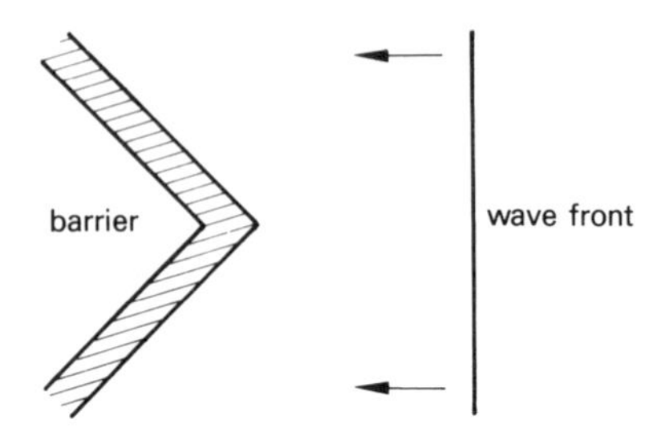

Figure 69 *Wave deflection diagram*

8 (*a*) Sound is a form of energy. TRUE/FALSE

(*b*) Any source of sound has some part which is vibrating. TRUE/FALSE

(*c*) Sound can be transmitted through gases, liquids and solids. TRUE/FALSE

(*d*) Sound can pass through a vacuum. TRUE/FALSE

(*e*) Sound waves are back-and-forth waves. TRUE/FALSE

(*f*) Sound is a pressure wave. TRUE/FALSE

(*g*) Sound can be reflected and refracted. TRUE/FALSE

(*h*) Sound always travels at the same speed, no matter what is the medium. TRUE/FALSE

(*i*) Hard materials are good reflectors of sound. TRUE/FALSE

9 With the aid of a diagram, explain the formation of echoes.

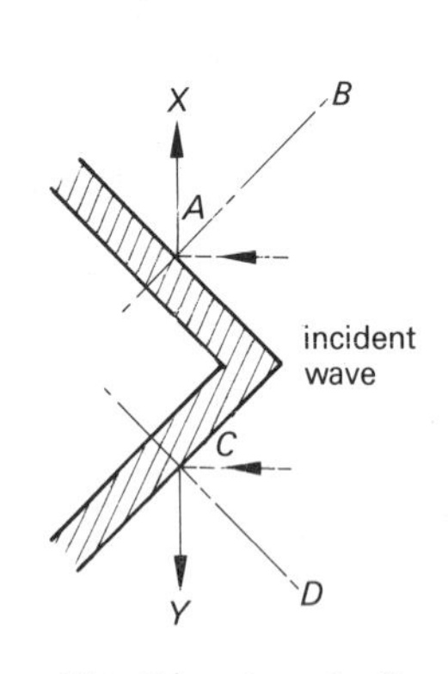

Figure 70 *Direction of reflected wave*

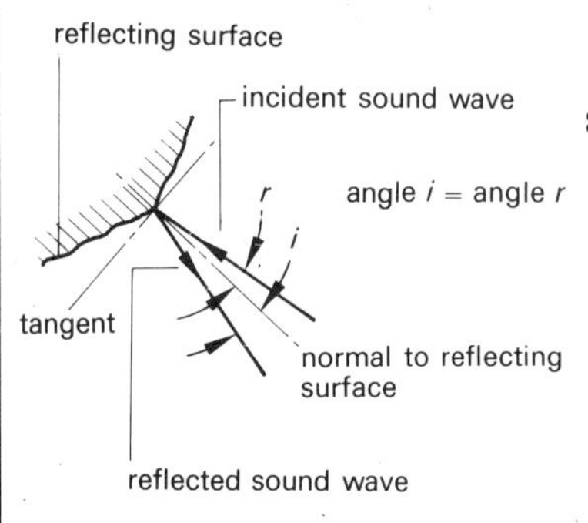

Figure 71 *Formation of echoes*

Solutions to self-assessment questions

1 Examples mentioned in the text are: water waves, sound waves, heat waves, light waves, radio waves, ultra-violet waves, X-rays, gamma rays, earthquake waves, infra-red waves, waves in a rope. There are other examples of wave motion.

2 (*a*) True.
(*b*) True.
(*c*) True.
(*d*) False.

3 Wave velocity = frequency × wavelength.

4

$$\text{In air: } v = f\lambda$$
$$\text{i.e. } 335 \text{ m/s} = 100 \frac{1}{s} \times \lambda \text{ m}$$
$$\therefore \underline{\text{Wavelength, } \lambda = 3.35 \text{ m}}$$
$$\text{In water: } v = f\lambda$$
$$\text{i.e. } 1\,360 = 100\,\lambda$$
$$\underline{\text{Wavelength, } \lambda = 13.6 \text{ m}}$$

5

$$v = f\lambda$$
$$300 \times 10^6 = f \times 150$$
$$\text{i.e. } f = \tfrac{300}{150} \times 10^6$$
$$\underline{\text{Frequency} = 2 \times 10^6 \text{ Hz}}$$

6 Diagram (*a*) represents reflection.
Diagram (*b*) represents refraction.

7 Figure 70 shows how the wave moves after reflection.

8 (*a*) True.
(*b*) True.
(*c*) True.
(*d*) False.
(*e*) True.
(*f*) True.
(*g*) True.
(*h*) False.
(*i*) True.

9 Figure 71 shows how the sound wave is reflected back. The reflecting surface must be hard and non-porous if the reflected sound is to be clear.

Topic area: Electricity

After reading the following material, the reader shall:

1 Know the preferred symbols used in circuit diagrams.
1.1 Identify the preferred symbols for electrical components.
1.2 Use the preferred symbols for electrical components when drawing circuit diagrams.

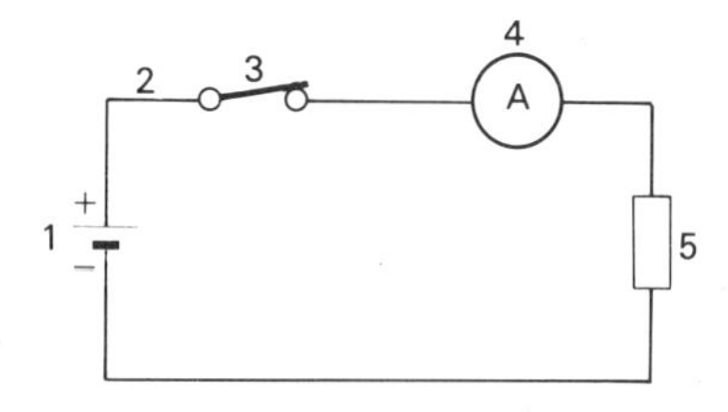

Figure 72 *British Standard graphical symbols for electrical components*

The circuit diagram shown in Figure 72 consists of symbols. These symbols are examples of *British Standard 3939 Graphical Symbols for Electrical Power Telecommunications and Electronic Diagrams.*

The circuit diagram shows a complete circuit which incorporates five symbols. The information contained in this circuit diagram may be summarized as follows: a single cell (1) is connected by means of a conductor (2) to a switch (3) which is shown in the closed position. The conductor is then connected to an ammeter (4) which is joined joined by the conductor to a resistor (5). The conductor then joins the resistor to the single cell. The short line in the symbol of the cell is marked negative (–) polarity, and the longer line positive (+) polarity.

If the switch was open circuited the circuit would not be complete. An incomplete circuit is referred to as an open circuit.

The symbols shown in the table are a selection of symbols in common use in electrical science.

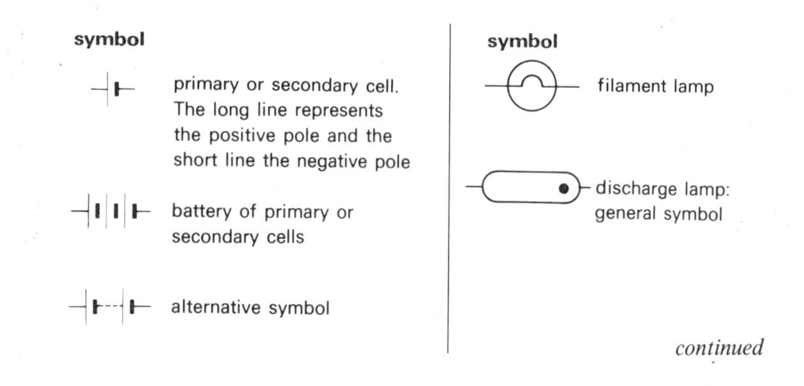

continued

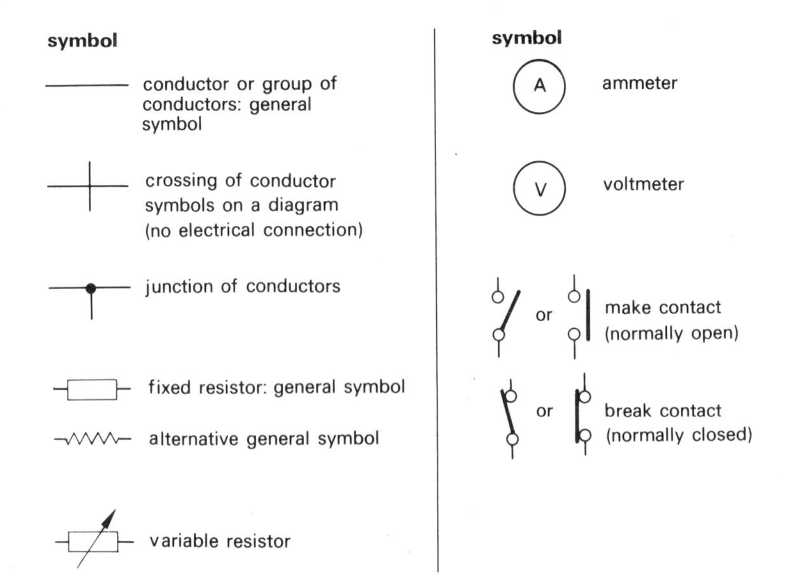

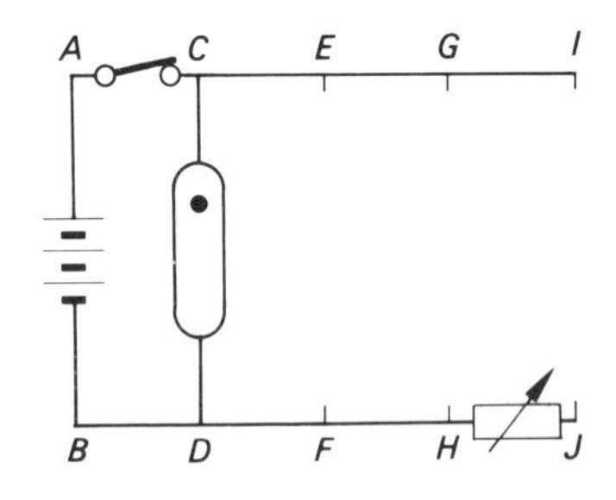

Figure 73

Self-assessment questions

The following six questions relate to Figure 73. In questions 1, 2 and 3 complete the statements.

1 The symbol between points A and B represents ______.

2 The symbol between points C and D represents ______.

3 The symbol between points H and J represents ______.

4 Draw the symbol for a voltmeter between the points E and F.

5 Draw the symbol for a filament lamp between the points G and H.

6 Draw the symbol for a fixed resistor between the points I and J.

After reading the following material, the reader shall:

2 Know the meaning of common electrical quantities.

2.1 Recognize that an electrical current is a unidirectional drift of free electrons.

2.2 State that free electrons tend to come from particular electron shells.

2.3 State that electrons have a constant negative charge.

2.4 State that like charges repel and that unlike charges attract.

2.5 Explain how current flows due to the existence of a potential difference (voltage) between two points in an electrical conductor.

2.6 Identify the coulomb as the unit of quantity of charge.
2.7 Identify the relationship between the coulomb and the ampere.
2.8 State that the unit of current is the ampere.
2.9 Identify the volt as the unit of electromotive force.
2.10 State that 1 V = 1 J/C.
2.11 State that the unit of potential difference is the volt.
2.12 State that for continuous current a complete circuit is necessary.
2.13 Distinguish between the conventional direction of current flow and the direction of the electron flow.

In order to explain how current flows it is necessary to be aware of some atomic theory and a little history.

Atomic theory

All matter, whether solid, liquid or gas, is made up of minute particles called molecules which can be subdivided into atoms. Until 1897 it was thought that the atom could not be further divided. In 1897 a particle called the *electron* was discovered in the atom, and in 1919 a further particle, the *proton*, was identified. The early discoveries, coupled with later researches, have produced evidence that all atoms are built up of particles some of which are electrically charged. A particle which is electrically charged exerts a force, the effect of which may be detected on another charged particle. These charged particles have been assigned polarities, i.e. they are called *negative* or *positive* charges. An important rule is that *like charges repel* and *unlike charges attract*.

In explaining how a current flows, the most important charged particles are the electron, which has a negative charge, and the proton, which has a positive charge. The charge on an electron and the charge on a proton are *equal* and of *opposite* polarities; between them, there is a force of attraction.

The early research on the atom produced a model of the atom, which helps to explain the interaction of the protons and the electrons.

The diagram in Figure 74 is a representation of a model of an atom. It shows a central nucleus which contains, along with other particles, the protons. Thus *the charge on the nucleus is positive*. This central nucleus is surrounded by a *cloud* of negatively charged electrons, which move in orbital paths determined by the forces of attraction and repulsion which exist between charged particles. It is known that each electron moves in an orbit around the nucleus, and that a number of such orbits constitute a *shell*. The shells are numbered outwards from the nucleus as shown in Figure 75. Shell one (the innermost) can contain a maximum of 2 electron orbits; shell two,

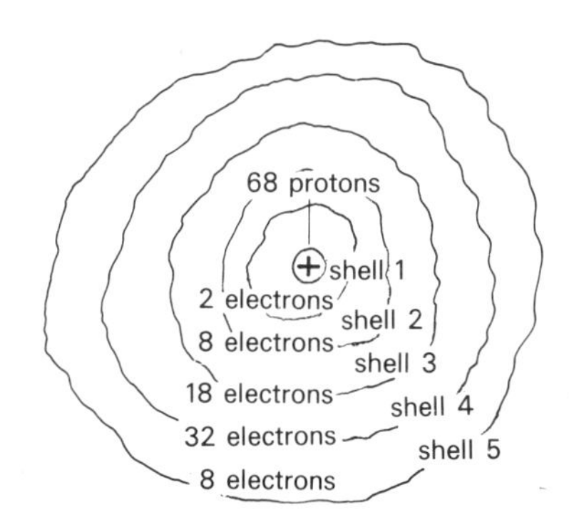

Figure 74 *A representation of a model of a neutron atom*

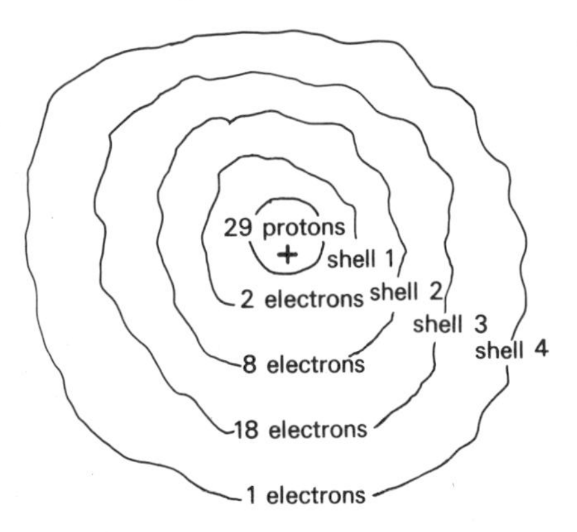

Figure 75 *A representation of a model of a neutral copper atom*

a maximum of 8 electron orbits; shell three, 18 electron orbits; shell four, 32 electron orbits. The outermost shell of all atoms can contain up to a maximum of 8 electron orbits. The electrons in the shells nearest to the nucleus are subject to relatively strong forces which bind them to the positive nucleus, but the electrons in the outermost shell are bound to the nucleus by relatively weak forces.

Figure 75 represents a copper atom. The nucleus contains 29 protons, i.e. 29 units of positive charge. When the atom is complete, they are exactly balanced by 29 electrons, i.e. 29 units of negative charge. In this condition the atom is electrically neutral. Figure 76 shows that the inner shell of the copper atom contains 2 electron orbits, shell two 8 electron orbits, shell three 18 electron orbits and the outer shell 1 electron orbit. The single outer electron is held in its shell by a comparatively weak force, and is easily detached from the parent atom. Electrons which have been detached from an atom are called *free electrons*. These electrons have the same quantity of negative charge as all other electrons. Free electrons are repelled by other negative charges, and are attracted by a positive charge. An atom with a full complement of electrons is neutral, i.e. it neither attracts nor repels other charges. When such an atom loses an electron from its outer shell it becomes positively charged, and is called a *positive ion*. A neutral atom which gains an electron becomes negatively charged, and is referred to as a *negative ion*.

On a larger scale, a substance whose atoms lose electrons is positively charged, and a substance whose atoms gain electrons is negatively charged. Suppose a substance which contains free electrons moving in a random motion between atoms (e.g. copper) is connected so that it joins a positively charged substance to a negatively charged substance. The electrons in the negatively charged substance repel the electrons which are in random motion in the copper. These in turn repel other free electrons in the copper, and produce a *unidirectional drift* of electrons through the copper towards the positively charged substance which is also exerting a force of attraction on the free electrons. *The unidirectional drift of electrons superimposed upon the random movement of the free electrons constitutes a current flow.*

The current flow is always from the more negatively charged substance towards the less negatively charged, i.e. from negative to positive.

The movement of negative charges (current) through the copper connecting the differently charged substances almost instantaneously reduces the difference between the charges to zero. When there is no difference in the state of charge, the unidirectional drift of electrons (current) ceases, and only the random movement of free electrons between atoms remains.

Solutions to self-assessment questions

1 A battery.
2 A discharge lamp.
3 A variable resistor.
4, 5, 6 The solutions are shown in Figure 76.

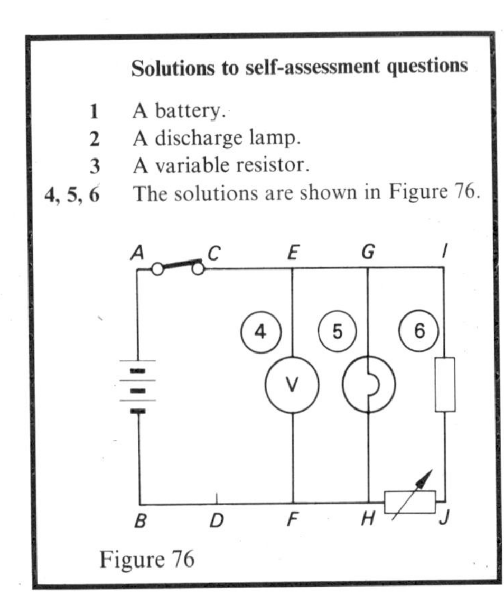

Figure 76

In order to maintain a unidirectional drift of electrons (i.e. a continuous current flow) it is essential that,

(*a*) there is a complete circuit through which current can flow,

and (*b*) the circuit contains a device which maintains a difference in the state of charges between its terminals.

Before 1799 the methods of intentionally producing differences in the electrical charge in substances were very limited. It had previously been discovered that rubbing a substance such as rubber, glass or amber with fur, cloth or silk produced differences in the electrical charge. It was also known that when differently charged substances were joined together by a metal wire, the difference in charge was rapidly reduced to zero. In some cases a spark could be observed, and in all cases the force between the charged substances disappeared. In 1799 a device, later called a primary cell, was invented. It was found that chemical action in the primary cell could,

(*a*) produce a difference in the state of charge between its two terminals,

and (*b*) maintain a difference in the state of charge for a considerable time.

The modern name for a difference in the state of charge between the terminals of a primary cell (or any other device which produces between its terminals a difference in the state of charge) is *electromotive force* (*e.m.f.*). The symbol for an e.m.f. is *E*. This is used to describe the difference in the state of charge between the terminals of a source of e.m.f. (e.g. the primary cell) when it is *not* supplying current to an external circuit. The unit of e.m.f. is the volt, (symbol V).

By about 1850, the principal sources of e.m.f. had been discovered. In addition to an e.m.f. produced by chemical action it was also found that relative movement between a magnetic field and a conductor produced an e.m.f. It was also discovered that when heat was applied to a joint between dissimilar metals, an e.m.f. was produced between the cold ends of the dissimilar metals. These discoveries had, by 1900, enabled the effects of a 'continuous' flow of current around a circuit to be classified, calculated and utilized by scientists and engineers.

The effects of a current flow in a circuit are,

(i) the temperature of the materials in the circuit is always raised.

(ii) the flow of current always produces a magnetic field.

(iii) a chemical reaction sometimes occurs.

The effects of a flow current indicate that energy from the source of e.m.f. is taken around the circuit, where it is converted into other forms of energy.

Engineers and scientists in the nineteenth and early twentieth century could readily produce sources of e.m.f., and set up conditions so that a current could flow; but they did not know what constituted a flow of current, or how the energy was transferred or transformed in the circuit.

Modern theories relating to the atomic structure of matter suggest that the energy from the source of e.m.f. is transferred to the free electrons which constitute the current flow around the circuit. The electrons in the current flow are now known,

(*a*) to carry a constant negative charge and to move away from negative charges towards a positive charge,

and (*b*) to have additional energy of motion transferred to them in the source of e.m.f.

This additional energy is called *electrical energy*.

When current flows around a circuit, the electrons in the current flow are involved in many millions of 'collisions' with electrons of the atoms in the materials which make up the circuit. These 'collisions' transfer energy from the electrons in the current flow to the atoms in the materials of the circuit. The electrical energy transferred is converted by the materials in the circuit into other forms of energy (e.g. heat, magnetic or chemical).

The magnitude of the charge on an individual electron is minute. Hence for practical purposes, a unit of charge is the charge on 6.29×10^{18} electrons. The unit is called a *coulomb* (C). The amount of electrical energy in joules which can be transferred to a coulomb of charge is determined by the e.m.f. of the source. A source of e.m.f. of 1 V means that 1 J is available in the source for transfer to each coulomb of charge.

$$E \text{ (volts)} = \frac{J \text{ (joules)}}{C \text{ (coulombs)}}$$

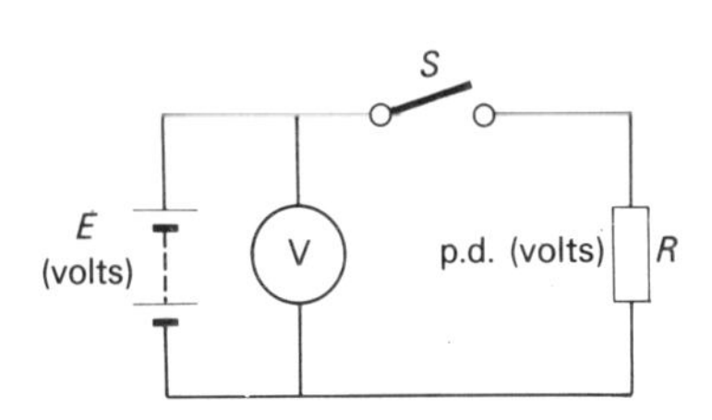

Figure 77 *An electric circuit*

The circuit in Figure 77 shows a battery as a source of electrical energy. A voltmeter (V) and a resistor (R) are connected between the terminals of the battery. The current flow in the circuit is controlled by the switch S.

In this theoretical circuit it is assumed that no energy is converted in the voltmeter or in the connecting conductors.

When the switch is open, the voltmeter indicates in volts the number of joules per coulomb which are available for transfer to the circuit. This is the e.m.f. of the battery. When the switch is closed, electrical energy is transferred in the source to the electrons which constitute the charge. The difference in the state of charge between the terminals of the battery moves the electrons around the circuit. The electrical energy transferred to the electrons in the battery is transformed in the resistor into (mainly) heat energy.

The joules per coulomb converted into other forms of energy in the resistor is recorded on the voltmeter in volts. This reading is referred to as a *potential difference* (p.d.). A p.d. of 1 V means 1 J/C of electrical energy has been converted into other forms of energy between the two points in the circuit across which the p.d. is measured.

The rate at which coulombs of charge are moved around a circuit is given a special name, the *ampere*. When a charge of 1 C passes a given point in a circuit in 1 s, an electrical current (I) of 1 ampere (A) is flowing in the circuit –

1 A = 1 coulomb per second

The coulomb is often referred to as the ampere-second and this produces the idea of a quantity of electricity (Q)

$$Q = I \times t$$

where Q is in coulombs,
I is in amperes,
t is in seconds.

Increased knowledge of the structure of the atom has revealed that the direction of a current flow is from a negative charge to a positive charge. This direction of current flow is referred to as the *electron flow*. The true direction of a current flow was determined long after the discovery of methods of producing a flow of current. The scientists who made the original discoveries *assumed* that a current flow was from a positive charge to a negative charge. This is referred to as *conventional current flow*. The convention is very widely used both in science and engineering; it is used in this text unless otherwise stated.

Self-assessment questions

7 (i) The charges on a proton and on an electron are of equal and opposite polarity.

TRUE/FALSE

(ii) An atom with a complete outer shell has several free electrons.

TRUE/FALSE

(iii) The random movement of free electrons constitutes a current flow.

TRUE/FALSE

(iv) A difference in the state of charge between the two ends of material which contains free electrons, will produce a unidirectional flow of electrons.

TRUE/FALSE

(v) One volt is equal to one joule per coulomb.

TRUE/FALSE

(vi) When current flows in a circuit, a potential difference exists between two points in the circuit.

TRUE/FALSE

(vii) Electron flow is from positive to negative.

TRUE/FALSE

(viii) The free electrons in a conductor originate from the incomplete electron shells.

TRUE/FALSE

(ix) A current of one ampere is equal to one coulomb per second passing a point in a circuit.

TRUE/FALSE

8 Complete each of the following statements by supplying the missing word:

(i) The charge on an electron is negative and is always ______.

(ii) The positively charged particles in the nucleus of an atom are called ______.

(iii) Between like charges there is a force of ______.

(iv) An electromotive force is represented by the symbol ______.

(v) The unit of potential difference is the ______.

(vi) Electromotive force and potential difference can be measured with an instrument called a ______.

(vii) For continuous current flow a circuit must contain a source of e.m.f. and be ______.

(viii) The symbol for an electric current is ______.

After reading the following material, the reader shall:

3 Know Ohm's law.

3.1 Identify the atomic structure of a material as the property which limits the flow of current.

3.2 Describe resistance as the property of a conductor that limits current.

3.3 Define resistance as the ratio of potential difference (voltage) across a resistor, to the current through it.

3.4 State Ohm's law in terms of the proportionality of current to potential difference.

3.5 Draw a graph of the relationship between potential difference and current using experimental data for

(*a*) a single resistor

(*b*) a non-linear component

3.6 Solve simple problems using Ohm's law.

An electric current is a flow of electrons through a material. The electrons do not move freely along the material because they are continuously being attracted or repelled by the atoms of the material.

Different materials oppose the passage of electrons by different amounts. The magnitude of the opposition to the flow of electrons in a material depends upon the atomic structure of the material. The atomic structures of some materials such as glass, porcelain, PVC, polystyrene, rubber, wood and paper, strongly oppose the passage of electrons, and a very considerable e.m.f. is required to produce a current. On the other hand, the atomic structure of metals such as copper, brass, aluminium, platinum, gold and silver offers very little opposition to a current. The magnitude of the opposition of the atomic structure of a material to a flow of electrons is called its *electrical resistance*.

The resistance (R), *unit the ohm* (Ω), *is measured in volts per ampere* (i.e. the number of joules per coulomb required to produce a flow of coulombs per second through a material) –

1 ohm = 1 volt per ampere

It is found that for a given solid conductor at a given temperature, the magnitude of the current flowing in the conductor is directly proportional to the potential difference between the ends of the conductor. Also, the current flowing through a conductor is inversely proportional to its resistance.

The relationships $V \propto I$ and $I \propto \frac{1}{R}$ were discovered by Dr Ohm in 1827 and are referred to as Ohm's law.
Ohm's law states:

The current in a circuit is proportional to the potential difference (*p.d.*) *across the circuit, and inversely proportional to the resistance of the circuit.*

The law can be written as

$$\text{current} = \frac{\text{potential difference}}{\text{resistance}}$$

Expressed in symbolic form, Ohm's law is:

$$I = \frac{V}{R}$$

This equation can be applied to any part of a circuit, or to a complete circuit.

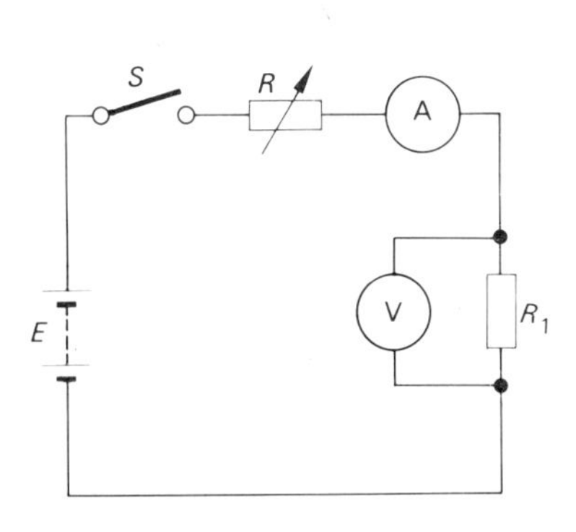

Figure 78 *Potential difference across a resistor*

In the circuit diagram in Figure 78 the battery provides an e.m.f. When the switch (S) is closed, a current flows in the circuit. The magnitude of the current can be controlled by the variable resistor (R), and measured by the ammeter (A). The current flow through the fixed resistor (R_1) produces a p.d. between the ends of the resistor. This p.d. is measured by the voltmeter (V).

Example 1

If the voltmeter in Figure 78 reads 12 V and the ammeter reads 4 A, calculate the resistance of R_1.

From Ohm's law $R = \frac{V}{I}$

when R is the resistance of the resistor R_1,
V is the p.d. across R_1,
I is the current through R_1.

The resistance of $R_1 = \frac{V}{I}$ ohms

$$\text{Resistance} = \frac{12}{4} = 3\ \Omega$$

Example 2

The rating plate on an electric kettle provides the following data: 240 V, 6 A. Calculate the resistance of the element of the kettle.

From Ohm's law $R = \frac{V}{I}$

when R is the resistance of the kettle element,
V is the p.d. across the element,
I is the current through the element.

$$\text{Resistance, } R = \frac{V}{I} = \frac{240}{6} = 40\ \Omega$$

Solutions to self-assessment questions

7 (i) True.
(ii) False.
(iii) False. A current flow is a unidirectional drift of electrons.
(iv) True.
(v) True.
(vi) True.
(vii) False. Electrons move from negative to positive.
(viii) True.
(ix) True.

8 (i) The charge on an electron is negative and is always *constant*.
(ii) The positively charged particles in the nucleus of an atom are called *protons*.
(iii) Between like charges there is a force of *repulsion*.
(iv) An electromotive force is represented by the symbol E.
(v) The unit of potential difference is the *volt*.
(vi) Electromotive force and potential difference can be measured with an instrument called a *voltmeter*.
(vii) For continuous current flow a circuit must contain a source of e.m.f. and be *complete*.
(viii) The symbol for an electric current is I.

Example 3
Calculate the current which flows through the element of the kettle in Example 2 when it is connected to a p.d. of 120 V.

From Ohm's law $I = \frac{V}{R}$

when I is the current through the element,
V is the p.d. across the element,
R is the resistance of the element.

$$\text{Current, } I = \frac{V}{R} = \frac{120}{40} = 3 \text{ A}$$

An alternative way of calculating the current in the element of the kettle in Examples 2 and 3 is to use the relationship

$V \propto I$.

This relationship means that when the p.d. across a circuit is increased, the current in the circuit increases in the same proportion.

Example 4
The rating plate on an electric kettle states 240 V, 6 A. If it is connected to a 120 V supply, calculate the current which flows in the element.

$$\frac{\text{low p.d.}}{\text{rated p.d.}} = \frac{\text{new current}}{\text{rated current}}$$

$$\frac{120 \text{ V}}{240 \text{ V}} = \frac{\text{new current}}{6 \text{ A}}$$

$$\therefore \text{ new current} = \frac{6 \times 120}{240}$$

$$\text{Current} = 3 \text{ A}$$

Self-assessment questions

9 The following data was obtained during an experiment to determine the relationship between p.d. and current in a circuit. The circuit diagram is shown in Figure 78 (page 115). The temperature of the resistor R_1 was maintained constant.

p.d. in volts across R_1	2	4	6	8	10	12	14	16
current flow in ampere in R_1	1	2	3	4	5	6	7	8

(*a*) Plot a graph of this relationship.
(*b*) What is the relationship between the p.d. across the resistor and the current in the circuit?
(*c*) What is known about the magnitude of the resistor R_1?

10 The following data was obtained during an experiment in which the p.d. across a filament lamp, rated at 250 V, was increased from 0 V to 250 V in increments of 50 V.

p.d. across lamp (volts)	50	100	150	200	250
current through lamp (amps)	0.18	0.27	0.35	0.4	0.46
resistance of lamp					

(*a*) Calculate the resistance of the lamp at each p.d.
(*b*) Plot a graph of the relationship between p.d. and the current in the circuit.
(*c*) Suggest a possible cause of the change in the resistance of the lamp.

11 (i) The passage of a unidirectional drift of electrons through a material is resisted by the atomic structure of the material.

TRUE/FALSE

(ii) In a circuit in which the resistance is maintained constant, a change in the p.d. across the circuit causes the current in the circuit to change.

TRUE/FALSE

(iii) Resistance can be defined as the ratio of the p.d. across a resistor to the current through the resistor.

TRUE/FALSE

12 Complete each of the following statements by supplying the missing word:
(i) The opposition to a flow of electrons in a material is called ______.
(ii) A device which is designed to oppose the flow of current is called a ______.
(iii) The unit of resistance is the ______.
(iv) 1 volt per ampere = ______.

After reading the following material, the reader shall:

4 Know the effect on electrical quantities of connecting electrical components in
(*a*) series
(*b*) parallel.
4.1 State that current is the same in all parts of a series circuit.

4.2 State that the sum of the voltages in a series circuit is equal to the total applied voltage.

4.3 Show that for resistors connected in series the equivalent resistance is given by: $R = r_1 + r_2 + r_3$

4.4 Solve simple problems involving up to three resistors connected in series, including the use of Ohm's law.

4.5 State that the sum of the current in resistors connected in parallel is equal to the current flowing into the parallel network.

4.6 State that the potential difference (voltage) is the same across resistors in parallel.

4.7 Show that for resistors connected in parallel the equivalent resistance is given by: $\frac{1}{R} = \frac{1}{r_1} + \frac{1}{r_2} + \frac{1}{r_3}$

4.8 Calculate the equivalent resistance of equal resistors in parallel.

4.9 Calculate the equivalent resistance of two unequal resistors connected in parallel.

4.10 Solve simple problems involving up to three resistors connected in parallel by the use of Ohm's law.

4.11 Describe the difference between series and parallel connections of resistors.

4.12 Identify in circuit diagrams the methods of connecting ammeters and voltmeters.

Resistors can be connected to sources of e.m.f. in three ways:

(i) series

(ii) parallel

(iii) series–parallel

At this stage, only the first two methods are considered; the third method is considered in a later unit.

The series arrangement of resistors is one in which *the current in the circuit flows through each resistor in turn.*

In the circuit diagram of Figure 79, the ammeter is connected so that it measures the current flow in the circuit. The current flows through the ammeter, and then through each resistor in turn. When resistors

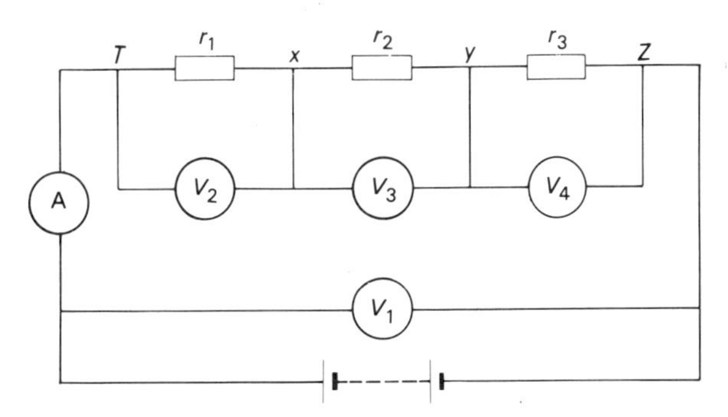

Figure 79 *Resistors connected in series*

or other items of electrical apparatus are connected so that the same current flows through each item in turn, the form of circuit is a series circuit and the items are connected in series. The ammeter in the circuit in Figure 81 is connected in series. The current an ammeter measures is often referred to as the 'load current'. A general name for electrical apparatus connected in a circuit is a 'load'. *Note that ammeters are always connected in series with the load.*

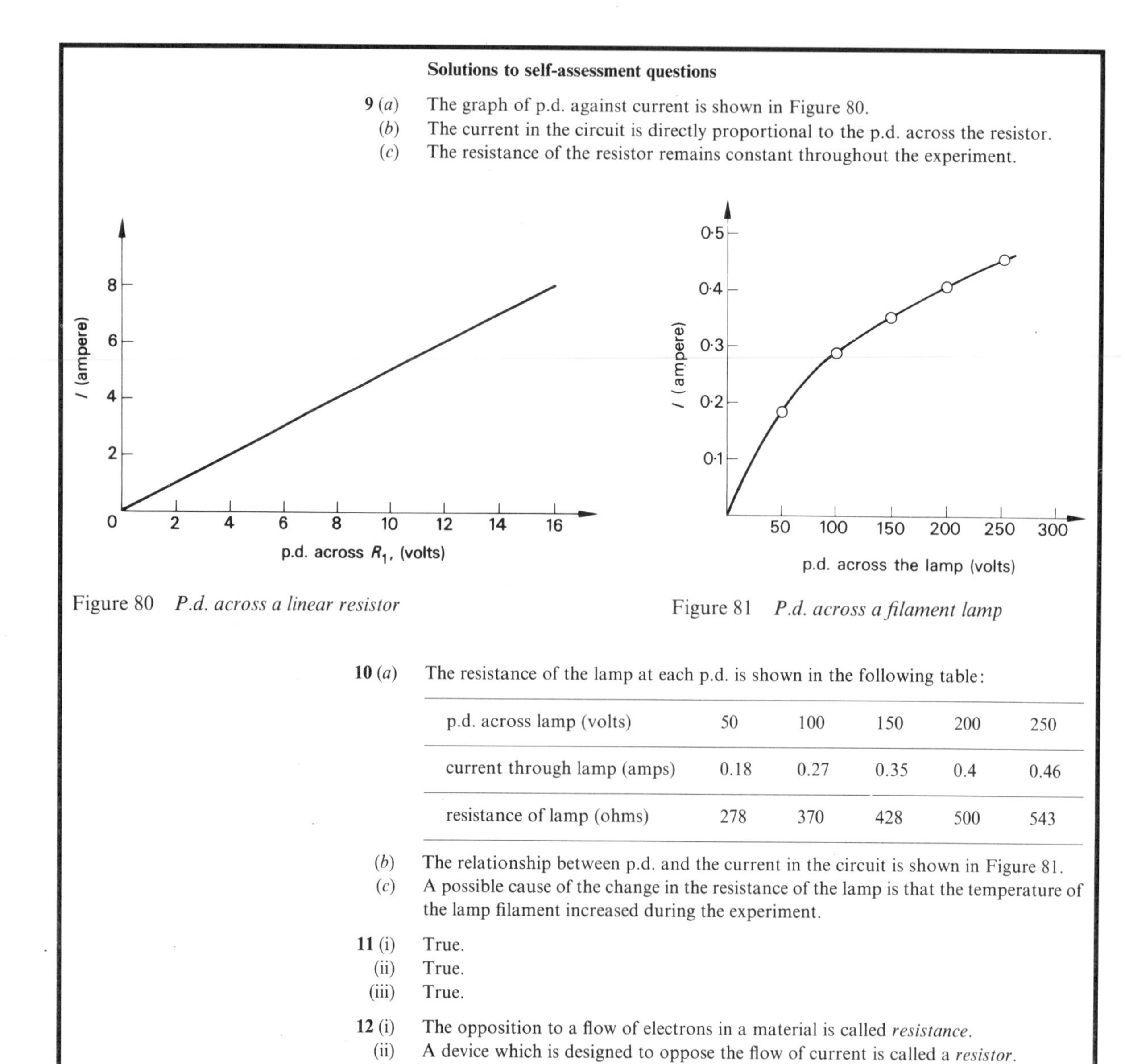

Solutions to self-assessment questions

9 (*a*) The graph of p.d. against current is shown in Figure 80.
(*b*) The current in the circuit is directly proportional to the p.d. across the resistor.
(*c*) The resistance of the resistor remains constant throughout the experiment.

Figure 80 *P.d. across a linear resistor*

Figure 81 *P.d. across a filament lamp*

10 (*a*) The resistance of the lamp at each p.d. is shown in the following table:

p.d. across lamp (volts)	50	100	150	200	250
current through lamp (amps)	0.18	0.27	0.35	0.4	0.46
resistance of lamp (ohms)	278	370	428	500	543

(*b*) The relationship between p.d. and the current in the circuit is shown in Figure 81.
(*c*) A possible cause of the change in the resistance of the lamp is that the temperature of the lamp filament increased during the experiment.

11 (i) True.
(ii) True.
(iii) True.

12 (i) The opposition to a flow of electrons in a material is called *resistance*.
(ii) A device which is designed to oppose the flow of current is called a *resistor*.
(iii) The unit of resistance is the *ohm*.
(iv) 1 volt per ampere = *1 ohm*.

In the circuit diagram in Figure 81, it is assumed that all the resistance R_T in the circuit is concentrated in the resistors r_1, r_2 and r_3. The ammeter, the four voltmeters and all the connecting conductors are regarded as not converting any of the electrical energy in the circuit into other forms of energy. In this strictly theoretical condition, the p.d. between T and Z measured on the voltmeter V_1 is equal to the current in the circuit multiplied by the total resistance of the three resistors r_1, r_2 and r_3

$$V_1 = I \times R_T$$
$$V_2 = I \times r_1$$
$$V_3 = I \times r_2$$
$$V_4 = I \times r_3$$

As it has been assumed that all the energy in joules per coulomb (i.e. the p.d. between T and Z) has been converted into other forms of energy in r_1, r_2 and r_3, then

$$V_1 = v_2 + v_3 + v_4$$
or $I \times R_T = I \times r_1 + I \times r_2 + I \times r_3$

Dividing each term by I

$$R_T = r_1 + r_2 + r_3$$

The total resistance (R_T) of a series circuit is the sum of the resistance of each resistor in the circuit.

Example 5

Using the data in the circuit in Figure 82 calculate:

(i) the resistance of the circuit
(ii) the resistance of r_3
and (iii) the potential difference across each resistor.

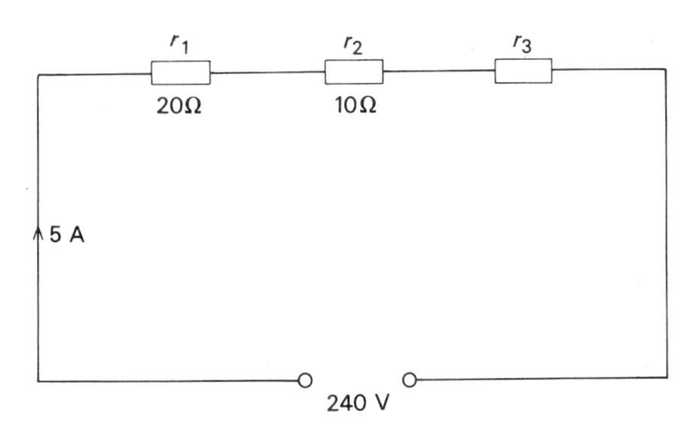

Figure 82 *Series circuit*

(i) From Ohm's law, $R = \frac{V}{I}$

when R_T is the resistance of the circuit
V is the potential difference across the circuit
I is the current in the circuit.

$$\text{Resistance, } R_T = \frac{V}{I} = \frac{240}{5} = 48\ \Omega$$

(ii)
$$\begin{aligned} \text{As } R_T &= r_1 + r_2 + r_3 \\ 48 &= 20 + 10 + r_3 \\ \therefore r_3 &= 48 - 30 \\ r_3 &= 18\ \Omega \end{aligned}$$

(iii)
$$\begin{aligned} \text{The p.d. across } r_1 &= I \times r_1 = 5 \times 20 = 100\ \text{V} \\ \text{The p.d. across } r_2 &= I \times r_2 = 5 \times 10 = 50\ \text{V} \\ \text{The p.d. across } r_3 &= I \times r_3 = 5 \times 18 = 90\ \text{V} \end{aligned}$$

An alternative method of calculating the p.d. across r_3 is

$$\begin{aligned} V &= \text{p.d.}_{r_1} + \text{p.d.}_{r_2} + \text{p.d.}_{r_3} \\ 240 &= 100 + 50 + \text{p.d.}_{r_3} \\ \therefore \text{p.d.}_{r_3} &= 240 - 150 \\ \text{P.d. across } r_3 &= 90\ \text{V} \end{aligned}$$

When resistors are connected in parallel:
(i) *the p.d. across each resistor in the parallel circuit is the same,*
and (ii) *each resistor provides a separate path for a part of the circuit current.*

The circuit diagram in Figure 83 shows three resistors connected in parallel. In series with each resistor is an ammeter. A_1 indicates the current i_1 flowing in resistor r_1; A_2 indicates the current i_2 flowing in resistor r_2; A_3 indicates the current i_3 flowing in resistor r_3. The ammeter A_4 indicates the total current I flowing in the circuit. The voltmeter measures the p.d. across the circuit. In order to measure a p.d., a voltmeter *must* be connected across a circuit or part of a circuit. It is assumed that no electrical energy is converted into other forms of energy by the instruments or the connecting conductors. In these theoretical conditions, the p.d. measured by the voltmeter is the same as that across each resistor.

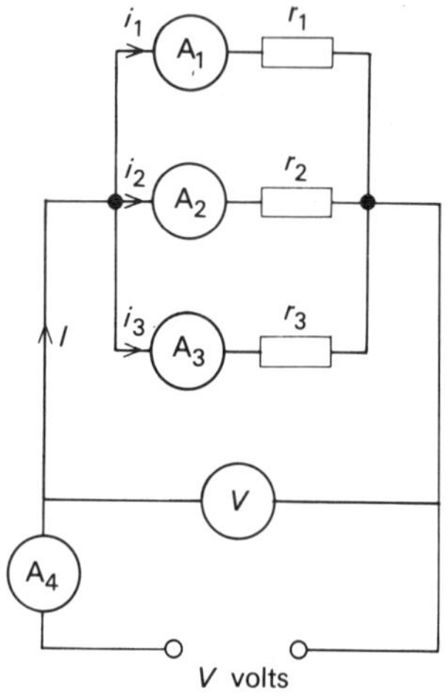

Figure 83 *Resistors connected in parallel*

The total current I = sum of the currents in each branch of the parallel circuit

$$I = i_1 + i_2 + i_3$$

Applying Ohm's law to each of the branches:

$$i_1 = \frac{V}{r_1},\ i_2 = \frac{V}{r_2},\ i_3 = \frac{V}{r_3}$$

Applying Ohm's law to the whole circuit,
when the resistance of the circuit is R_T,

then $R_T = \frac{V}{I}$ and $I = \frac{V}{R_T}$

Hence $I = \frac{V}{R_T} = \frac{V}{r_1} + \frac{V}{r_2} + \frac{V}{r_3}$

Dividing each term by V:

$$\frac{1}{R_T} = \frac{1}{r_1} + \frac{1}{r_2} + \frac{1}{r_3}$$

In electrical circuits it is common practice
(i) to connect equal resistors in parallel,
or (ii) to connect two unequal resistors in parallel,
or (iii) to connect more than two unequal resistors in parallel.

The circuit diagram in Figure 84 shows three equal resistors r_1, r_2 and r_3 connected in parallel.

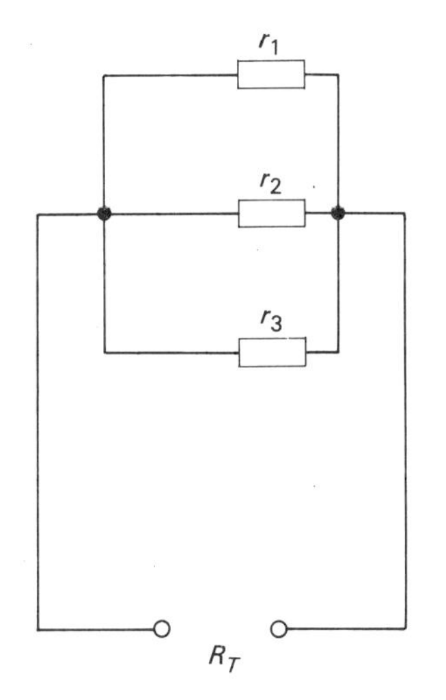

Figure 84 *Equal resistors connected in parallel*

Then as $\frac{1}{R_T} = \frac{1}{r_1} + \frac{1}{r_2} + \frac{1}{r_3}$

$\frac{1}{R_T} = \frac{3}{r}$ where $r = r_1 = r_2 = r_3$

$\therefore R_T = \frac{r}{3}$

R_T is therefore $\frac{1}{3}$ of the resistance of one of the equal resistors.

Example 6
If each of the resistors in Figure 84 is of 9 Ω, calculate the total resistance of the circuit.

Since $\frac{1}{R_T} = \frac{1}{r_1} + \frac{1}{r_2} + \frac{1}{r_3}$

$\frac{1}{R_T} = \frac{1}{9} + \frac{1}{9} + \frac{1}{9} = \frac{3}{9}$

$\therefore$ Total resistance, $R_T = \frac{9}{3}$ ohms $= 3\ \Omega$

R_T is $\frac{1}{3}$ of the value of one of the equal resistors.

If n equal resistors are connected in parallel, their equivalent resistance R_T is equal to $\frac{1}{n}$-th the resistance of one of the resistors, where n is the number of resistors.

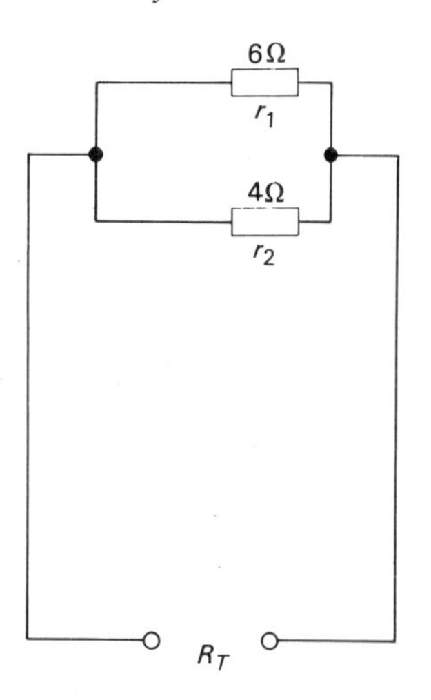

Figure 85 *Unequal resistors connected in parallel*

In the circuit diagram of Figure 85, two parallel resistors r_1 and r_2 have different values of resistance.

$$\text{As } \frac{1}{R_T} = \frac{1}{r_1} + \frac{1}{r_2}$$

$$\text{then } \frac{1}{R_T} = \frac{r_1 + r_2}{r_1 \times r_2}$$

$$R_T = \frac{r_1 \times r_2}{r_1 + r_2}$$

Example 7
In Figure 85, if $r_1 = 6\ \Omega$ and $r_2 = 4\ \Omega$, calculate the total resistance of the circuit.

$$\text{As } R_T = \frac{r_1 \times r_2}{r_1 + r_2}$$

$$R_T = \frac{6 \times 4}{6 + 4}$$

$$= \frac{24}{10}\ \Omega$$

$\therefore$ Total resistance, $R_T = 2.4\ \Omega$

Example 8
The circuit diagram in Figure 86 shows three resistors which are connected in parallel. The resistances of the resistors are $r_1 = 4\ \Omega$, $r_2 = 6\ \Omega$, $r_3 = 12\ \Omega$. The current in the circuit is 12 A.
Calculate
(i) the equivalent resistance of the circuit
(ii) the p.d. across the circuit
and (iii) the current in each branch of the circuit.

(i)
$$\text{As } \frac{1}{R_T} = \frac{1}{r_1} + \frac{1}{r_2} + \frac{1}{r_3}$$

$$\frac{1}{R_T} = \frac{1}{4} + \frac{1}{6} + \frac{1}{12}$$

$$= \frac{3 + 2 + 1}{12} = \frac{6}{12}$$

Resistance, $R_T = \frac{12}{6}$ ohms $= 2\ \Omega$

(ii) From Ohm's law the p.d. across the circuit is

$$V = I \times R_T$$

P.d. $= 12 \times 2 = 24$ V

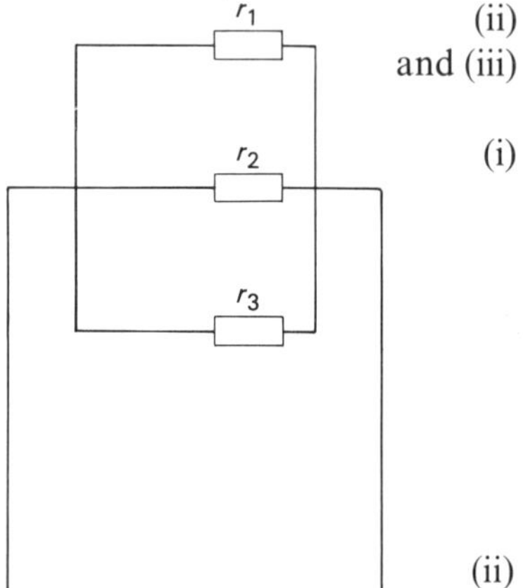

Figure 86 *Parallel circuit*

(iii) From Ohm's law, the current in the 4 Ω resistor is

$$i_1 = \frac{V}{r_1} = \frac{24}{4} = 6 \text{ A}$$

The current in the 6 Ω resistor is

$$i_2 = \frac{V}{r_2} = \frac{24}{6} = 4 \text{ A}$$

The current in the 12 Ω resistor is

$$i_3 = \frac{V}{r_3} = \frac{24}{12} = 2 \text{ A}$$

In each of the preceding examples the parallel resistors can be replaced by an *equivalent* resistor, which always has a value of resistance smaller than the smallest resistor in the circuit. Hence, when replacing parallel resistors by an equivalent resistor, care must be taken to ensure that the current will not damage the equivalent resistor.

Self-assessment questions

13 Complete each of the following statements by supplying the missing word(s):

(i) In a circuit containing resistors connected in series, the sum of the potential differences across each resistor is equal to the ______.

(ii) A series circuit contains two resistors. One resistor has a value of 10 Ω, and the total resistance of the circuit is 25 Ω. The resistance of the second resistor is ______.

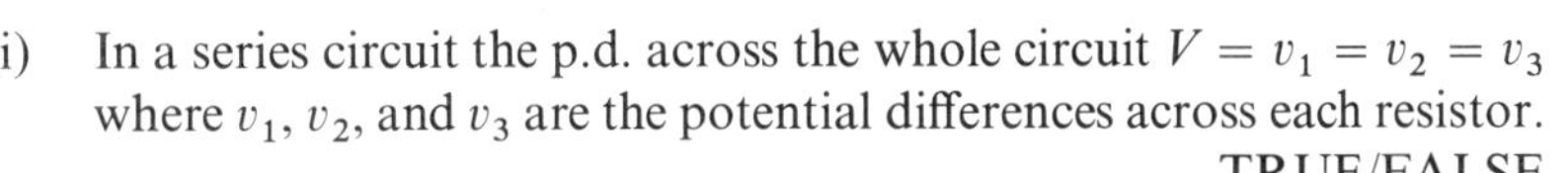

14 (i) In a series circuit the p.d. across the whole circuit $V = v_1 = v_2 = v_3$ where v_1, v_2, and v_3 are the potential differences across each resistor.

TRUE/FALSE

(ii) In a parallel circuit with three branches, the total current $I = i_1 + i_2 + i_3$.

TRUE/FALSE

(iii) In a parallel circuit with three branches, the p.d. across the circuit is the same as the p.d. across each branch.

TRUE/FALSE

(iv) In a parallel circuit containing two branches, the total current $I = i_1 = i_2$.

TRUE/FALSE

(v) In the circuit diagram in Figure 87, the ammeters A_1, A_2 and A_3 are each connected in series with a resistor.

TRUE/FALSE

(vi) In the circuit diagram in Figure 87, the voltmeter V is connected in parallel with the resistors.

TRUE/FALSE

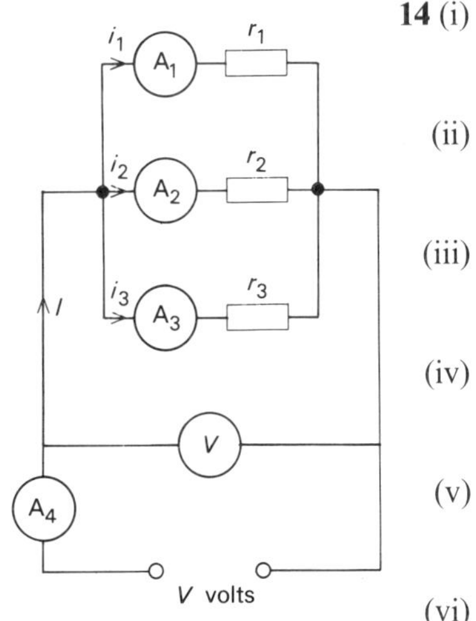

Figure 87 *Resistors connected in parallel*

15 In the circuit diagram in Figure 88, each instrument is numbered, and measures a particular quantity in the circuit. Identify each instrument as an ammeter or voltmeter by ticking the appropriate column.

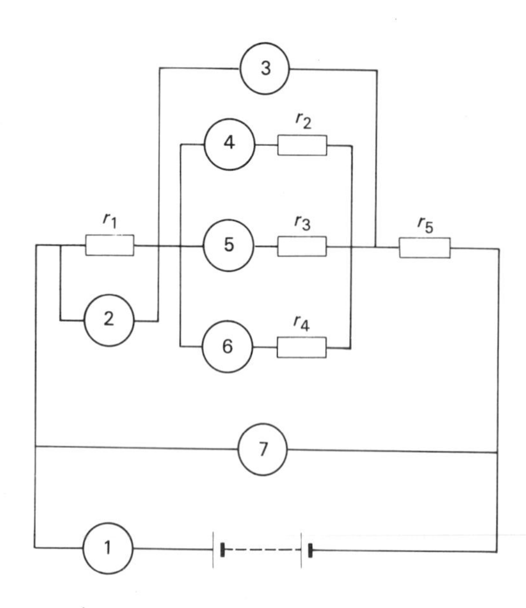

instrument number	ammeter	voltmeter
1		
2		
3		
4		
5		
6		
7		

Figure 88 *Instruments in a series parallel circuit*

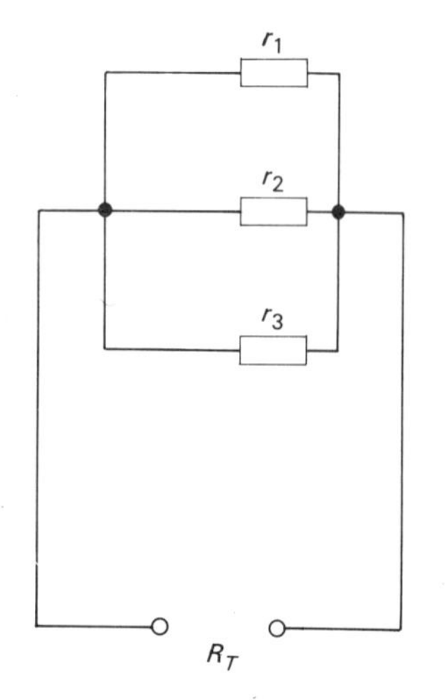

Figure 89

16 Three resistors each with a resistance of 210 Ω are connected in parallel to a 140 V supply. Calculate the current in the circuit.

17 Three resistors each with a resistance of 10 Ω are connected in series to a 90 V supply. Calculate

(*a*) the current in the circuit

(*b*) the p.d. across one resistor

18 A 10 Ω and a 5 Ω resistor are connected in parallel to a 100 V supply. Calculate the total current in the circuit.

19 In the circuit diagram in Figure 89 resistor $r_1 = 4\ \Omega$, $r_2 = 10\ \Omega$ and $r_3 = 20\ \Omega$. When the circuit is connected to a 25 V supply calculate

(i) the current in the circuit,

(ii) the current in each resistor.

After reading the following material, the reader shall:

5 Know the factors which determine the resistance of conductors and insulators.

5.1 List the factors which affect resistance.

5.2 State the relationship between the resistance of a conductor and its length, cross-sectional area and material.

5.3 State that resistance varies with temperature.

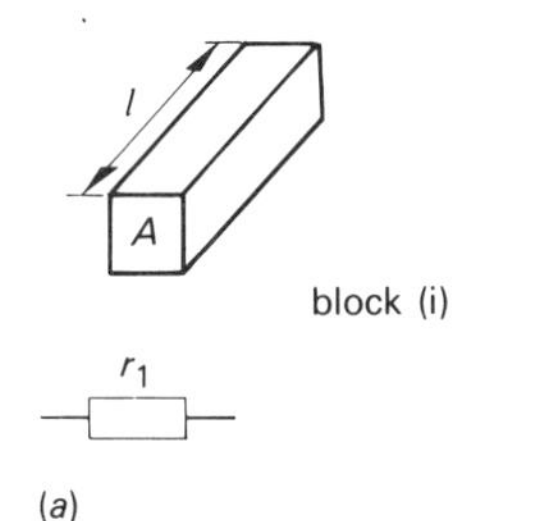

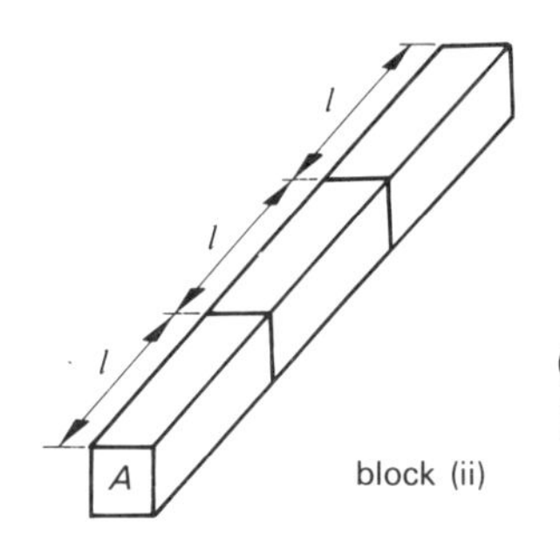

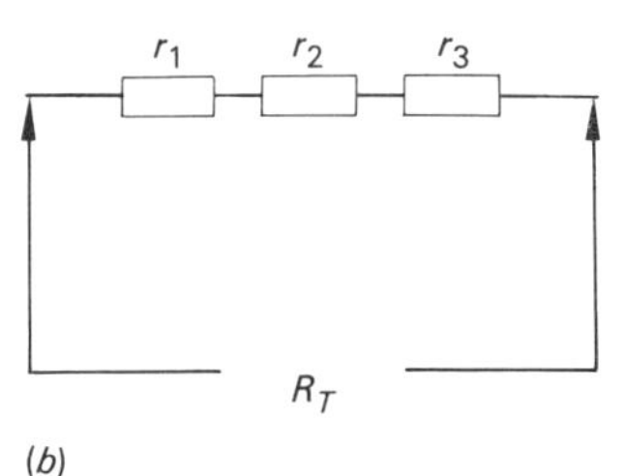

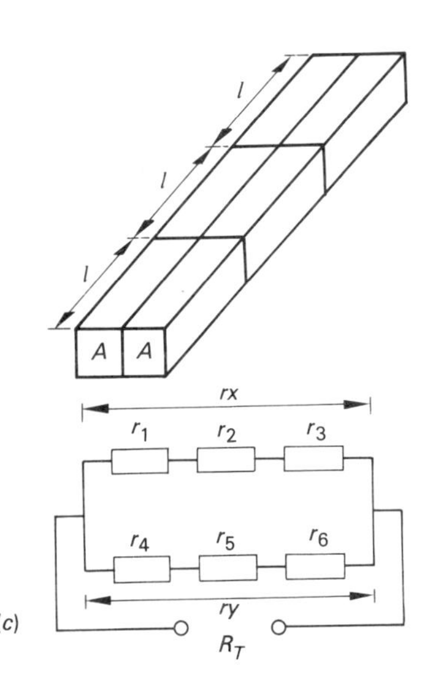

Figure 90 *Relationship between resistance, length and cross-sectional area*

All materials offer resistance to a flow of current. The resistance to a unidirectional drift of electrons through a material depends upon four factors:

(i) the length of the material
(ii) the cross-sectional area of the material
(iii) the type of material
(iv) the temperature of the material

Each of these factors is taken into account when electrical apparatus is designed.

The block of material in Figure 90(*a*) has a length l and a uniform cross-sectional area A. At a specified temperature this material offers a particular resistance to a current flow along its length. This resistance is represented in Figure 90(*a*) by the resistor r_1.

Figure 90(*b*) represents three blocks of the same material joined in series. This block has the same uniform cross-sectional area but is three times longer than one block. A given flow of electrons through this length of material meets with three times the resistance encountered when the same current flows through one-third of the length of the material. This resistance is represented by R_T, which is the sum of the three resistors r_1, r_2, and r_3, connected in series as Figure 90(*b*).

It is found that for a material with a uniform cross-sectional area and temperature, *the resistance measured between the ends of the material is directly proportional to the length of the material.*

Two samples of the same material are joined in parallel as shown in Figure 90(*c*). Compared with Figure 90(*b*), the cross-sectional area has been doubled, but the length is the same. The circuit diagram in Figure 90(*c*) shows the arrangement of the resistors. Since each resistor has the same resistance

$$r_x = r_y$$

where r_x = the resistance of $r_1+r_2+r_3$
and r_y = the resistance of $r_4+r_5+r_6$

As r_x and r_y represent equal equivalent resistors which are connected in parallel, when n is the number of equal resistors in parallel

$$R_T = \frac{1}{n}\text{th of the resistance of one of the resistors}$$

$$= \frac{r_x}{2}\text{, i.e. } \tfrac{1}{2} \text{ of the resistance of one branch}$$

Hence it follows that, if the cross-sectional area of a specified length of a particular material at a uniform temperature is doubled, the resistance of the material is halved, i.e. *the resistance is inversely proportional to the cross-sectional area.*

The relationship between resistance and length, and resistance and cross-sectional area is illustrated in the following examples.

Solutions to self-assessment questions

13 (i) *Potential difference across the circuit.*

(ii) The resistance of the second resistor is *15 ohms*. Since

$$R_T = r_1 + r_2$$
$$25 = 10 + r_2$$
$$\therefore r_2 = 25 - 10 = 15\ \Omega$$

14 (i) False. $V_1 = v_1 + v_2 + v_3$.

(ii) True. The circuit current is equal to the sum of the current in the branches.

(iii) True.

(iv) False. The total current $I = i_1 + i_2$.

(v) True.

(vi) True.

15 The table should be completed as follows:

instrument number	ammeter	voltmeter
1	√	
2		√
3		√
4	√	
5	√	
6	√	
7		√

16 In a parallel circuit in which all of the resistors have equal resistance, and n is the number of resistors, the resistance of the circuit, R_T, is given by

$$R_T = \frac{1}{n}\text{th the resistance of one of the resistors}$$

$$= \tfrac{210}{3} = 70\ \Omega$$

From Ohm's law

$$\text{Current, } I = \frac{V}{R} = \frac{140}{70} = 2\ \text{A}$$

Example 9

The resistance of a length of wire, 3 m long, is 30 Ω at a particular temperature. Calculate the resistance of 100 m of the same wire at the same temperature.

Resistance of 1 m of wire $= \frac{30}{3} = 10\ \Omega$

Since resistance is directly proportional to length, the resistance of 100 m of the wire $= 10 \times 100$ ohm $= 1$ kΩ.

Example 10

A length of conducting material has a resistance of 20 Ω at a particular temperature. The same material is used to make a conductor to operate at the same temperature as the original conductor, but which is twice as long and has half the cross-sectional area. Calculate the resistance of the new conductor.

17

$$\text{As } R_T = r_1 + r_2 + r_3$$

$$R_T = 10 + 10 + 10 = 30\ \Omega$$

$$\text{From Ohm's law } I = \frac{V}{R_T}$$

$$\therefore \text{ Current, } I = \frac{90}{30} = 3\text{ A}$$

$$\text{The p.d. across a } 10\ \Omega \text{ resistor} = I \times r$$

$$\text{p.d.} = 3 \times 10 = 30\text{ V}$$

18

$$\text{As } R_T = \frac{r_1 \times r_2}{r_1 + r_2} = \frac{5 \times 10}{5 + 10}$$

$$R_T = \frac{50}{15} = 3.33\ \Omega$$

$$\text{From Ohm's law } I = \frac{V}{R_T}$$

$$\text{Current} = \frac{100}{3.33} = 30\text{ A}$$

19

$$\text{As } \frac{1}{R_T} = \frac{1}{r_1} + \frac{1}{r_2} + \frac{1}{r_3} = \frac{1}{4} + \frac{1}{10} + \frac{1}{20}$$

$$\frac{1}{R_T} = \frac{5 + 2 + 1}{20} = \frac{8}{20}$$

$$\therefore R_T = \frac{20}{8} = 2.5\ \Omega$$

$$\text{From Ohm's law, } I = \frac{V}{R} = \frac{25}{2.5} = 10\text{ A}$$

$$\text{Current } (i_1) \text{ in } r_1 = \frac{V}{r_1} = \frac{25}{4} = 6.25\text{ A}$$

$$\text{Current } (i_2) \text{ in } r_2 = \frac{V}{r_2} = \frac{25}{10} = 2.5\text{ A}$$

$$\text{Current } (i_3) \text{ in } r_3 = \frac{V}{r_3} = \frac{25}{20} = 1.25\text{ A}$$

As $R \propto l$, when the length of the conductor is doubled, the resistance will double.

$\therefore$ R is increased to $2 \times 20\ \Omega = 40\ \Omega$

As $R \propto \frac{1}{A}$, when the cross-sectional area is halved, the resistance doubles. Therefore the resistance of the new conductor is $2 \times 40\ \Omega = 80\ \Omega$.

At a specified temperature, the resistance of a material is dependent upon the physical dimensions of the material and upon the atomic structure of the material. The atomic structure of the material determines the availability of free electrons to carry current, and the magnitude of the resistance to a flow of electrons. In order to compare the resistance of materials, the resistance between opposite faces of a unit cube of the material at a stated uniform temperature is measured. The result of this measurement is called the *resistivity* of the material. The symbol used for resistivity is ρ (rho). *The fundamental unit of resistivity is the ohm metre* in which the ohm is the unit of resistance, and the metre the length of the side of the unit cube.

$$R = \frac{\rho l}{A} \text{ ohms}$$

Materials used as conductors (e.g. copper, aluminium, brass) have low resistivities. For conductors, the preferred unit of resistivity is the microhm millimetre (μΩ mm). Materials which have very high resistivities (e.g. glass, rubber, PVC) are used to limit the flow of current. These materials are called electrical insulators. The preferred unit of resistivity for insulators is the megohm millimetre (MΩ mm).

The resistivity of a particular material at a specific temperature is a constant. Therefore the resistance of a material at a particular temperature is dependent upon its physical dimensions and the resistivity of the material at that same temperature –

The resistivity of an insulator *decreases* with increasing temperature. The resistivity of a conductor *increases* with increasing temperature. This phenomenon may be explained in the following manner.

Consider an insulator subjected to an e.m.f. and to heat. The heat releases electrons from their parent atoms, and because of the e.m.f., these electrons constitute a current flow. Hence the resistivity is decreased.

If a conductor, subjected to an e.m.f., is heated, the random movement of the free electrons is increased. This increases the number of collisions between electrons, and hence the resistivity is increased. Thus, the resistance of both insulators and conductors is dependent upon temperature.

Self-assessment questions

20 List the four factors which affect the resistance of a material.

21 State the formula which relates physical dimensions and the type of material to the resistance of a material. State the units of each symbol in the formula.

22 Complete each of the following statements by supplying the missing word(s):

(i) A temperature rise in a material with very few free electrons causes the resistivity of the material to ______.

(ii) At a specific temperature the resistivity of a material is ______.

(iii) A change in the resistance of most materials occurs when the material is subjected to (*a*) a change in physical dimensions, and (*b*) ______.

23 If the length of a conductor is doubled, what happens to its value of resistance?

24 If the cross-sectional area of a conductor is doubled, what happens to its value of resistance?

After reading the following material, the reader shall:

6 Know the effects of changes in the p.d. across filament lamps.
6.1 Determine the p.d. across lamps connected in series.
6.2 Compare the merits of wiring lamps in –
(*a*) series
(*b*) parallel.

Filament lamps are used in domestic, commercial and industrial electrical installations to provide illumination. The filament in an electric lamp is designed to operate at temperatures up to about 3 000 °C; at around this temperature the filament glows white hot and emits light. The quantity of heat energy converted into light depends upon the resistance of the filament at its working temperature, and the current flowing through the filament at this temperature. The product of this current and resistance produces a particular p.d. across the lamp. This p.d. is the 'designed operating p.d.' of the lamp. When a lamp is connected to a supply which maintains the required operating p.d. across its terminals, the lamp works at its rated temperature, i.e. is at full brilliance. Should a lamp be connected to a p.d. below its designed operating p.d., the lamp does not reach its correct operating temperature and is below full brilliance. If a lamp is connected to a p.d. above its designed operating p.d., the lamp exceeds its design temperature and 'burns out'.

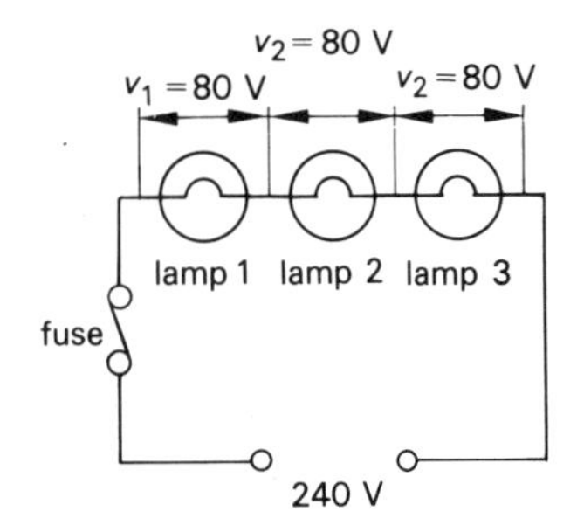

Figure 91 *Lamps connected in series*

The circuit diagram in Figure 91 shows three filament lamps connected in series. Each lamp is designed to work efficiently when a p.d. of 240 V is maintained between its terminals; also each lamp is designed to convert electrical energy into heat energy at the same rate. The p.d. across the circuit is maintained at 240 V.

In this circuit when V is the p.d. across the circuit and the p.d. across lamp 1 is v_1, across lamp 2 is v_2, and across lamp 3 is v_3, then

$$V = v_1 + v_2 + v_3$$

In this case, as the lamps are identical, the p.d. across each lamp is the circuit p.d. divided by the number of lamps.

$$\text{p.d. across a lamp} = \frac{V}{\text{number of lamps}} \text{ volts}$$

$$= \frac{240}{3} = 80 \text{ volts}$$

This much lower than normal p.d. across each lamp prevents each lamp reaching its design temperature. Instead of being white hot the filaments of the lamps glow red hot. They would last a very long time, but provide little illumination. Should one of the lamps develop a fault which reduces its resistance to around zero ohms, this lamp goes out and the current in the circuit would increase. The results of this are that the p.d. across each remaining lamp would rise, and the amount of heat energy converted into light in these lamps would increase. Should one of the lamps 'open circuit', i.e. the filament break, all the lamps in the circuit go out. It follows that when lamps are connected in series in a circuit connected to a constant p.d., any change in the resistance of the circuit affects all the lamps in the circuit.

Solutions to self-assessment questions

20 The four factors which affect the resistance of a material are length, cross-sectional area, material and temperature.

21 $R = \dfrac{\rho l}{A}$ ohms

The units of ρ are ohm metre; the units of l are metres; the units of A are m^2.

22 (i) A temperature rise in a material with very few free electrons causes the resistivity of the material to *decrease*. The resistivity may change from megohm millimetre to ohm millimetre.

(ii) At a specific temperature the resistivity of a material is *constant*.

(iii) A change in the resistance of most materials occurs when the material is subjected to (*a*) a change in physical dimensions, and (*b*) a change in *temperature*.

23 Its resistance is doubled, since R is proportional to l.

24 Its resistance is halved, since R is proportional to $1/A$.

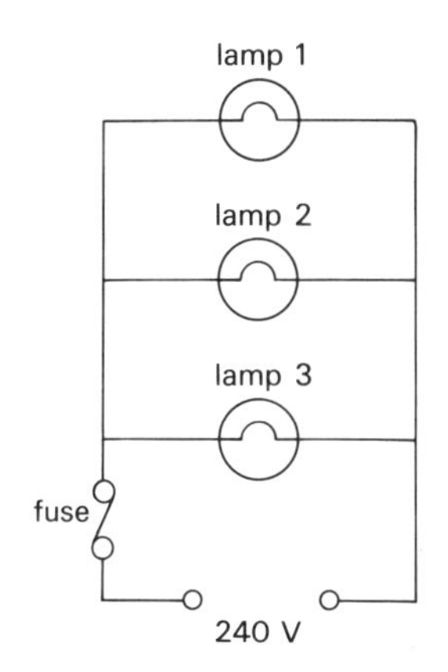

Figure 92 *Lamps connected in parallel*

In a parallel circuit, shown in Figure 92, each lamp is connected across a constant p.d. Provided the lamps are designed to operate on this p.d., each lamp can be a different size, i.e. have a different resistance, without affecting the other lamps in the circuit. One lamp may be switched off, and the others continue to operate at their normal efficiency. Should one lamp 'short circuit', then the current increases, but the circuit protection device, in this case a fuse, operates to interrupt the flow of current in the whole circuit.

The flexibility afforded by the connection of lamps in parallel makes this the most widely used form of connection in electrical installations. The series form of connecting lamps is limited to decorative lighting, such as the lights on a Christmas tree.

Self-assessment questions

25 A set of decorative lighting consists of twelve identical lamps connected in series. What is the potential difference across each lamp when the set is connected to a 240 V supply?

26 In a set of twelve lamps connected in series and operating on a 240 V supply, two of the lamps have been 'short circuited'. If the supply is maintained at 240 V, what is the p.d. across each remaining lamp?

27 Six 240 V lamps, each having the same filament resistance, are connected to a 240 V supply, first in series, then in parallel. Each circuit is protected by a fuse. The current in each circuit is measured with an ammeter. Assume that the resistance of the filaments remain constant.

Compare the effects of each form of connection by selecting, from the following statements, those which refer to the series circuit (*S*) and those which refer to the parallel circuit (*P*):

(i) The p.d. across each lamp is 40 V.

S/P

(ii) The lamps operate at full brilliance.

S/P

(iii) The p.d. across each lamp is 240 V.

S/P

(iv) Each lamp is at about $\frac{1}{6}$th of full brilliance.

S/P

(v) Each lamp operates at full rated current.

S/P

(vi) The circuit current is lowest when the lamps are connected in ______.

S/P

(vii) The current in one circuit is about six times greater than in the ______ circuit.

S/P

(viii) The lamps in this circuit last a very long time compared with the other circuit.

S/P

(ix) A 'short' across one lamp may cause the circuit fuse to interrupt the supply.

S/P

(x) An open circuit in one lamp has no effect on the other lamps.

S/P

After reading the following material, the reader shall:

7 Apply the power concept to electrical circuits.
7.1 State that the power produced in a circuit is equal to the product of potential difference and current.
7.2 Use Ohm's law to show that $P = I^2R$.
7.3 Calculate the power dissipated in simple circuits.

When a current flows in a circuit, electrical energy is converted into other forms of energy. The rate at which this happens is very important. The rate of conversion of electrical energy is called *power*.

$$\text{Power} = \text{rate of energy conversion} = \frac{\text{energy converted in joules}}{\text{time in seconds for conversion to take place}}$$

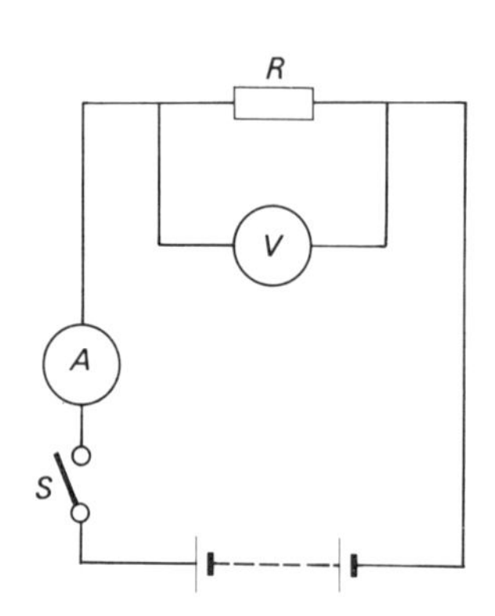

Figure 93 *Power in a circuit*

The circuit diagram in Figure 93 shows a resistor (R) across which a voltmeter (V) measures the p.d. The circuit contains an ammeter (A) which measures the current in the circuit, and a battery which supplies the current (electrical energy) when the switch (S) is closed. When the source of electrical energy is connected to the load, all the components in the circuit convert electrical energy into other forms of energy; for the purposes of the present analysis it is assumed that all the electrical energy is converted in the resistor. Note that this is not the case in practice.

Let the e.m.f. of the battery be sufficient to maintain a p.d. of 100 V across the resistor when a current of 6 A is flowing in the circuit. A current of 6 A means that 6 coulombs of charge pass a point in the resistor in every second. The p.d. of 100 V across the resistor means that each coulomb of charge passing through the resistor transfers 100 joules of electrical energy to the resistor. In the resistor, this electrical energy is converted mainly into heat energy, but there is some magnetic energy and possibly a chemical effect.

The energy converted in each second in the resistor

$$= V \times I \times t$$
$$= 100 \times 6 \times 1$$
$$= 600 \text{ J}$$

$$\text{As power} = \frac{\text{energy converted in joules}}{\text{time in seconds}}$$

$$\text{power} = \frac{600}{1} \left[\frac{V \times I \times t}{t}\right] = 600 \text{ J/s}$$

The joule per second is called the watt (W)

$$1 \text{ W} = 1 \text{ J/s}$$
$$\text{Power, } P = VI \text{ watts}$$

The power rating of many items of electrical apparatus is displayed on the apparatus, e.g. electric lamps in domestic situations used for room illumination are rated at 40 W, 60 W, 100 W, etc. This information, and the p.d. at which the apparatus is designed to operate, indicate the rate at which electrical energy is converted.

Example 11

A battery supplies a current of 20 A to a heater which operates at a p.d. of 100 V. Calculate the power rating of the heater.

$$\text{Power rating } P = V \times I \text{ watts}$$
$$= 100 \times 20$$
$$= 2\,000 \text{ W} = 2 \text{ kW}$$

This means that the heater, when connected across a supply which can maintain a p.d. between the heater terminals of 100 V, will convert electrical energy at the rate of 2 000 J/s.

In an electrical circuit the current can be determined from Ohm's law: $I = \dfrac{V}{R}$ amperes. Combining this equation with the equation, $P = VI$ watts produces two other equations:

$$\text{From Ohm's law, } I = \frac{V}{R} \text{ amperes}$$
$$\text{also } P = VI \text{ watts}$$

By substituting for I in the power equation,

$$P = V \times \frac{V}{R} = \frac{V^2}{R} \text{ watts}$$

$$\text{Again from Ohm's law, } V = I \times R$$
$$\text{also } P = V \times I$$

substituting for V in the power equation,

$$P = I^2R \text{ watts}$$

Hence the three power equations are

$$P = VI$$
$$P = \frac{V^2}{R}$$
$$P = I^2R$$

Example 12
A soldering iron is rated at 100 W when connected to a p.d. of 240 V. Calculate the resistance of the soldering iron when energy is being converted at this rate.

Power, $P = \frac{V^2}{R}$ with power in watts, p.d. in volts and resistance in ohms

$$100 = \frac{(240)^2}{R}$$

$$R = \frac{(240)^2}{100}$$

Resistance, $R = 576\ \Omega$

Example 13
A resistor has a resistance of 10 ohms. Calculate the heat generated per second when a current of 5 A is flowing.

Heat generated per second is in joules per second, and

1 J/s = 1 watt

Heat generated per second,

i.e. power, $P = I^2R$
$= (5)^2 10$
Power $= 250$ W

Example 14
Three resistors with resistances of 2 Ω, 4 Ω and 6 Ω are connected in series to a supply which maintains a p.d. of 24 V across the circuit. Calculate the minimum power rating of each resistor.

$$R_T = r_1 + r_2 + r_3$$
$$= 2+4+6 = 12\ \Omega$$

From Ohm's law, $I = \frac{V}{R}$

$\therefore$ Circuit current $= \frac{24}{12} = 2$ A

Solutions to self-assessment questions

25	20 V.
26	24 V.
27 (i)	S
(ii)	P
(iii)	P
(iv)	S
(v)	P
(vi)	S
(vii)	S
(viii)	S
(ix)	P
(x)	P

In a series circuit, the current is the same throughout the circuit, and hence the current passing through each resistor is 2 A. Hence:

$$\text{Power rating of resistor } r_1 = I^2 r_1 = 2^2 \times 2 = \underline{8\ \text{W}}$$

$$\text{Power rating of resistor } r_2 = I^2 r_2 = 2^2 \times 4 = \underline{16\ \text{W}}$$

$$\text{Power rating of resistor } r_3 = I^2 r_3 = 2^2 \times 6 = \underline{24\ \text{W}}$$

Example 15

Three resistors with resistance of 2 Ω, 4 Ω and 6 Ω are connected in parallel to a supply which maintains a p.d. of 24 V across the circuit. Calculate the minimum power rating of each resistor.

In a parallel circuit, the current is equal to the sum of the current in each branch, and the p.d. across each resistor is the same.

The p.d. across resistors r_1, r_2 and r_3 = 24 V

$$\therefore \text{Current in } r_1 = i_1 = \frac{V}{r_1} = \frac{24}{2} = 12\ \text{A}$$

$$\text{Current in } r_2 = i_2 = \frac{V}{r_2} = \frac{24}{4} = 6\ \text{A}$$

$$\text{Current in } r_3 = i_3 = \frac{V}{r_3} = \frac{24}{6} = 4\ \text{A}$$

$$\text{Power rating of resistor } r_1 = {i_1}^2 \times r_1 = 12^2 \times 2 = 288\ \text{W}$$

$$\text{Power rating of resistor } r_2 = {i_2}^2 \times r_2 = 6^2 \times 4 = 144\ \text{W}$$

$$\text{Power rating of resistor } r_3 = {i_3}^2 \times r_3 = 4^2 \times 6 = 96\ \text{W}$$

All electrical apparatus is designed to allow for the rise in temperature which occurs when a current flows. The temperature rise permitted is limited by the safe operating temperature of the resistor, and the rate at which heat energy is dissipated. To ensure that the temperature rise stays within foreseen limits, each resistor is rated in watts. In Example 14, the 2 ohm resistor would have a rating such that it can convert a minimum of 8 J/s into heat energy without causing damage. A 2 ohm resistor with a rating of 10 W would be used in practice. However, if such a resistor with a 10 W power rating is connected in parallel with a 24 V supply as in Example 15, it would certainly overheat and possibly melt, unless special precautions are taken to dissipate the excess heat energy.

Self-assessment questions

28 The p.d. across a resistor is 200 V, and the current flowing through the resistor is 5 A. What is the minimum power rating of the resistor?

29 A current of 8 A flows through a resistor of 10 ohms. Calculate the energy converted in joules per second.

30 Two resistors $r_1 = 10\ \Omega$, $r_2 = 4\ \Omega$ are connected in parallel to a 40 V supply. Calculate the power rating of each resistor.

31 Using Ohm's law, show that
$P = I^2R$.

After reading the following material, the reader shall:

8 Describe the magnetic field concept and the relationships between magnetic fields and electric currents.
8.1 State that a magnet experiences a force when in a magnetic field.
8.2 State that a current carrying conductor produces a magnetic field.
8.3 Describe the type of magnetic field pattern produced by
(*a*) a bar magnet
(*b*) a solenoid.

Around 600 BC a mineral called lodestone was found to have strange properties. These properties are
(i) Lodestone can attract or repel other pieces of lodestone.
(ii) Lodestone, when freely suspended, aligns itself in a north and south direction.

This second property was exploited as a navigational aid. The part of the lodestone which points towards the north pole is called a *north-seeking pole*; the opposite end is called a *south-seeking pole*. A modern version of this direction-locating device is called a magnetic compass.

The mineral lodestone was discovered in a place in Turkey called Magnesia, and the properties of lodestone were named after this place. A substance which can exert a force of attraction and repulsion on another substance, and aligns itself in a north and south direction is said to possess *magnetism*, and is referred to as a *magnet*.

Through the centuries it was discovered that magnetism could be produced in short lengths of iron bar. It was also found that the effects of magnetization were most noticeable at the ends of the iron

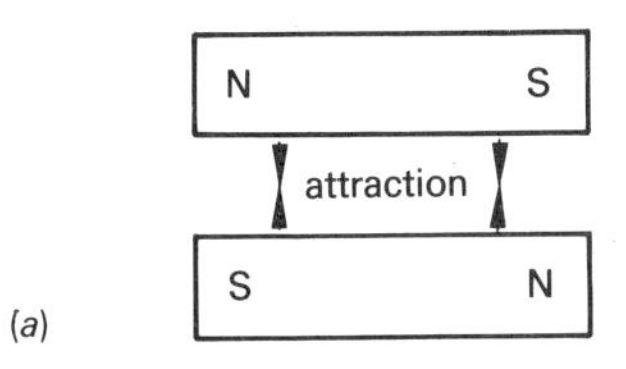

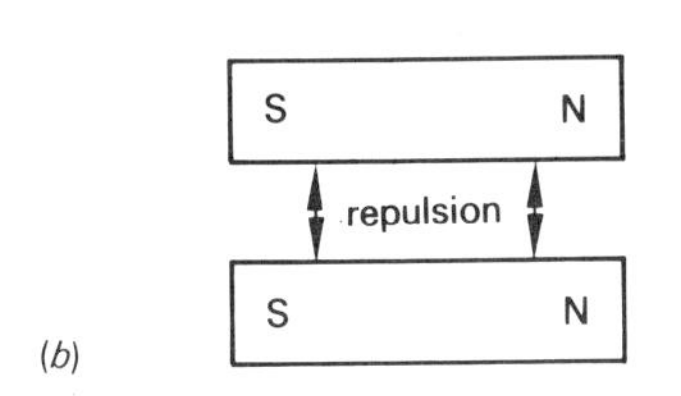

Figure 94 (a) *Unlike poles attract;* (*b*) *like poles repel*

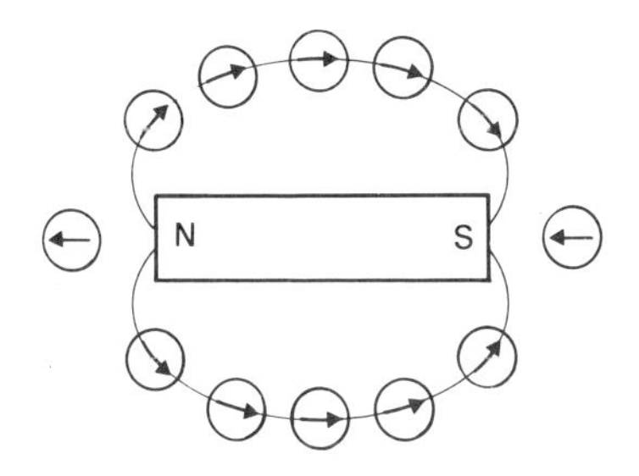

Figure 95 *Direction of magnetic field*

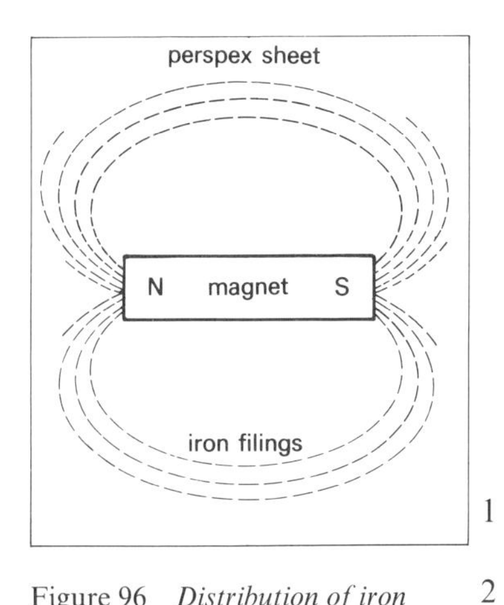

Figure 96 *Distribution of iron filings in a magnetic field*

bar. Magnets of this type are called *bar magnets* and their ends *poles*. One pole is a north-seeking pole, the other is a south-seeking pole. The force of attraction or repulsion between the poles of bar magnets gives rise to a fundamental rule.
This rule is:

Unlike poles attract, like poles repel.

In Figure 94(*a*) two bar magnets are located so that unlike poles are close together. In this condition, a force of attraction in the space between the two magnets will tend to draw the magnets together.

In Figure 94(*b*) the bar magnets are located so that there is a force of repulsion in the space between the two magnets.

The force between the two magnets is produced by the *magnetic flux* (Φ) possessed by each magnet. The space in which a magnetic effect (i.e. a force of attraction, repulsion) can be detected is called a magnetic field.

The rule, *unlike poles attract and like poles repel*, enables magnetic fields to be mapped with the aid of a small magnet, i.e. a magnetic compass.

Figure 95 shows a bar magnet with the north-seeking pole marked N, and the south-seeking pole marked S. Around the bar magnet are located small magnetic compasses. The north-seeking pole of each compass needle is marked with an arrow head. The arrow heads point away from the 'N' pole of the magnet because of the force of repulsion between like poles. Each compass needle indicates the direction of the magnetic field at that point, which is produced by the magnetic flux possessed by the bar magnet. A line drawn through each compass pivot point is called a *magnetic line of flux or force*. Note that lines of magnetic flux do not exist in a physical sense; they are a concept, first introduced by Faraday, to represent the distribution and direction of a magnetic field.

Iron filings, sprinkled on a sheet of perspex placed over a magnet, may also be used to obtain a pictorial representation of the shape and distribution of a magnetic field. This is shown in Figure 96. The filings settle in chains, which form loops between the 'N' pole and the 'S' pole of a bar magnet.

Characteristics are attributed to the imaginary lines of magnetic flux or force, which help to explain magnetic phenomena.

There are five characteristics:

1 the direction of a line of magnetic flux or force in a magnetic field is that of the north-seeking pole of a compass needle at that point.
2 the lines form closed loops.
3 magnetic forces repel each other when lines are in the same direction.

4 the lines behave like stretched elastic, in that they tend to shorten themselves.

5 the lines never intersect.

Up to 1820 the only magnetic properties used were those relating to the magnetic compass. In 1820, about twenty years after the invention of the electric battery, a discovery was made which enabled magnetism to be utilized widely.

Experiments with batteries revealed that when a conductor was connected between the terminals of a battery, the current which flowed in the conductor caused a compass needle to be deflected. The experiments revealed that the current in the conductor was producing a magnetic field which interacted with the magnetic field of the compass needle. This caused a force which deflected the needle from its normal alignment.

Solutions to self-assessment questions

28 The minimum power rating of the resistor, P, is given by:

$$P = V \times I \text{ watts}$$
$$= 200 \times 5$$

Power rating $= 1\,000$ W $= 1$ kW

29 The energy converted by the resistor in joules per second $= I^2R$ watts

$$= 8^2 \times 10$$

Energy converted $= 640$ J/s

30 The p.d. across each resistor is 40 V.

Therefore the minimum power rating of resistor $r_1 = \frac{V^2}{r_1}$ watts

$$= \frac{40^2}{10}$$

Power rating of $r_1 = \frac{1\,600}{10} = 160$ W

The minimum power rating of resistor $r_2 = \frac{V^2}{r_2}$ watts

$$= \frac{40^2}{4}$$

Power rating of $r_2 = \frac{1\,600}{4} = 400$ W

31 From Ohm's law

$$V = I \times R \text{ volts}$$

and power $P = V \times I$ watts

Substituting for V in $P = V \times I$

$$P = (I \times R) \times I$$
$$\therefore P = I^2R \text{ watts}$$

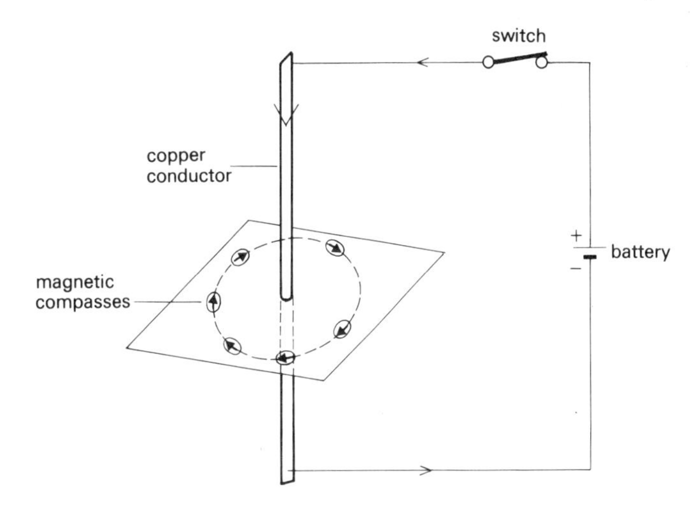

Figure 97 *The direction of the magnetic field around a current-carrying conductor*

The sketch in Figure 97 shows a copper conductor around which small magnetic compasses are arranged in a circle. Current is supplied to this conductor by a battery, and the current flow controlled by a switch. It is usual to represent the current as flowing from the positive terminal of the battery, and to call this direction *conventional current flow*.

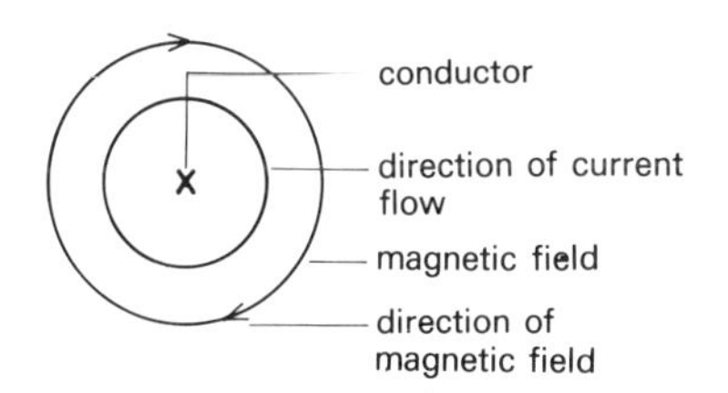

Figure 98 *Conventional representation of a conductor, current and magnetic field*

When the switch is closed, a current flows in the copper conductor and the compass needles align themselves in a circle around the conductor, as shown in the sketch. When the switch is opened, the compass needles re-align to point north and south. When the direction of current flow is reversed, the compass needles re-align into a circle, the north-seeking poles of the compasses pointing in the reverse direction.

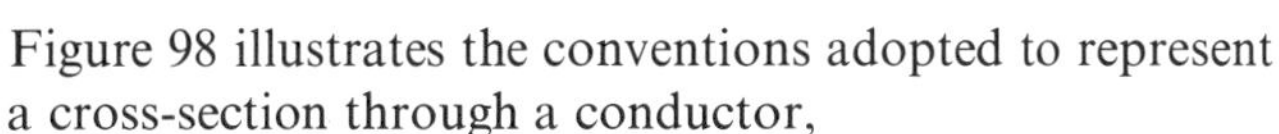

Figure 98 illustrates the conventions adopted to represent

(i) a cross-section through a conductor,
(ii) the direction of current flow,
(iii) the magnetic field,
and (iv) the direction of the magnetic field

for the direction of the current flow shown. A cross is used to indicate that the current is flowing away from the observer.

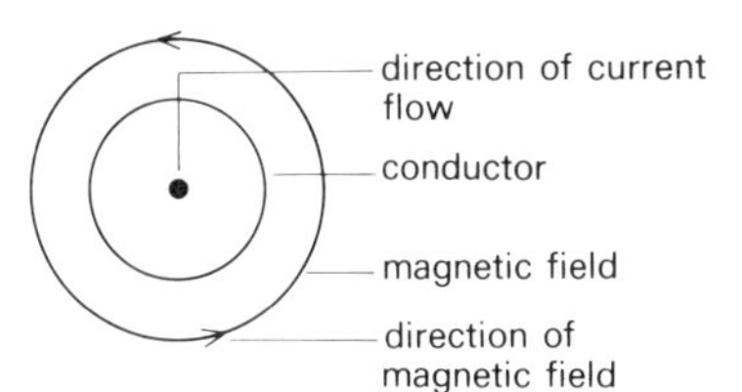

Figure 99 *Representation of a conductor, current and magnetic field*

Figure 99 shows a conductor with the current flowing towards the observer. This direction of current flow is represented by a dot; the direction of the magnetic field is anticlockwise.

The magnetic effect produced by a current can be concentrated by winding the conductor into a coil.

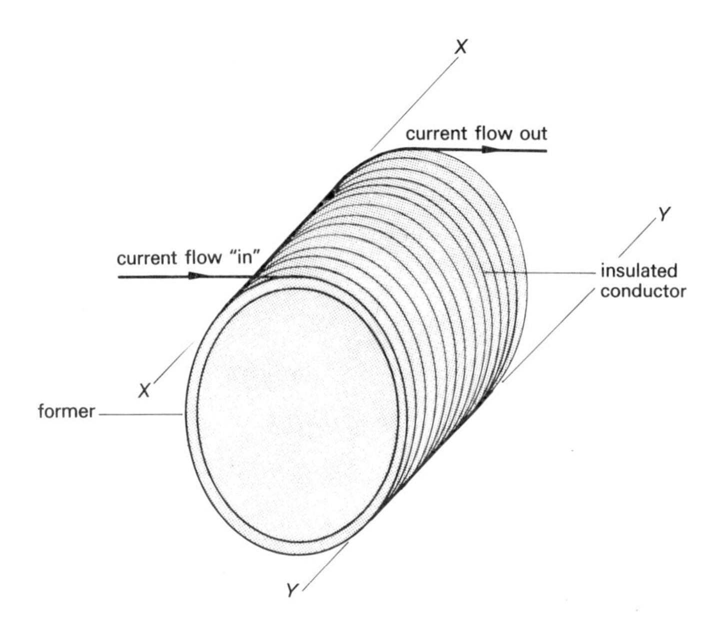

Figure 100 *A coil on a former*

Figure 100 shows an insulated conductor wound around a former to produce a coil.

Figure 101 shows the coil sectioned along the planes *X–X* and *Y–Y*. In this figure, each half turn of the coil is represented by a circle with a cross to show the current flowing inwards at one end of the half turn, and a circle with a dot to show the current flowing outwards at the other end of the half turn. The magnetic field, and the direction of the magnetic field for each half turn, is indicated by a circle and an arrowhead. The compass needles indicate that one end of the coil is a north-seeking pole, and the other a south-seeking pole.

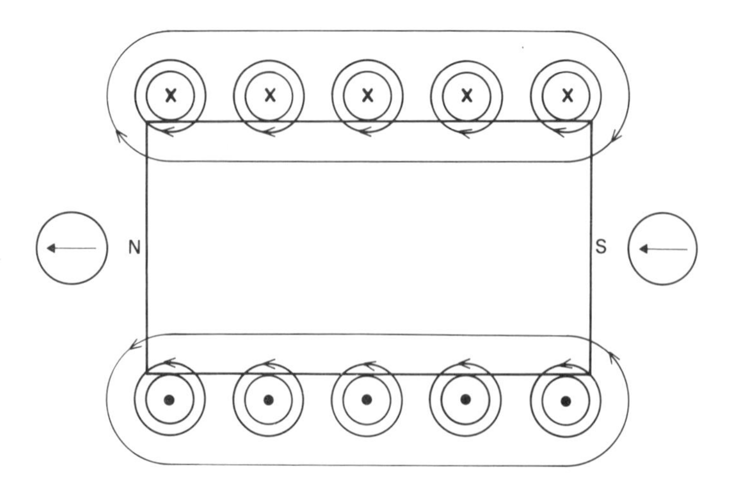

Figure 101 *Representation of the magnetic field produced by a current in a coil*

The magnetic effect of a current in a coil is greatly increased if an iron core is introduced into the coil. This arrangement is called an *electro-magnet* or *solenoid*, and is very widely used in electrical apparatus.

Figure 102 shows an application of a solenoid. The current in the coil sets up a magnetic field in the iron core, and in the ferrous frame. This magnetic field attracts the moving armature towards the iron core. To the moving armature is secured a contact which closes the bell and battery circuit. This circuit operates until the current in the coil of the solenoid is switched off.

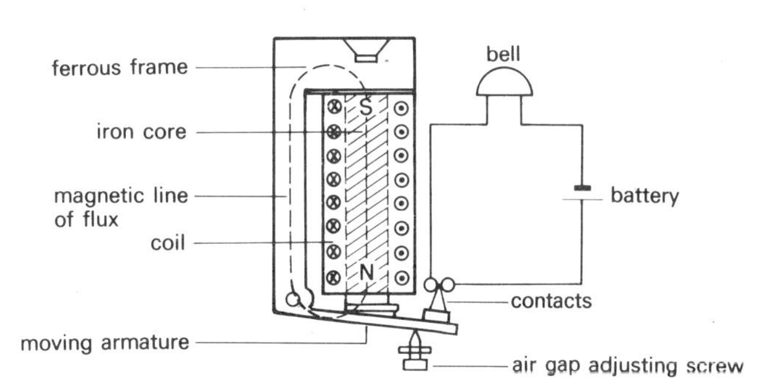

Figure 102 *A solenoid controlling a bell circuit*

The discovery that current flowing in a coil could be used to magnetize ferrous materials enabled materials to be classified as *low retentivity* or *high retentivity* magnetic materials. Low retentivity materials lose most of their magnetism when removed from a magnetic field, and are used for electro-magnets. High retentivity materials retain their magnetism, and are used in very great numbers in small electric motors, telecommunications, etc.

Self-assessment questions

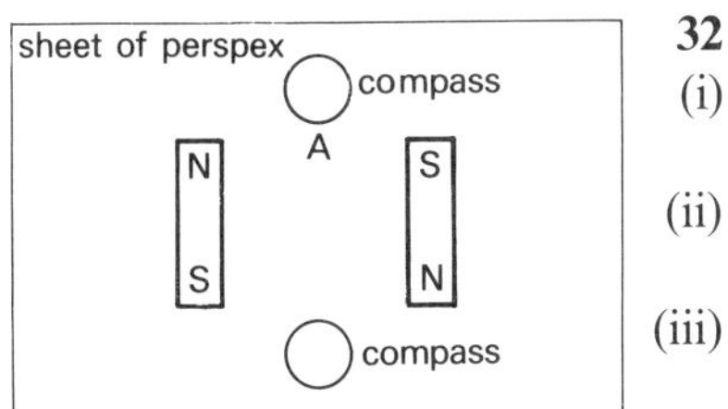

Figure 103 *Representation of the magnetic field due to bar magnets*

32 In Figure 103:

(i) Draw the pattern produced by iron filings on a sheet of perspex placed above two magnets.

(ii) Draw in the compass needles, and indicate their direction with arrow heads at their north-seeking pole.

(iii) State the type of effect (repulsion or attraction) experienced by the magnets in the area marked *A*.

33 Complete the following statements by supplying the missing word(s):

(i) A magnet when placed in a magnetic field is subjected to ______.

(ii) Lines of magnetic flux form continuous ______ ______.

(iii) When current flows in a conductor it produces a heating effect, a chemical effect and a ______ ______.

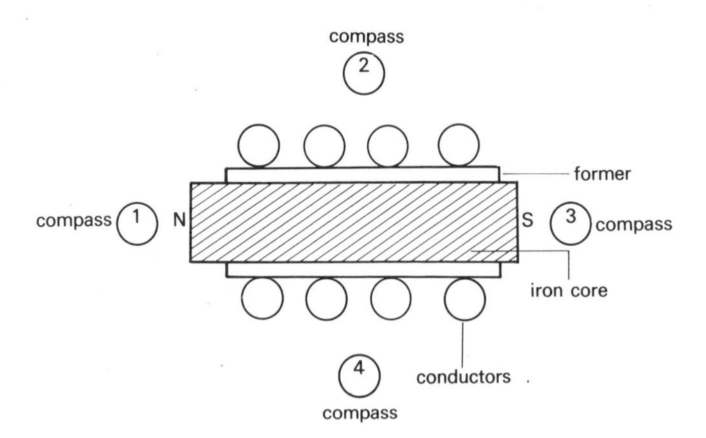

Figure 104 *Representation of a section through a solenoid*

34 The sketch in Figure 104 represents a section through a solenoid.

(i) Indicate the direction of the current flow in each turn of the coil to produce the magnetic polarity shown at the ends of the iron core.

(ii) Draw the compass needles in each of the four compasses, and indicate their direction with arrow heads at their north-seeking poles.

(iii) Draw the lines of magnetic force indicated by the compass needles 2 and 4.

After reading the following material, the reader shall:

8.4 State that a current carrying conductor experiences a force when in a magnetic field.

8.5 Indicate the direction of the force in given conditions.

8.6 Explain, the basic operation of a moving coil meter.

8.7 Explain, the basic operation of a d.c. motor.

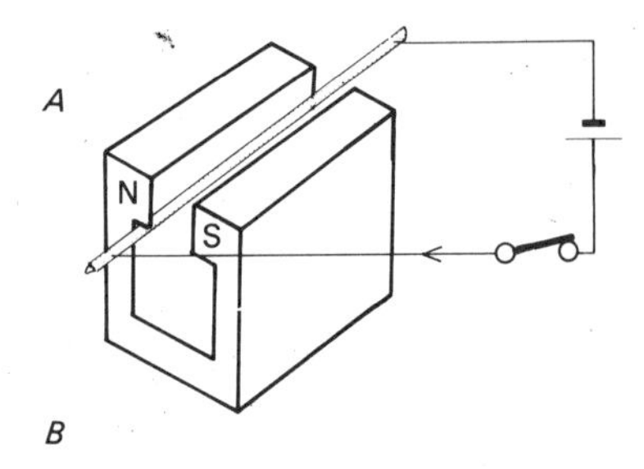

Figure 105 *Conductor in a magnetic field*

The diagram in Figure 105 shows a magnet with a conductor between the poles. The conductor is arranged so that it is free to move in any direction; and is connected to a battery. The current flow is controlled by a switch.

The diagram in Figure 106(*a*) shows an endview of the magnet, and the lines of magnetic force between the poles of the magnet. The conductor is stationary, and is shown with no current flowing in the circuit.

Figure 106(*b*) shows the magnetic field due to the magnet, and the field due to the current in the conductor. The directions of both fields are indicated in the diagram. A characteristic attributed to lines of magnetic force is that when they are in the same direction, they repel each other. In the area between the poles marked *A*, the

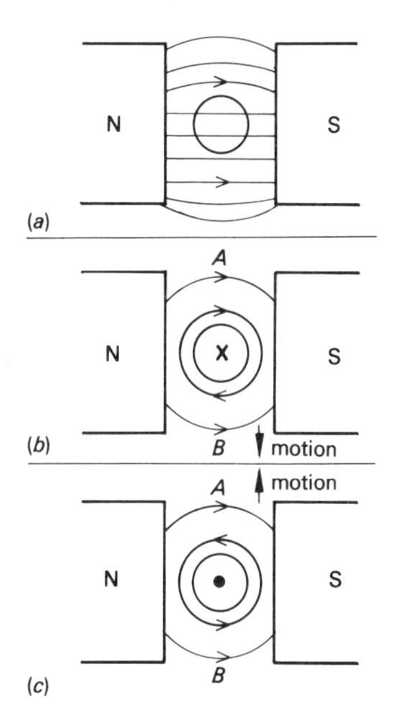

Figure 106 *The interaction of magnetic fields*

lines are in the same direction. In this area there is a force of repulsion. If the force were large enough and the conductor were free to move, the conductor would move down towards the area marked *B*.

In Figure 106(*c*), the direction of the current flow in the conductor has been changed. The lines of magnetic force are in the same direction in the area marked *B*. If the conductor were free to move, and the force large enough, the conductor would move up towards the are marked *A*.

Self-assessment questions

35 Interpret the data in Figure 107(*a*), and state which way the conductor would tend to move.

36 Interpret the data in Figure 107(*b*), and indicate the polarity of the magnet.

37 Interpret the data in Figure 107(*c*), and indicate the direction of the current flow.

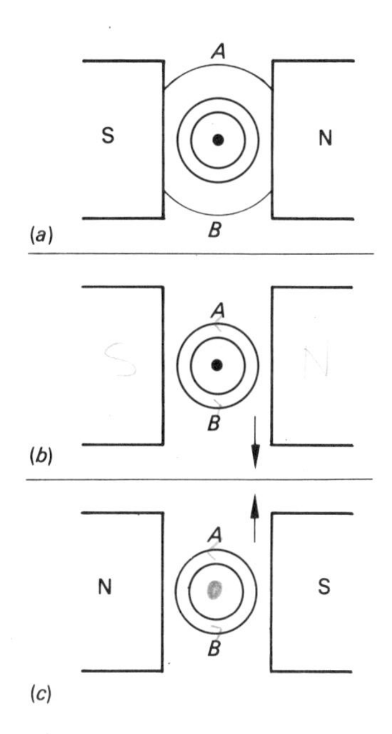

Figure 107

In each of the cases in Figures 106 and 107, interaction between the field due to the conductor and the field due to the magnet, has produced a force which would tend to move the conductor.

The interaction of the magnetic fields, due to (i) a magnet and (ii) a current flowing in a conductor, is used to provide the deflecting force on the pointers of instruments. A very common instrument which uses this type of interaction of magnetic fields is a *moving coil instrument*. Its scales can be calibrated to read current, potential difference, temperature, wind speed, etc.

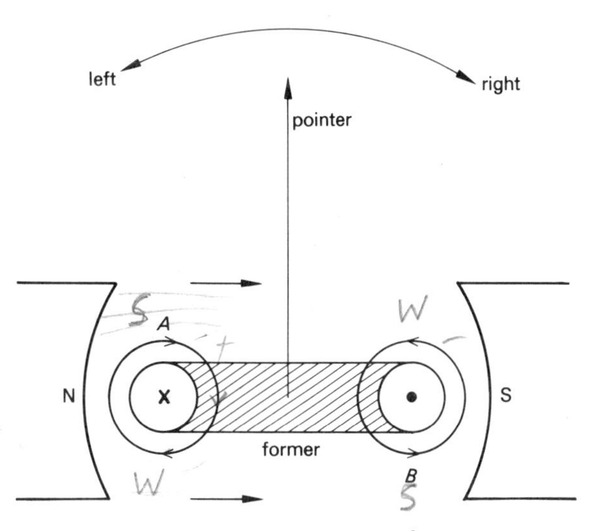

Figure 108 *Magnetic interactions in a moving coil instrument*

Figure 108 is a simplified diagram of a section through the coil of a moving coil instrument, and shows the former, the pointer and the permanent magnet. The direction of current flow, and the direction of the magnetic field produced by the current flow, are shown in the diagram. The direction of the permanent magnetic field is also indicated. An area is marked A and an area marked B.

In area A, the lines of magnetic force produced by the magnet and by the current in the side of the coil are in the same direction. Since lines of magnetic force in the same direction repel, there is a force on the left hand side of the coil which tends to move this part of the coil downwards. In area B the lines of magnetic force due to both the magnet and the current, are in the same direction. Hence there is a force of repulsion, which tends to move this side of the coil away from area B. As the coil is on a former which is free to rotate, the result of these interactions between the magnetic fields is that the pointer moves towards the left.

Solutions to self-assessment questions

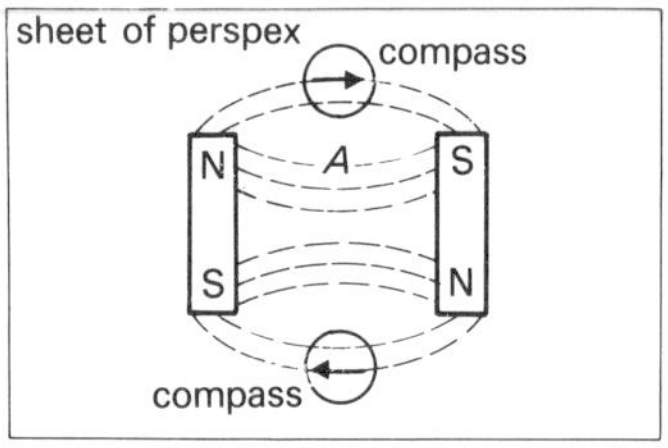

Figure 109

32 (i) The pattern produced by the iron filings should be similar to that shown in Figure 109.
(ii) The directions are as shown in Figure 109.
(iii) The force in area A is a force of attraction between the two magnets.

33 (i) A magnet when placed in a magnetic field is subjected to *a force*.
(ii) Lines of magnetic flux form continuous *loops or chains*.
(iii) When current flows in a conductor it produces a heating effect, a chemical effect and a *magnetic effect*.

34 (i) The direction of current is indicated in Figure 110.
(ii) The directions are shown in Figure 110.
(iii) The lines of magnetic force are shown in Figure 110.

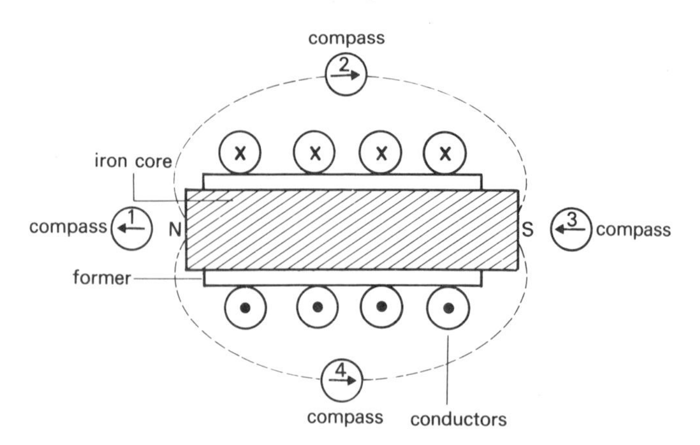

Figure 110 *Representation of the magnetic field of a solenoid*

35 Downwards, i.e. towards area B.

36 The polarity of the magnet is an 'S' on the left hand pole and an 'N' on the right hand pole.

37 The direction of the current flow is outwards and is represented by a dot.

The magnitude of the force on a side of the coil is directly proportional to the current in the coil; hence the deflection of the pointer is also directly proportional to the current. The scale in this case is therefore uniform.

When the current in the coil is reversed, the direction of the lines of force in the magnetic field produced by the current is also reversed. In these conditions the instrument illustrated in Figure 110 would indicate to the right.

If the direction of the current is continually changed, the pointer moves first in one direction, then in the other. If the rate of reversals is very fast, the pointer vibrates around the zero position. This effect means that, in its basic form, a moving coil instrument may only be used with a unidirectional current, i.e. a direct current (d.c.).

The interaction between the magnetic field caused by a current flowing through a conductor and a second magnetic field is the principle on which a d.c. motor operates.

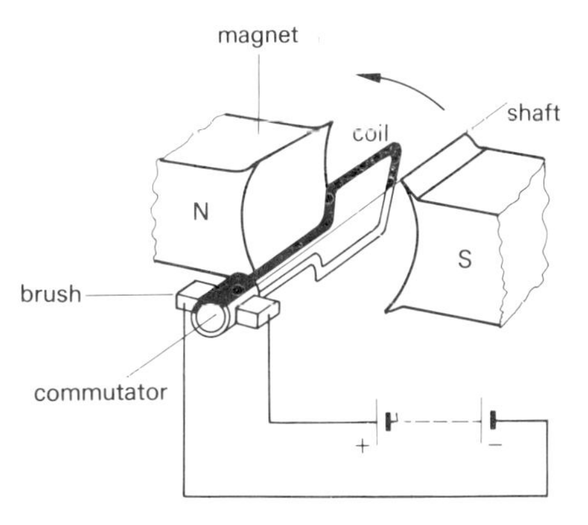

Figure 111 *Simplified d.c. motor*

The sketch in Figure 111 shows a simplified d.c. motor. Current from the battery is supplied to the coil through carbon *brushes*, which are mounted in the motor frame. The fixed carbon brushes form a sliding contact with contacts which are directly connected to each end of the coil. The current from the battery flows through one carbon brush on to the contact to which one end of the coil is connected. The current then flows through the coil to the second contact, and through the second carbon brush to return to the battery. The contacts connected to the ends of the coil are referred to as commutator segments and form a part of the *commutator* as shown in Figure 111. The coil and the commutator are fixed to a shaft which is free to rotate.

When a current flows through the coil, it sets up a magnetic field which interacts with the magnetic field due to the magnets, and causes the coil to rotate.

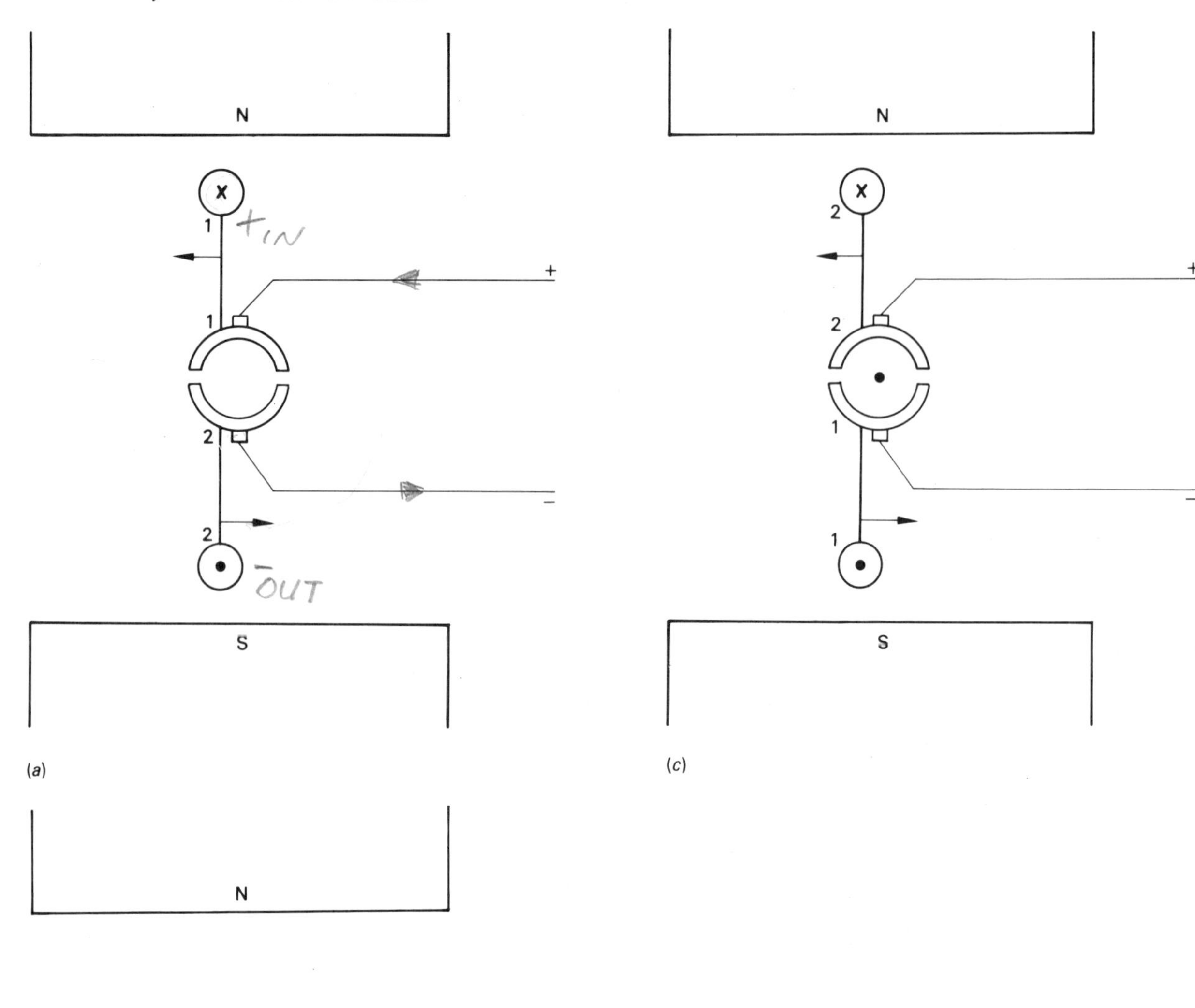

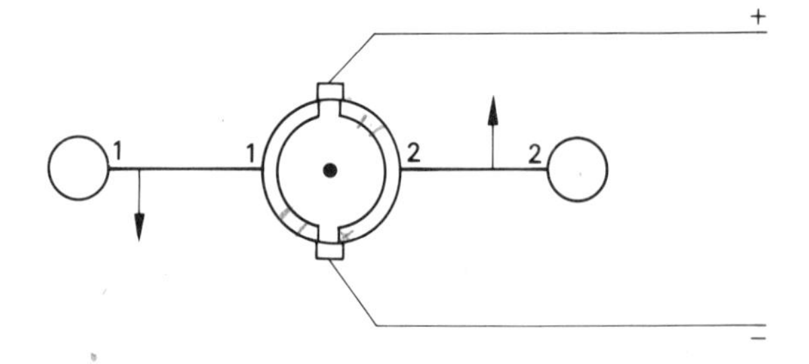

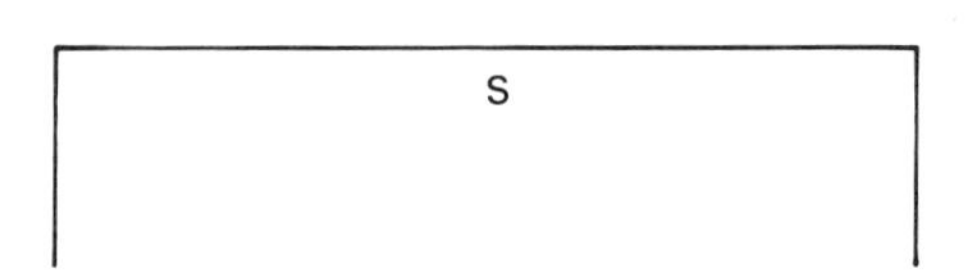

Figure 112 *Principle of a d.c. motor*

Figure 112 illustrates the principle of a d.c. motor. In Figure 112(*a*) the coil sides (numbered 1 and 2) carry a current in the directions shown. They cause a magnetic field, which interacts with the field of the magnet to produce the direction of rotation shown. The commutator segments connected to each coil side are also numbered 1 and 2.

In Figure 112(*b*) the coil sides are shown after a rotation of 90°. At this point the commutator segment connected to side 2 is just breaking contact with one brush, and just making contact with the second brush. The commutator segment connected to coil side 1 is also just making contact with one brush, and just breaking contact with the second brush. At this instant no current is flowing in the coil sides. A fraction of a second later the commutator segments are in full contact with the brushes and current is flowing in the coil sides, but in the reverse direction. The reversal of the direction of the current is referred to as *commutation*.

In Figure 112(*c*) the coil has rotated a further 90°. The directions of the current in the coil sides, the field caused by the current and the field of the magnet are shown.

The process is repeated to complete the revolution, and is continued until the current is switched off.

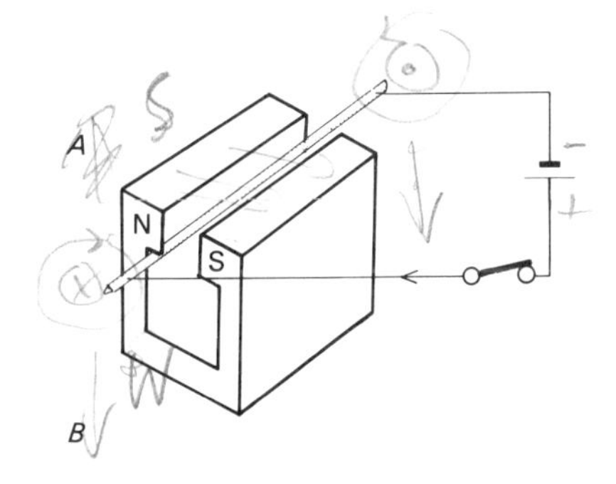

Figure 113

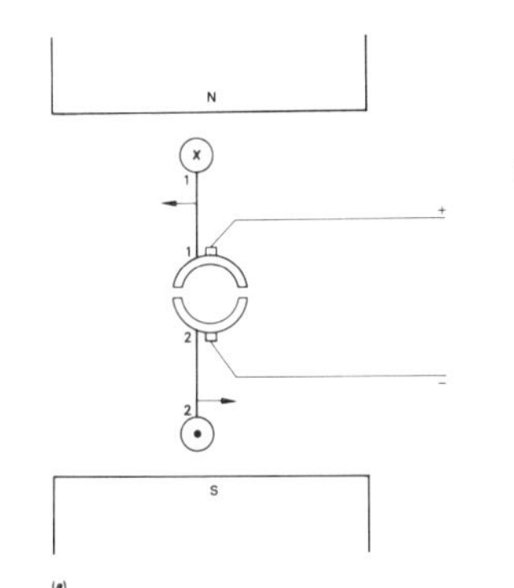

Figure 114

Self-assessment questions

38 For the conditions illustrated in Figure 113 draw,
(i) the direction of the field of the magnet
(ii) the direction of the field due to the current
(iii) an arrow showing the direction in which the conductor tends to move.

39 In Figure 110 the conditions illustrated produce a deflection of the pointer to the left.
(*a*) State the reason for the deflection.
(*b*) State what must happen to cause a deflection to the right.

40 What is the principle on which the operation of a d.c. motor depends?

41 The d.c. motor in Figure 114 is rotating in an anti-clockwise direction. State how it can be made to rotate in a clockwise direction.

After reading the following material, the reader shall:

8.8 Describe what happens when a permanent magnet is moved in a coil of wire connected to a galvanometer.

8.9 Sketch the wave form of the current produced by a magnet moving into and away from a coil.

8.10 Explain in terms of 8.8, the basic operation of an a.c. generator.

8.11 Use Lenz's law to determine the direction of an induced current.

The discovery in 1820 that an electric current in a conductor produces a magnetic flux around the conductor posed a question: can a magnetic flux produce an electric current? The answer to this question was published in England and America in 1832. Two scientists, Faraday in England and Henry in America, found that a magnetic flux could produce an electric current in a conductor. The discovery is credited to Faraday who discovered the phenonemon, called *electromagnetic induction*, on 29 August 1831.

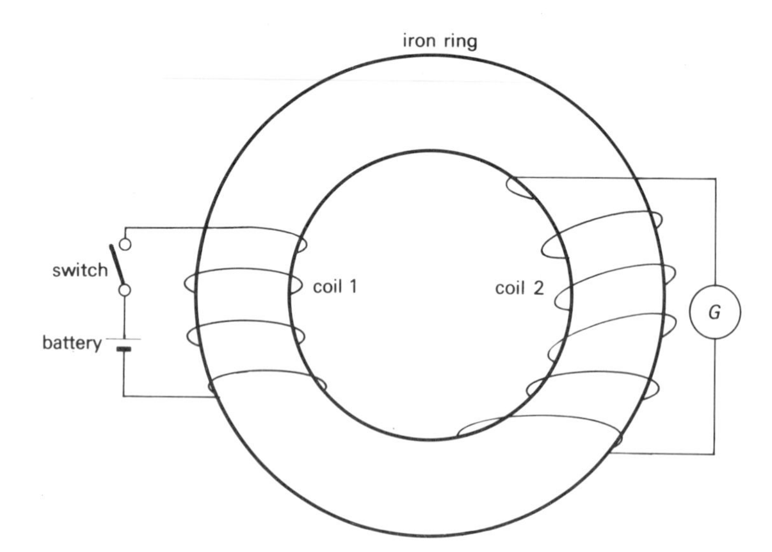

Figure 115 *The production of an induction current*

Figure 115 shows a coil wound on an iron ring and connected to a battery. The current in this coil is controlled by a switch. A second coil is wound on the iron ring. It is connected to a moving coil instrument, which indicates the direction and magnitude of a current flowing in the second coil. The instrument is called a *galvanometer*. This arrangement is similar to that used by Faraday in 1831.

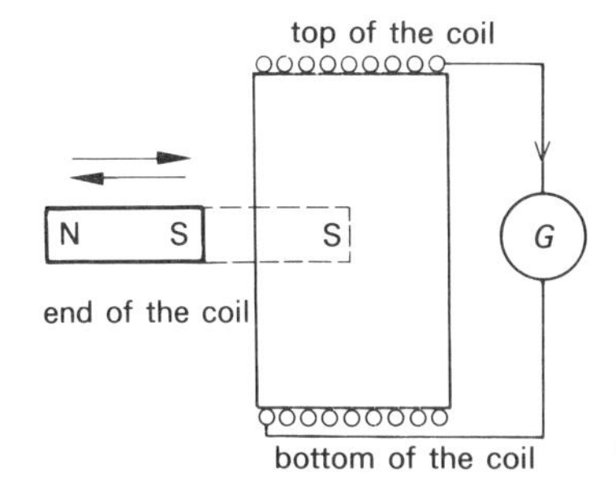

Figure 116 *The production of an induced current moving a magnet*

When the switch is closed, current flows in coil 1. Almost instantaneously the galvanometer needle is deflected, indicating a flow of current in coil 2. The galvanometer needle just as quickly returns to zero, indicating that the current flow in coil 2 has ceased. When the switch is opened to interrupt the current flow in coil 1, the galvanometer again almost instantaneously indicates a current flow in coil 2, which just as quickly ceases. In this experiment Faraday demonstrated that magnetism (electromagnetism) could produce an e.m.f. The name of the phenonemon, electromagnetic induction, is derived from this experiment.

Continuing his research into electromagnetic induction, Faraday found that when a permanent magnet was moved backwards and forwards in a coil, a galvanometer connected to the coil indicated a current flow in the coil. The diagram in Figure 116 shows an arrangement similar to that used by Faraday. He found that the direction of the current flow was dependent upon

(*a*) the direction in which the magnet was moving (towards or away from the end of the coil),
(*b*) whether an 'N' pole or an 'S' pole was nearer to the end of the coil.

Faraday also found that the magnitude of the current flow was dependent upon

(*a*) the speed at which the magnet was moved towards or away from the coil,
(*b*) the number of turns in the coil.

Faraday needed a galvanometer to establish the direction of the current in a coil. In 1834 a scientist called Lenz suggested a method by which the direction of an induced e.m.f., and the current it produces in a circuit, could be forecast.

The method, referred to as Lenz's law, can be expressed as follows: *An induced e.m.f. tends to cause a current to flow in such a direction as to oppose the cause of the induced e.m.f.*

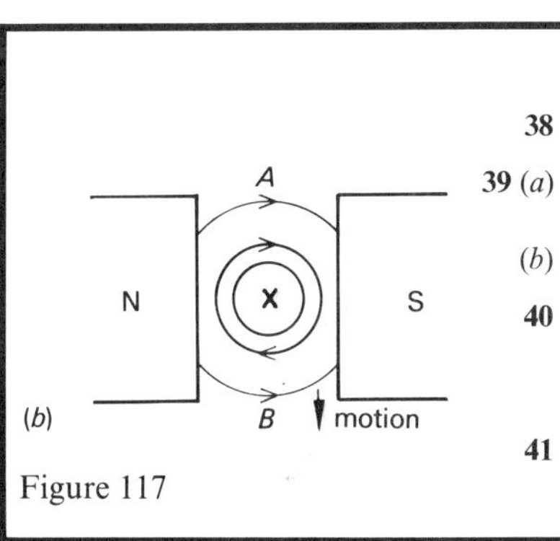

Figure 117

Solutions to self-assessment questions

38 The solution is shown in Figure 117.

39 (*a*) The deflection is caused by the interaction between the magnetic field around a current carrying conductor, and the magnetic field of the magnet.
(*b*) To reverse the direction of the deflection the current in the coil must be reversed.

40 The principle on which the operation of a d.c. motor depends is the interaction between the magnetic field around a current carrying conductor and a second magnetic field.

41 To make the d.c. motor, illustrated in Figure 114, rotate in a clockwise direction, connections to the coil must be reversed.

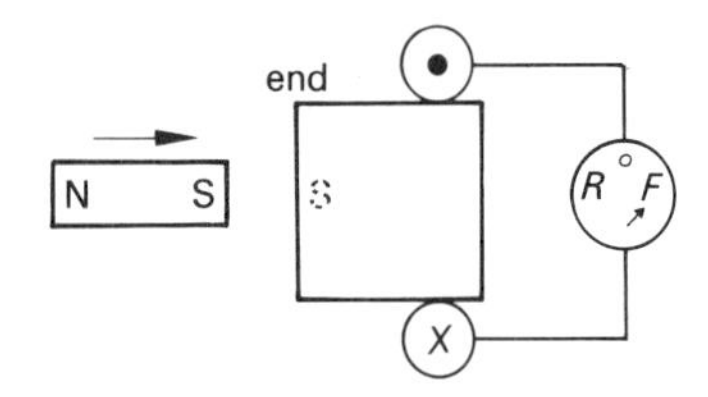

Figure 118 *Lenz's law*

In Figure 118 the magnet is moving towards the end of the coil. Lenz's law requires that the current in the coil produces a magnetic flux in the coil which opposes the movement of the magnet. As the magnet is moving towards the end of the coil, a force of repulsion is required at this end of the coil. The pole of the magnet nearest to this end of the coil is an 'S' pole. It is therefore repelled by an 'S' pole at this end of the coil. In order to produce an 'S' pole at this end of the coil, the coil turns must carry a current in the direction indicated.

The key to understanding electromagnetic induction is to recognize that *the magnetic flux interlinking with conductors must be changing* for an e.m.f. to be induced in the conductors.

In Faraday's initial experiment, illustrated in Figure 115, the magnetic flux interlinking the conductors in coil 2 changes when the current is switched on and off. When the flux is constant, no e.m.f. is induced in coil 2. In the experiment illustrated in Figure 116, the magnetic flux interlinking the conductors in the coil is changing, because the magnet is being moved towards or away from the coil. In this experiment the same effects can be achieved by moving the coil towards and away from a stationary magnet.

Industrial applications of the principle of electromagnetic induction have produced more efficient methods of generating electrical energy than Faraday used in his initial experiments. The diagram in Figure 119 shows a simplified arrangement of a generator. This generator produces an *alternating current* (*a.c.*) and is referred to as an *alternator*.

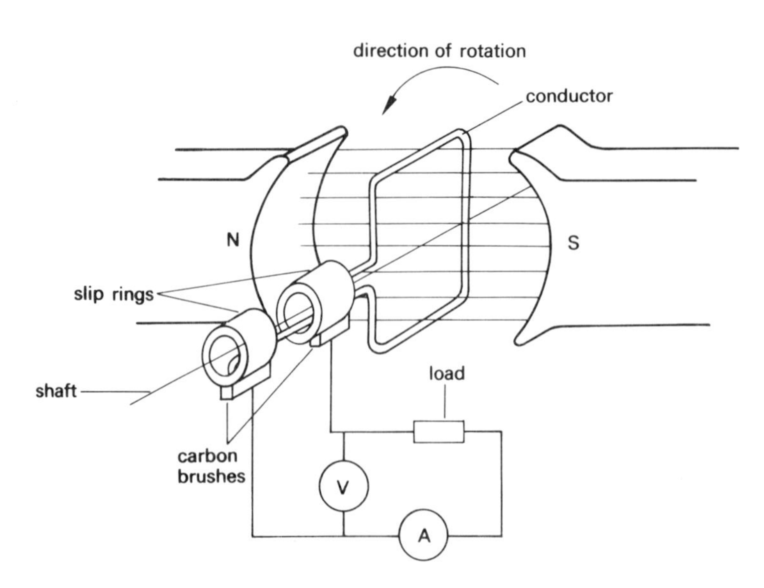

Figure 119 *A simplified generator*

In Figure 119 a coil, represented as a single conductor, is mounted on a shaft; when the shaft is rotated, the coil also rotates between the poles of a magnetic field.

The magnetic field is designed to ensure a uniform distribution of the magnetic flux in the air gap between the poles and the conductors. The ends of the coil are connected to conducting rings, referred to as *slip rings*, mounted on the shaft. The slip rings rotate with the shaft. Each slip ring is connected to the external circuit by a block of carbon called a carbon brush. The carbon brushes are stationary, and contact with the slip rings is maintained by springs located so that they exert a force which pushes the brushes on to the slip ring. When the coil rotates, a sliding contact is maintained between the slip rings and the brushes. The carbon brushes are conductors, and conduct the current generated in the coil to the external circuit. The external circuit contains a resistor, a voltmeter which measures the p.d. across the resistor and an ammeter which measures the current flow in the resistor at any instant in time.

In Figure 120(*a*) the machine is shown in a simplified form as a coil represented by two conductors in between the poles of a uniform magnetic field. The direction of the field is from the pole marked N towards the pole marked S. The coil is rotating. At the instant in time illustrated in Figure 120(*a*) the conductors are moving parallel to the direction of the lines of force in the magnetic field. At this instant, no lines of magnetic force are being 'cut' or are 'cutting conductors'. An alternative way of describing this condition is to state that there are no interlinkages between the conductor and the flux. In this condition, no current will flow in the external circuit. Thus the current and the p.d. are at zero as shown in Figure 120(*a*).

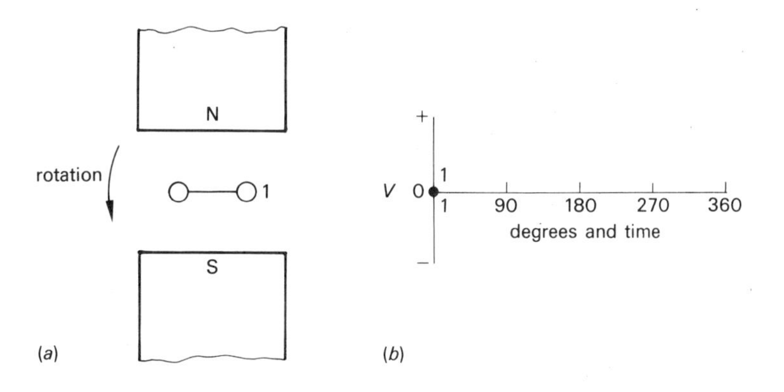

Figure 120

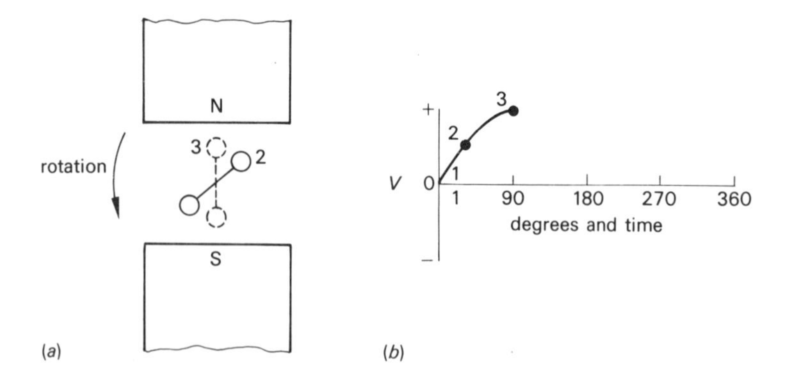

Figure 121

Figure 121(*a*) shows the conductor first after a rotation of 45°, and later after a rotation of 90°. At the 45° position the conductors are 'cutting lines of magnetic force'; therefore a current is flowing in the external circuit. This is represented on the graph in Figure 121(*b*) by point 2. At the 90° position the conductor is 'cutting lines of magnetic force' at right angles to their direction. At this instant the interlinkages between the magnetic fields are at a maximum; the current in the resistor is therefore also at a maximum. This is represented on the graph in Figure 121(*b*) by point 3.

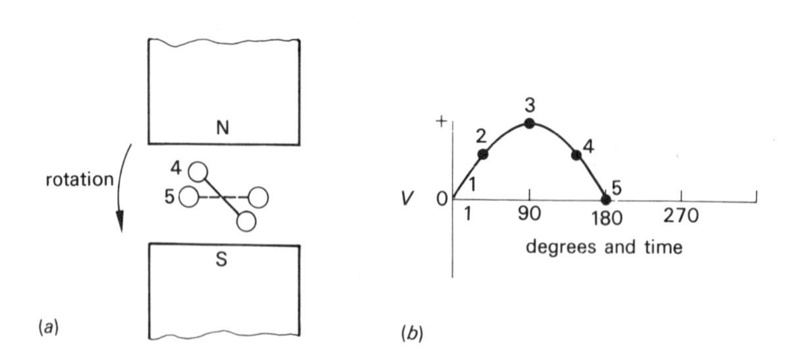

Figure 122

Figure 122(*a*) shows the conductors at 135°, and later at 180° from the start of the rotation. At 135° the rate at which the conductors are interlinking with the magnetic lines of force is reduced, as the conductors are moving in a direction more parallel to the direction of the lines of magnetic force than earlier in the revolution. The decreasing number of interlinkages produces a decreasing flow of current. The current in the external circuit at 135° of rotation is shown in Figure 122(*b*) by point 4. After 180° of rotation, the conductors are moving parallel to the direction of the lines of magnetic force. At this instant there are no flux interlinkages, and therefore there is no current flow in the external circuit. This is represented in Figure 122(*b*) by point 5.

During the 180° of rotation described, the current has changed in magnitude from zero to a maximum value and returned to zero. In the next half revolution the magnitude of the current changes from zero to a maximum value and returns to zero, but acts in the opposite direction.

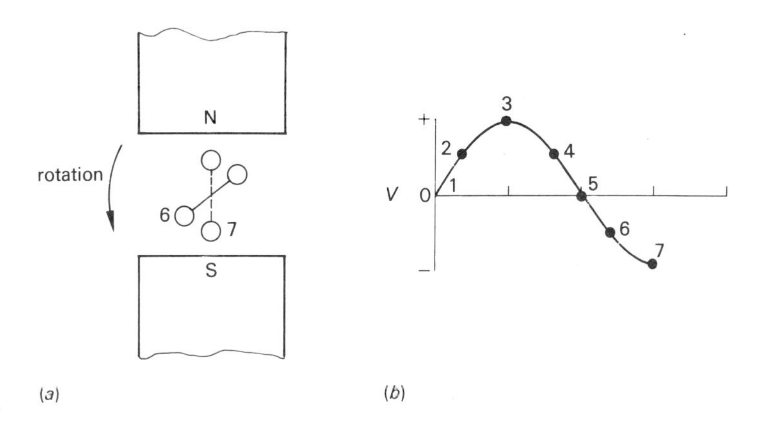

Figure 123 (*a*) (*b*)

Figure 123(*a*) shows the conductors at the 225° position, and later at the 270° position from the start of the rotation. The magnitudes and the direction of the current are shown in Figure 123(*b*) at points 6 and 7.

Figure 124(*a*) shows the conductors at the 315° position, and later at the 360° position from the start of the rotation. The magnitudes and the direction of the current are shown in Figure 124(*b*) at points 8 and 9.

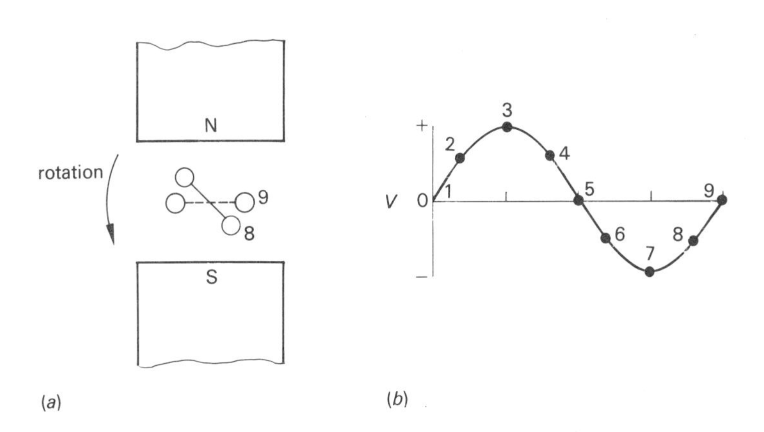

Figure 124 (*a*) (*b*)

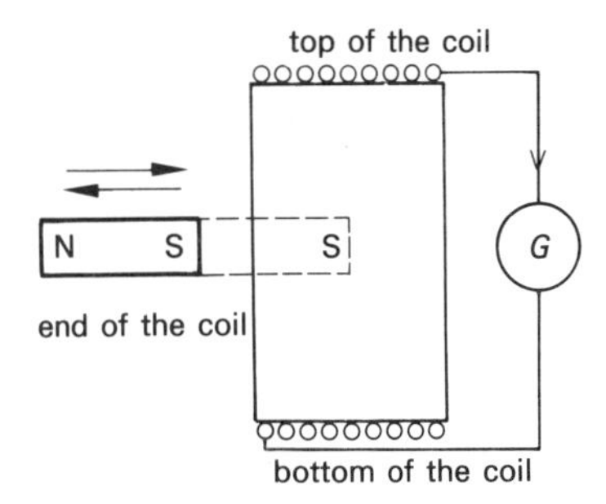

Figure 125

The graph of the variations of the current produced when a coil is rotated in a uniform magnetic field is called an *alternating waveform*. In one revolution the value of the current rises from zero to a maximum, returns to zero, increases to a maximum in the reverse direction, then returns to zero. A current with an alternating wave form is an *alternating current* (*a.c.*). Alternating current is the type of current most commonly used.

Self-assessment questions

42 For the conditions illustrated in Figure 125, two factors determine the direction of the current flow in the galvanometer. State these factors.

43 For the conditions illustrated in Figure 125, two factors determine the magnitude of the current flow in the galvanometer. State these factors.

44 Sketch the waveform of the current flow in the galvanometer in Figure 125 when the magnet is moved towards the coil, and then immediately moved away from the coil.

45 Place a dot or a cross in a conductor on top of the coil shown in Figure 125 to indicate the direction of the current in the coil when the magnet is being moved towards the end of the coil.

46 Complete the following statements by supplying the missing word(s). The questions all refer to the simple a.c. generator shown in Figure 126.

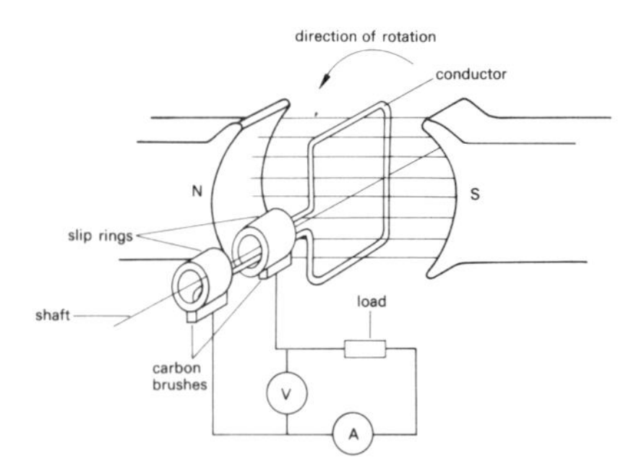

Figure 126

(i) The coil sides are connected to ______.

(ii) The carbon brushes provide a sliding contact between the external circuit and the ______.

(iii) The coil sides rotate in a magnetic field designed to be ______.

(iv) If the flux in the magnetic field is halved, but the speed at which the coil rotates is maintained constant, the maximum current in the external circuit is ______.

(v) The current waveform produced in a 360° rotation is ______.

After reading the following material, the reader shall:

9 Be aware of the three main effects of an electric current.
9.1 State examples of electric current being used for its magnetic effect.
9.2 State examples of electric current being used for its chemical effect.
9.3 State examples of an electric current being used for its heating effect.
9.4 Identify the effect being made use of in given specific cases e.g. electro-magnet, electroplating, electric fire, fuse.

The modern world contains many examples of the utilization of the effects of an electric current. The observable effects of an electric current in a circuit depend to a large extent on the materials in the circuit. Circuits are designed so that electrical energy is converted predominantly into heat energy or magnetic energy or chemical energy. Examples of electrical energy being converted into predominantly heat energy include resistors, filament lamps, electric heaters (e.g. radiant, convection, storage heaters, for domestic and industrial use), resistance furnaces, immersion heaters. Examples of the conversion of electrical energy into mainly magnetic energy include solenoids, moving coil instruments, electromagnets, electric motors and generators. Examples of the conversion of electrical energy into chemical energy are the electric battery and electro-plating.

In all circuits, whenever an electric current flows, there are a heating, a magnetic and a chemical effect, although the chemical may not be obvious unless the circuit is designed so that this effect is predominant.

Self-assessment questions

47 List three applications of the heating effect of an electric current.

48 List three applications of the magnetic effect of an electric current.

49 List two applications of the chemical effect of an electric current.

50 In a heating element which is being used for 15 minutes in every hour
(*a*) There is a magnetic effect.
TRUE/FALSE
(*b*) There is a chemical effect.
TRUE/FALSE

51 In the conductors of a coil of an electric solenoid, which is in continual use
(*a*) There is a heating effect.
TRUE/FALSE
(*b*) There is a chemical effect.
TRUE/FALSE

52 In a battery which is supplying current

(*a*) There is a heating effect.

TRUE/FALSE

(*b*) There is a magnetic effect.

TRUE/FALSE

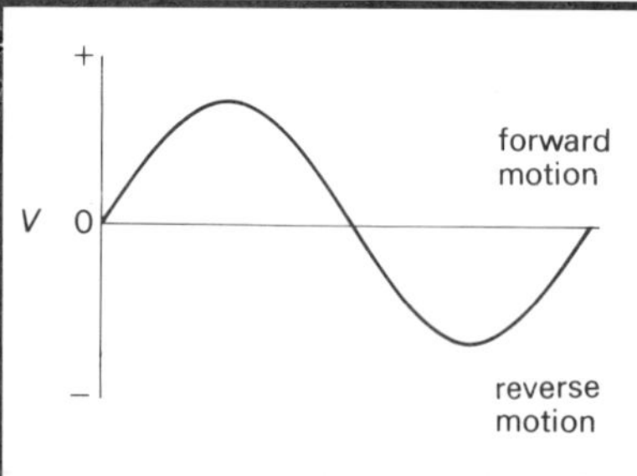

Figure 127 *An alternating current waveform*

Solutions to self-assessment questions

42 The two factors which determine the direction of the current in the coil are,
(*a*) the direction in which the magnet is moving
(*b*) whether an 'N' pole or an 'S' pole is nearer to the end of the coil

43 The two factors which determine the magnitude of the current in the coil are,
(*a*) the speed at which the magnet is moving
(*b*) the number of turns in the coil

44 The wave form of the current flow in the galvanometer in Figure 125 is shown in Figure 127.

45 The direction of the current flow is indicated by a dot in the conductors at the top of the coil.

46 (i) The coil sides are connected to *slip rings*.
(ii) The carbon brushes provide a sliding contact between the external circuit and the *slip rings*.
(iii) The coil sides rotate in a magnetic field designed to be *uniform*.
(iv) The maximum current in the external circuit is *halved*.
(v) The current waveform produced in a 360° rotation is *an alternating waveform*.

Solutions to self-assessment questions

47 The list may include: lamps, electric fires, electric soldering irons, domestic electric irons, electric cookers, immersion heaters, resistance furnaces.

48 The list may include: solenoids, moving coil instruments, various electric motor applications, electric generators.

49 The list may include: electro-plating, battery charging, metal purification (a special form of plating), corrosion, the separation of water into oxygen and hydrogen.

50 (*a*) True. There is a magnetic effect because of current flow.
(*b*) True. Each time the heater cools, the results of a chemical change may show as scale around the element. The heater element may eventually 'corrode' away.

51 (*a*) True. The predominant effect is magnetic, but when current flows through a conductor there is a temperature rise.
(*b*) False. Provided the temperature does not rise high enough to change the characteristics of the conductor, there is little if any chemical effect due to the current.

52 (*a*) True. The flow of current in the battery is resisted; therefore there is a temperature rise.
(*b*) True. Whenever current flows, a magnetic effect is produced, but this effect is very small and difficult to detect inside the battery.

Topic area: Dynamics

After reading the following material, the reader shall:

1 Understand the relationship between displacement and velocity, under uniform motion conditions.
1.1 Define speed.
1.2 Identify the recommended SI unit for speed.
1.3 Calculate speeds from given time and distance data.
1.4 Plot distance–time graphs.
1.5 Calculate the slope of such graphs, and interpret the slope as speed.
1.6 Calculate average speeds from given numerical and graphical data.

Suppose a motor car travels 3 km in 2 minutes. Its speed is 1.5 km per minute, and if this speed is converted to km per hour, it becomes

$$1.5 \times 60 \text{ km per hour} = 90 \text{ km/hour}$$

In general, the speed of a body is the distance travelled in unit time, i.e. the *rate* at which distance is covered. The distance can be expressed in any convenient unit such as metre or kilometre, and the time expressed in hours, minutes or seconds. However, the recommended SI units are the metre, kilometre or millimetre for distance, and the second for time.

Example 1
An aircraft, flying at a constant speed, covers a distance of 9 000 km in 10 hours. Find
(i) its speed in m/s
(ii) the number of kilometres covered in 75 minutes
(iii) the time taken to travel 6 500 km

(i)
$$9\,000 \text{ km} = 9\,000 \times 10^3 \text{ m} = 9 \times 10^6 \text{ m}$$
$$10 \text{ hours} = 10 \times 3\,600 \text{ s} = 36 \times 10^3 \text{ s}$$
$$\therefore \text{Speed in m/s} = \frac{9 \times 10^6}{36 \times 10^3} = 250 \text{ m/s}$$

(ii) The aircraft takes 10 hours to travel 9 000 km.
Hence in 1 hour, it travels $\frac{9\,000}{10}$ km
$= 900$ km

In 15 minutes (i.e. a quarter of an hour), it travels a quarter of 900 km.

$\therefore$ In 15 minutes, distance covered $= \frac{900}{4} = 225$ km
$\therefore$ In 75 minutes, distance covered $= (900+225)$ km $= 1\,125$ km

(iii) Since the aircraft travels 9 000 km in 10 hours, it covers 1 km in $\frac{10}{9\,000}$ hours.

$\therefore$ It covers 6 500 km in $\frac{10}{9\,000} \times 6\,500$ hours $= 7.22$ hours

The information given in the above example can be presented in a rather different fashion, which leads to a different, and perhaps easier, method of solution. For example, suppose the distance travelled by the aircraft in given intervals of time is measured, the results of the measurement being as follows:

time (hours)	1	3	5	7	9
distance travelled (km)	900	2 700	4 500	6 300	8 100

Assignment

(i) Plot the above results on graph paper, with distance travelled on the vertical axis, and time on the horizontal axis.
(ii) The graph should appear as a straight line. Calculate the slope of the straight line.
(iii) Read off from the graph
(*a*) the distance travelled in 75 minutes
(*b*) the time taken to travel 6 500 km

The graphical method of the assignment involved less calculation than the method of example 1. This was concerned with an aircraft flying at constant speed. Now suppose the speed changes during a journey.

Example 2
A motor car travels 80 km in 2 hours at constant speed, and then travels a further 90 km in 1.5 hours, again at constant speed. Find
(i) the speed in km/hour for the first 80 km
(ii) the speed in km/hour for the second 90 km
(iii) the average speed for the total 170 km
(iv) the distance travelled after 3 hours
(v) the time taken to travel 160 km

If the information is plotted graphically, it appears as in Figure 128. The line $0A$ represents the first part of the journey, and the line AB the second half of the journey. Note that the graph consists of straight lines, because the speeds are constant in the two separate parts of the journey.

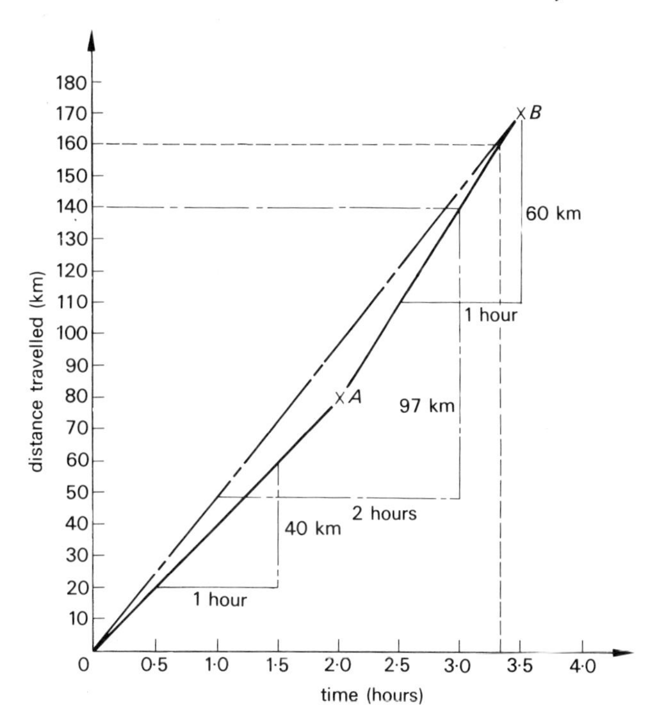

Figure 128 *Distance travelled–time (change in speed)*

(i) Slope of $0A = \dfrac{40}{1}\ \dfrac{\text{km}}{\text{hours}} = 40$ km/hour

$\therefore$ Speed for first 80 km is 40 km/hour.

(ii) Slope of $AB = \dfrac{60}{1}\ \dfrac{\text{km}}{\text{hours}} = 60$ km/hour

$\therefore$ Speed for final 90 km is 60 km/hour.

(iii) Slope of $0B = \dfrac{97}{2}\ \dfrac{\text{km}}{\text{hours}} = 48.5$ km/hour

$\therefore$ Average speed for whole journey = 48.5 km/hour.

(iv) Reading from Figure 128,
Distance travelled after 3 hours = 140 km.

(v) Reading from Figure 128,
Time taken to travel 160 km = 3.33 hours.

All the above answers could have been derived without reference to graphical methods, perhaps using methods similar to those of example 1. However, the opportunities for error in calculation, particularly in parts (iv) and (v), are less with graphical methods.

Results of assignment

(i) The graph of time against distance travelled should be that of Figure 129.

(ii) The value of the slope of the graph is

$$\frac{3\,600}{4}\ \frac{\text{km}}{\text{hour}} = 900 \text{ km/hour}$$

Converting this value to m/s:

$$900 \text{ km/hour} = 900 \times 10^3 \text{ m/hour}$$

$$= \frac{900 \times 10^3}{3\,600} \text{ m/s} = 250 \text{ m/s}$$

Compare this answer with the answer to part (i) of the example. Since the answers are the same, it would seem that the slope of the distance–time graph represents the speed of the aircraft. This conclusion is correct: in general *the slope of any distance–time graph represents speed.*

(iii) Using Figure 129:
Distance travelled in 75 minutes is 1 000 km.
Time taken to travel 6 500 km is 7.2 hours.

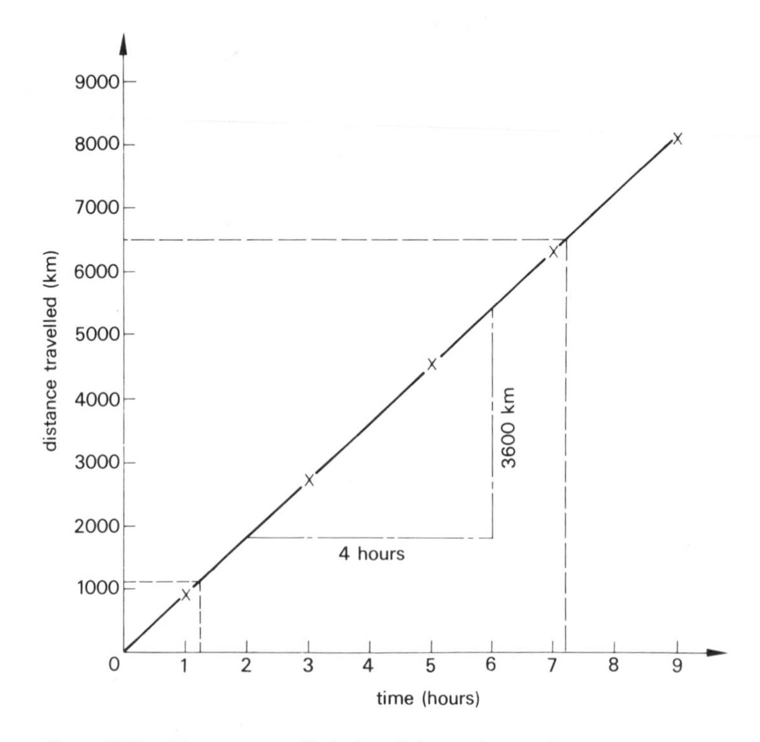

Figure 129 *Distance travelled–time (change in speed)*

Self-assessment questions

1 The table below shows the relationship between time and distance covered by a motorcycle.

time (seconds)	0	0.5	1.5	2.5	3.5	4.5	5.5	6.5	7.5
distance (metres)	0	8	24	40	56	72	88	104	120

Plot a graph of distance against time, and find:

(i) the speed of the motorcycle
(ii) the distance covered after 5 seconds
(iii) the time to cover a distance of 20 m

2 An estimate is made of the distance covered by a motor car at 10 second intervals. The car travels at a constant speed for the first 40 seconds, and at a different constant speed for the next 40 seconds. The results are given in the table below:

time (seconds)	0	10	20	30	40	50	60	70	80
estimated distance (km)	0	0.16	0.325	0.485	0.65	0.71	0.775	0.835	0.90

Plot the distance–time graph, and find:

(i) the speed in m/s over the first 40 seconds
(ii) the speed in m/s over the second 40 seconds
(iii) the average speed in m/s for the whole 80 seconds
(iv) the distance covered after 53 seconds
(v) the time taken to cover 0.8 km

After reading the following material, the reader shall:

1.7 Distinguish between distance and displacement.
1.8 State the difference between speed and velocity.

Suppose a person walks a distance of 14 km. Does that mean that the distance between his initial and final positions is 14 km? The answer is 'No, not necessarily:' he could be at any position within a radius of 14 km from his starting point, dependent upon how straight a path he takes. In fact he could have returned to his starting point. Just one of the almost infinite number of routes he could have taken is shown in Figure 130, where '*a*' represents the initial position, and '*c*' the final position.

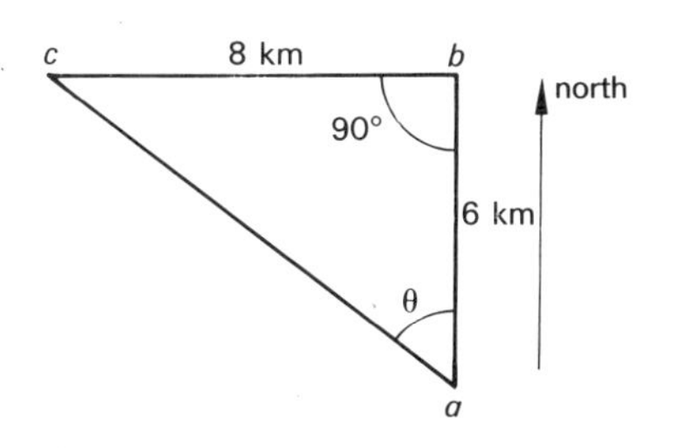

Figure 130 *Vector diagram*

Solutions to self-assessment questions

1 The data are plotted in Figure 131.

From this figure, slope $= \dfrac{112-16}{6}\,\dfrac{\text{m}}{\text{s}} = \dfrac{96}{6}$ m/s

$\therefore$ Speed = 16 m/s

Also, from Figure 131,

Distance covered after 5 s is 80 m.

Time to cover a distance of 20 m is 1.25 s.

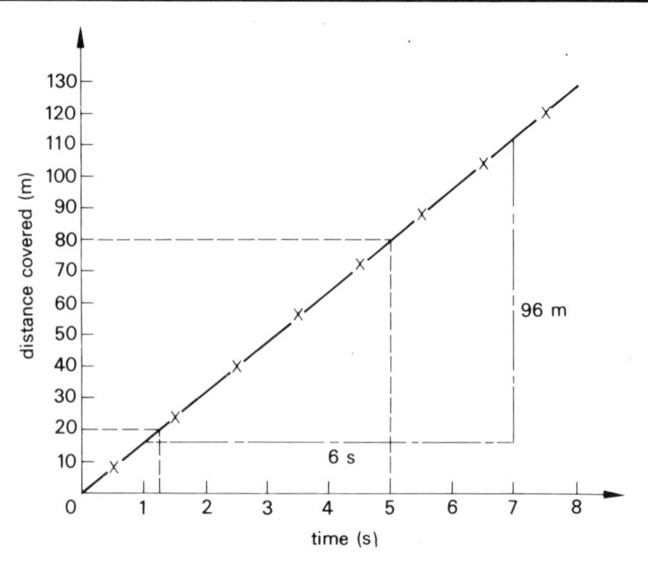

Figure 131 *Time–distance relationship*

2 The data are plotted on Figure 132.

(i) Speed over first 40 s is $\dfrac{(0.565-0.24)}{20}\,\dfrac{\text{km}}{\text{s}}$

Speed $= \dfrac{0.325\times10^3}{20}\,\dfrac{\text{m}}{\text{s}} = 16.25$ m/s

(ii) Speed over second 40 s is $\dfrac{(0.865-0.74)}{20}\,\dfrac{\text{km}}{\text{s}}$

Speed $= \dfrac{0.125\times10^3}{20}\,\dfrac{\text{m}}{\text{s}} = 6.25$ m/s

(iii) Average speed over 80 s is $\dfrac{(0.672-0.112)}{50}\,\dfrac{\text{km}}{\text{s}}$

Average speed $= \dfrac{0.56\times10^3}{50}\,\dfrac{\text{m}}{\text{s}} = 11.2$ m/s

(iv) Distance covered after 53 s is 0.73 km or 730 m.

(v) Time taken to cover 0.8 km is 64.5 s.

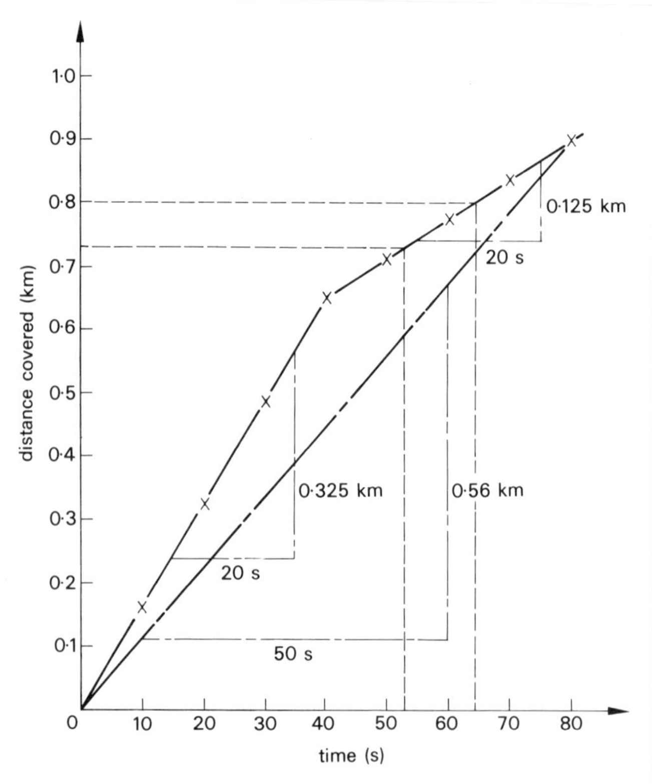

Figure 132 *Estimated time–distance relationship*

Example 3
Calculate or find graphically (by drawing Figure 130 to scale) the distance between 'a' and 'c', and the angle ac makes with the northerly direction.

If Figure 130 is drawn to scale, then ac is equivalent to 10 km, and the angle between ac and the north is approximately 53°.

Alternatively, using the theorem of Pythagoras:

$$(ac)^2 = (ab)^2 + (bc)^2$$
$$(ac)^2 = [6^2 + 8^2]\ \text{km}^2$$
$$(ac)^2 = [36 + 64]\ \text{km}^2 = 100\ \text{km}^2$$
$$\therefore \underline{ac = 10\ \text{km}}$$

Also, $\sin\theta = \frac{8}{10} = 0.8$

$$\therefore \underline{\theta = 53^\circ 6'}$$

Since 'c' represents the final position of the walker, and 'a' the initial position, ac represents the *change in position* of the walker, and is termed the *displacement*. Note that the displacement not only has magnitude (in example 3, this is 10 km), but also direction (53°6′ from north).

The distinction between distance and displacement is that distance has magnitude only, whereas displacement has both magnitude and direction.

Essentially the same distinction exists between *speed* and *velocity*. Speed has magnitude only. If the direction of motion is specified as well as the magnitude, then the quantity is termed the velocity of the body. Thus the units of speed and velocity are the same, and the magnitude of the two quantities is the same. They differ because velocity also specifies the direction of motion.

Speed has been defined as the distance travelled in unit time, or *the rate at which distance is covered with respect to time.* In a similar manner, velocity is defined as the change in displacement in unit time or *the rate of change of displacement with respect to time.* In symbol form, if the displacement of a body is s in time t, then the average velocity v is given by:

$$v = \frac{s}{t}$$

Self-assessment questions

3 A car travels 3 km southwards, and then travels 4 km westwards.
(i) What distance has the car travelled?
(ii) What is the displacement of the car from the initial position?

4 An aeroplane leaves an airport and flies due south for a distance of 180 km. It lands, refuels and returns to the original airport.

(i) What distance has the aeroplane travelled?

(ii) What is its displacement from the original position?

5 Write down the SI unit of velocity.

6 Most countries impose 'speed limits' upon motor vehicles, a typical value being approximately 50 km/hour. Is this strictly a speed limit or a velocity limit?

7 Three statements are written below, and are followed by two questions. Answer the questions by selecting the appropriate statements.

Statement (*a*): A motor vehicle travels 30 metres in a direction due south.

Statement (*b*): A motor vehicle travels at 30 m/s in a direction due south.

Statement (*c*): A motor vehicle travels at 30 m/s.

(i) Which statement describes the velocity of the motor vehicle?

(ii) Which statement describes the speed of the motor vehicle?

After reading the following material, the reader shall:

2.9 Solve problems using the equation
s = average velocity × time.

In the previous section, the average velocity v of a body which experiences a displacement s in time t was shown to be

$$\frac{s}{t} = v$$

Multiplying both sides of this equation by t:

$$s = vt$$

i.e., displacement = average velocity × time

Suppose a motor vehicle changes its velocity from 12 m/s to 24 m/s in a given time. Its average velocity is then

$$\left(\frac{12+24}{2}\right) \text{ m/s} = 18 \text{ m/s}$$

Expressed in symbol form:
If v_1 is the initial velocity and v_2 is the final velocity

$$\text{average velocity } v = \frac{v_1+v_2}{2}$$

$$\text{But } s = vt$$

$$\therefore s = \left(\frac{v_1+v_2}{2}\right) t$$

Example 4

A car accelerates uniformly from rest to a velocity of 50 km/hour in 10 seconds. Find the distance travelled.

$$\text{Using the equation } s = \left(\frac{v_1+v_2}{2}\right)t$$

$$v_1 = 0,\ v_2 = 50 \text{ km/hour}$$

$$= 50 \times 10^3 \text{ m/hour}$$

$$= \frac{50 \times 10^3}{3\,600} \text{ m/s} = 1.389 \text{ m/s}$$

$$\text{time, } t = 10 \text{ seconds}$$

$$\therefore s = \left(\frac{0+1.389}{2}\right) \times 10$$

$$\therefore \text{Distance } s = \underline{69.4 \text{ m}}$$

Self-assessment questions

8 A cage is lowered down a vertical pit shaft, 800 m deep, in 50 s. Find its average velocity.

9 A car travels at 50 km/hour for half an hour, and then at 80 km/hour for the next hour. Calculate
(i) the total distance travelled in km
(ii) the average speed of the car in km/hour.

10 A motorcycle accelerates uniformly from rest to a velocity of 10 m/s in 4 seconds. At this instant, the throttle is opened wider so that the velocity increases uniformly to 20 m/s in 2 seconds. Find:
(i) the average velocity during the first 4 seconds
(ii) the average velocity during the next 2 seconds
(iii) the distance travelled during the first 4 seconds
(iv) the distance travelled during the total 6 seconds

After reading the following material, the reader shall:

2 Understand the relationship between velocity and acceleration when the velocity is increasing steadily.
2.1 Define acceleration.
2.2 Identify the recommended SI unit for acceleration.
2.3 Plot velocity-time graphs for motion in a straight line.
2.4 Calculate the slope of such graphs and interpret the slope as acceleration.

When driving a car, velocity (or speed) can be increased by pressing the accelerator pedal; the car is then said to *accelerate*. Conversely, if the accelerator pedal is released, or the brake pedal is pressed, the car *decelerates*, which can be considered negative acceleration. Thus, acceleration of the car (or indeed of any object) is concerned with a change in velocity over a period of time.

The magnitude of an acceleration is dependent upon two factors:

(i) The change in velocity. The greater the change in velocity, the greater will be the acceleration.

(ii) The time taken to change the velocity. The shorter the time interval to change velocity, the greater the acceleration.

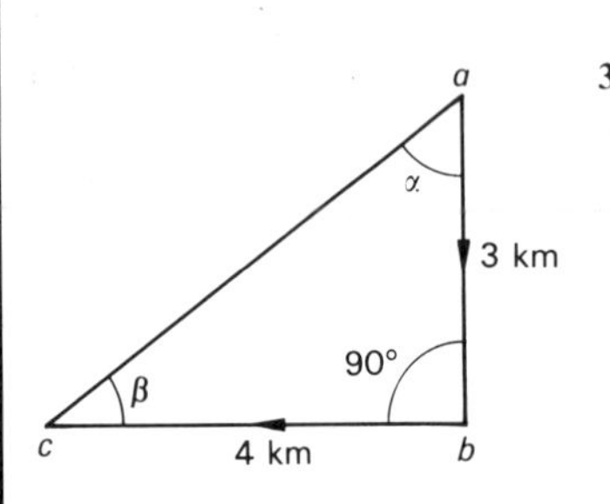

Figure 133

Solutions to self-assessment questions

3 (i) The car has travelled a distance of $(3+4)$ km $= 7$ km.

(ii) The travel path of the car is shown in Figure 133, '*a*' being the initial position and '*c*' the final position. The displacement *ac* can be found either graphically or by calculation, using the theorem of Pythagoras.

$$\text{Thus } (ac)^2 = [3^2+4^2] \text{ km}^2$$
$$(ac)^2 = [9+16] \text{ km}^2 = 25 \text{ km}^2$$
$$\therefore ac = 5 \text{ km}$$
$$\text{Sin } \alpha = \tfrac{4}{5} = 0.8$$
$$\therefore \alpha = 53°6'$$
$$\text{and } \beta = 90° - \alpha = 90° - 53°6' = 36°54'$$

Thus, the displacement of the car from the initial position is 5 km, at an angle of 53°6′ to the north–south direction, or 36°54′ to the east–west direction.

4 (i) The distance the aeroplane has travelled is 180 km + 180 km = 360 km.

(ii) Since the aeroplane has returned to its original position, its displacement from the original position is zero.

5 Average velocity has been defined as

$$v = \frac{s}{t}$$

and the unit of *s* is metre, and of *t* is second, then the unit of velocity is m/s. In SI, the unit of velocity could also be mm/s or km/s.

6 A 'speed limit' such as 50 km/hour does not specify the direction in which a motor vehicle is travelling: indeed the direction of the vehicle can change frequently while it is travelling at 50 km/hour. Under these conditions, the vehicle's velocity changes every time its direction changes, whereas its speed remains constant.
Thus, the 50 km/hour limit is a speed limit, and not a velocity limit.

7 Statement (*b*) describes the velocity of the vehicle, since it quotes the direction of motion as well as the distance covered in unit time.
Statement (*c*) describes the speed of the vehicle, because it quotes the distance covered in unit time, but not the direction of motion.
Statement (*a*) describes neither the velocity nor the speed of the vehicle. It states only the displacement of the vehicle.

Self-assessment question

11 Three cars each increase their velocity uniformly from zero to 35 m/s. To produce this change in velocity, car X takes 10 seconds, car Y takes 20 seconds and car Z takes 30 seconds. Which car has the greatest acceleration, and which car has the least acceleration?

Acceleration therefore depends upon the change in velocity, and the time taken to produce that change. The strict definition of acceleration is *the rate of change of velocity with respect to time*, but for the present, it is sufficient to define acceleration as the change in velocity in unit time, i.e.

$$\text{acceleration} = \frac{\text{change in velocity}}{\text{time taken}}$$

Suppose a motorcycle increases its velocity from 6 m/s to 16 m/s in a period of 5 seconds. Then:

$$\text{acceleration} = \frac{\text{10 metre per second}}{\text{5 seconds}}$$
$$= \text{2 metre per second per second.}$$

To write out these units is rather cumbersome and time consuming, and in SI, they are abbreviated to m/s^2. Thus the acceleration of the motorcycle is 2 m/s^2.

Solutions to self-assessment questions

8 Average velocity $= \dfrac{\text{displacement}}{\text{time}}$

$= \left[\dfrac{800}{50}\right]\dfrac{\text{m}}{\text{s}}$

Average velocity = 16 m/s, vertically down.

9 (i) Distance covered in first part of journey

$= [50 \times 0.5]\dfrac{\text{km}}{\text{hour}} \times \text{hour}$

$= 25$ km

Distance covered in second part of journey

$= [80 \times 1.0]\dfrac{\text{km}}{\text{hour}} \times \text{hour}$

$= 80$ km

$\therefore$ Total distance covered $= [25+80]$ km $= 105$ km

(ii) Average speed $= \dfrac{\text{distance}}{\text{time}}$

$= \left[\dfrac{105}{1.5}\right]\dfrac{\text{km}}{\text{hour}}$

Average speed = 70 km/hour

10 (i) During first 4 seconds:

Average velocity $= \dfrac{v_1+v_2}{2} = \left[\dfrac{0+10}{2}\right]$ m/s

Average velocity = 5 m/s

(ii) During final 2 seconds:

Average velocity $= \left[\dfrac{10+20}{2}\right]$ m/s $= 15$ m/s

(iii) During first 4 seconds:

Distance travelled = average velocity × time

$= [5 \times 4]\dfrac{\text{m}}{\text{s}} \times \text{s}$

Distance travelled = 20 m

(iv) During final 2 seconds:

Distance travelled $= [15 \times 2]\dfrac{\text{m}}{\text{s}} \times \text{s}$

$= 30$ m

$\therefore$ Total distance travelled $= [20+30]$ m $= 50$ m

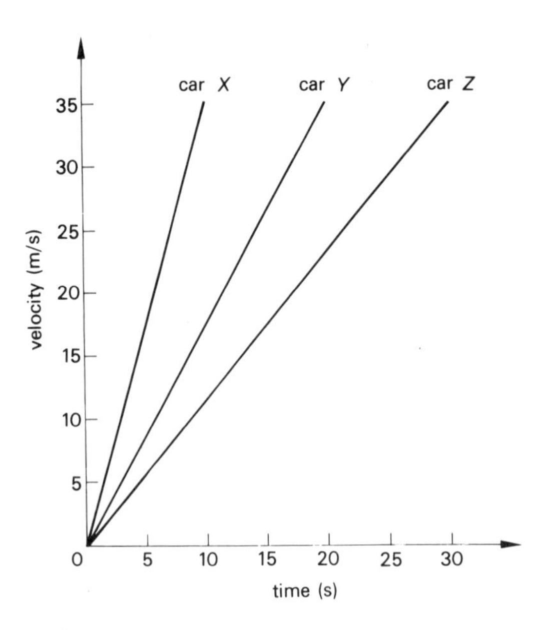

Figure 134 *Velocity–time*

Self-assessment question 11 was concerned with three cars, each of which increased their velocity from 0 to 35 m/s. Car X took 10 seconds, car Y took 20 seconds and car Z took 30 seconds. This information is presented graphically in Figure 134, assuming that the velocities increase uniformly with time.

Assignment

Calculate the slopes of the three straight lines in Figure 134. Suggest the units for the answers.

If the accelerations of the three cars are calculated using the expression:

$$\text{acceleration} = \frac{\text{change in velocity}}{\text{time taken}}$$

$$\text{For car } X\text{, acceleration} = \frac{35}{10}\frac{\text{m/s}}{\text{s}} = 3.5 \text{ m/s}^2$$

$$\text{For car } Y\text{, acceleration} = \frac{35}{20}\frac{\text{m/s}}{\text{s}} = 1.75 \text{ m/s}^2$$

$$\text{For car } Z\text{, acceleration} = \frac{35}{30}\frac{\text{m/s}}{\text{s}} = 1.167 \text{ m/s}^2$$

Solution to self-assessment question

11 Car X has the greatest acceleration since it took the least time, and car Z the smallest acceleration, since it took the most time.

Comparing the calculated values with those derived graphically, it can be seen that *the slope of a velocity–time graph gives the acceleration*. Note that this statement is exactly analogous to that concerning distance, time and speed on pages 161 and 162.

Example 5

A train makes a journey between two stations in 8 minutes. The velocity of the train at 1 minute intervals is given in the following table:

time (seconds)	0	60	120	180	240	300	360	420	480
velocity (m/s)	0	3.5	6.9	10.5	14.0	14.0	9.3	4.6	0

Assuming the graph to be three straight lines, draw the velocity–time graph, and determine:

(i) the acceleration between 0 and 4 minutes

(ii) the deceleration between 5 and 8 minutes.

The velocity–time graph is shown in Figure 135.

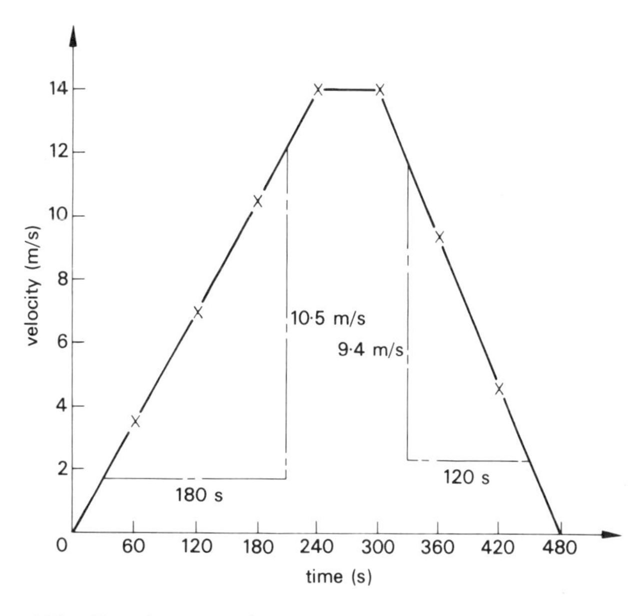

Figure 135 *Travel pattern of a train*

From the figure,

Slope between 0 and 4 minutes is

$$\frac{[12.2 - 1.7]}{180} \; \frac{\text{m/s}}{\text{seconds}}$$

$$= \frac{10.5}{180} \frac{\text{m/s}}{\text{s}} = 0.058 \text{ m/s}^2$$

Slope between 5 and 8 minutes is

$$-\frac{[11.7-2.3]}{120}\ \frac{\text{m/s}}{\text{seconds}}$$

$$=\frac{-9.4}{120}\ \frac{\text{m/s}}{\text{s}} = -0.078\ \text{m/s}^2$$

Thus, the acceleration between 0 and 4 minutes is 0.058 m/s^2, and the deceleration between 5 and 8 minutes is 0.078 m/s^2. Note:

(i) The slope of the straight line between 5 and 8 minutes is, in a mathematical sense, negative. The negative sign indicates negative acceleration, i.e. deceleration.

(ii) The slope of the straight line between 4 and 5 minutes is zero, which means that the acceleration is zero. This is confirmed by inspecting the table, since the velocity is constant between the relevant times.

Self-assessment questions

12 The table below shows the variation of velocity with time for a motor car. Plot velocity against time, and determine:

(i) the acceleration of the motor car

(ii) its velocity after 6.5 seconds

(iii) the time to reach a velocity of 14 m/s

velocity (m/s)	0	1.5	3.0	4.5	6.0	7.5	9.0	10.5	12.0
time(s)	0	1	2	3	4	5	6	7	8

velocity (m/s)	13.5	15.0	16.5	18.0
time(s)	9	10	11	12

13 A motorcycle accelerates uniformly for 2.5 s. At this instant, its rate of change of velocity decreases. Estimates are made of the velocity of the motorcycle during the initial 2.5 s, and during a further 7.5 s, the results being shown below.

time(s)	0	1	2	2.5	4	5	6	7	8	9	10
velocity (km/hour)	0	18	36.5	45	49	51.7	54.4	57	59.6	62.3	65

Plot a graph of velocity against time, and find:

(i) the acceleration during the first 2.5 s

(ii) the acceleration between 2.5 s and 10 s

(iii) the velocity after 7.5 s

(iv) the time to reach a velocity of 50 km/hour

14 The table below shows measurements of time and velocity taken during the performance testing of a high-powered motor car:

time(s)	0	5	11.5	20	25	30	35	40	45	50
velocity (m/s)	0	6.5	15.0	26.5	33.0	40.0	40.0	40.0	40.0	48.5
time(s)	54	57	60	62	66	70	72	75	77	80
velocity (m/s)	55.0	60.0	65.0	58.0	45.5	32.5	26.0	16.0	9.5	0

Plot a graph of velocity against time, assuming that the points are joined by a series of straight lines. Hence determine:

(i) two values of acceleration and one value of deceleration of the car
(ii) the time for the velocity to reach 50 m/s
(iii) the velocity of the car, 3 seconds after it has begun to brake

After reading the following material, the reader shall:

2.5 Calculate quantities involved in the equation $v = u+at$, excluding motion under gravity.

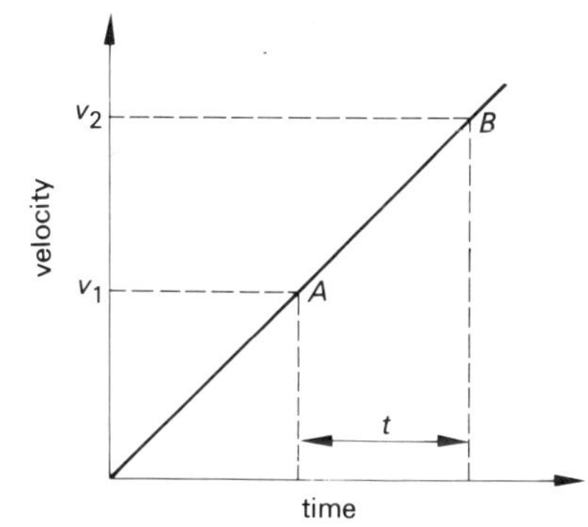

Figure 136 *Velocity–time relationship*

Consider the velocity–time relationship shown in Figure 136. The slope of the line AB is

$$\frac{v_2 - v_1}{t}$$

But the slope of a velocity–time graph is acceleration, and denoting acceleration by the symbol a, then

$$a = \frac{v_2 - v_1}{t}$$

Multiplying both sides by t

$$at = v_2 - v_1$$

or $v_2 = v_1 + at$

Results of assignment

The slopes of the straight lines are found by dividing velocity by time, and hence the units of the slope are $\frac{\text{m/s}}{\text{s}}$, i.e. m/s^2. The values of the slopes are:

Car X – 3.5 m/s^2
Car Y – 1.75 m/s^2
Car Z – 1.17 m/s^2 (approximately)

Solutions to self-assessment questions

12 The graph of velocity against time is shown in Figure 137. From the graph:

(i) Acceleration $= \dfrac{(12.75-2.2)}{7} \dfrac{\text{m/s}}{\text{s}}$

Acceleration $= \dfrac{10.55}{7} \dfrac{\text{m/s}}{\text{s}} = 1.5 \text{ m/s}^2$

(ii) Velocity after 6.5 s is 9.75 m/s.

(iii) Time to reach a velocity of 14 m/s is 9.33 s.

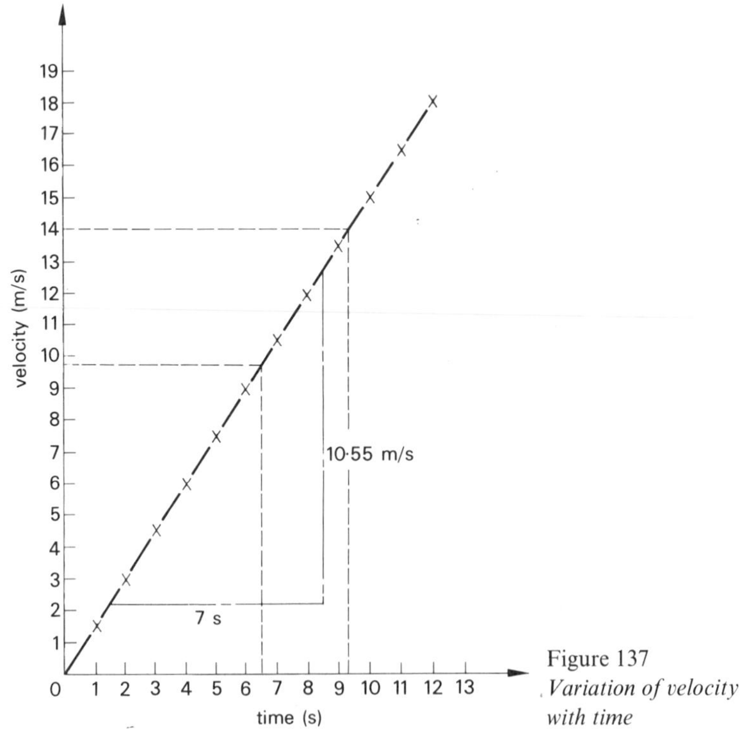

Figure 137
Variation of velocity with time

13 The graph of velocity against time is shown in Figure 138. From the graph:

(i) Acceleration during first 2.5 s is

$\dfrac{(27-8.8)}{1} \dfrac{\text{km/hour}}{\text{s}}$

$= \dfrac{18.2}{1} \dfrac{\text{km/hour}}{\text{s}} = \dfrac{18.2 \times 10^3}{\frac{3\,600}{1}} \dfrac{\text{m/s}}{\text{s}}$

Acceleration $= 5.05 \text{ m/s}^2$

(ii) Acceleration between 2.5 s and 10 s is

$\dfrac{(63.5-47.5)}{6} \dfrac{\text{km/hour}}{\text{s}}$

$= \dfrac{16}{6} \dfrac{\text{km/hour}}{\text{s}} = \dfrac{16 \times 10^3}{\frac{3\,600}{6}} \dfrac{\text{m/s}}{\text{s}}$

Acceleration $= 0.74 \text{ m/s}^2$

(iii) Velocity after 7.5 s is 58.3 m/s.

(iv) Time to reach a velocity of 50 km/hour is 4.4 s.

Example 6

A car starting from rest accelerates uniformly at 2 m/s² for 10 seconds. Find the final velocity.

$$v_1 = 0$$
$$a = 2 \text{ m/s}^2$$
$$t = 10 \text{ s}$$
$$\therefore v_2 = 0+(2\times 10)\frac{\text{m}}{\text{s}^2}\text{s}$$

$\therefore$ Final velocity, $v_2 = 20$ m/s

14 The graph of velocity against time is shown in Figure 139. From the graph:

(i) Acceleration between 0 and 30 s is

$$\frac{(36.8-10)}{20}\frac{\text{m/s}}{\text{s}} = \frac{26.8}{20}\frac{\text{m/s}}{\text{s}}$$

First acceleration $= 1.34$ m/s²

Acceleration between 45 s and 60 s is

$$\frac{(61-44.3)}{10}\frac{\text{m/s}}{\text{s}} = \frac{16.7}{10}\frac{\text{m/s}}{\text{s}}$$

Second acceleration $= 1.67$ m/s²

Deceleration between 60 s and 80 s is

$$-\frac{(48.5-8)}{12.5}\frac{\text{m/s}}{\text{s}} = -\frac{40.5}{12.5}\frac{\text{m/s}}{\text{s}}$$

Deceleration $= -3.24$ m/s²

(ii) Times for velocity to reach 50 m/s are 51 s and 64.5 s.

(iii) Velocity 3 s after breaking begins is 55 m/s.

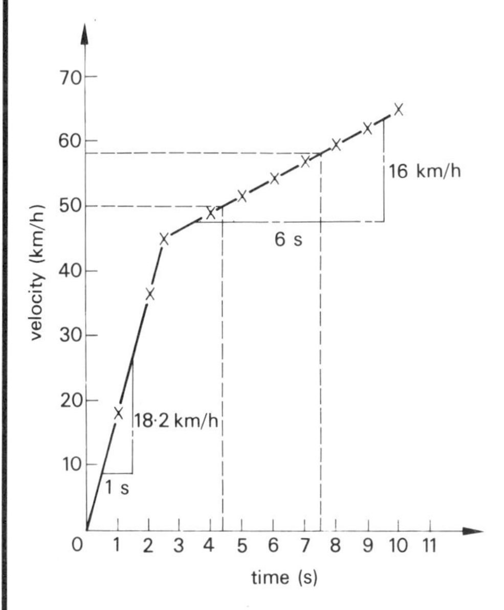

Figure 138 *Estimated velocity–time relationship*

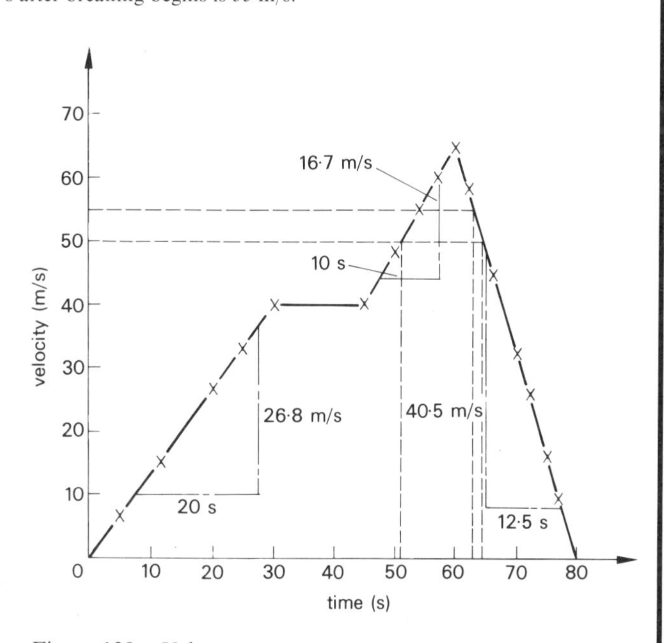

Figure 139 *Velocity–time measurements*

Example 7
The velocity of a motorcycle falls uniformly from 72 km/hour to 18 km/hour at a rate of 3 m/s². Find the time taken.

$$v_1 = 72 \text{ km/hour} = \frac{72 \times 10^3}{3\,600} \text{ m/s} = 20 \text{ m/s}$$

$$v_2 = 18 \text{ km/hour} = \frac{18 \times 10^3}{3\,600} \text{ m/s} = 5 \text{ m/s}$$

$$a = -3 \text{ m/s}^2 \text{ (negative because the motorcycle is decelerating)}$$

Using $v_2 = v_1 + at$

$$5 = 20 + (-3) \times t$$

$$-15 = -3t$$

$$t = 5s$$

Note that in examples 6 and 7, v_1 was assumed to be the initial velocity and v_2 the velocity at a later instant in time. The reader should adhere to this convention; if it is used with the correct sign for acceleration or deceleration, the solution of velocity–time equations is greatly simplified.

Self-assessment questions

15 A train starting from rest, reaches a velocity of 20 m/s in 60 seconds. Assuming the acceleration to be uniform, calculate the value of the acceleration.

16 A motor car accelerates uniformly at 2 m/s² from a velocity of 50 km/hour. Find the velocity of the car
(i) after 10 seconds
(ii) after 15 seconds

17 The landing velocity of an aeroplane is 250 km/hour, and it takes 2 minutes for the brakes to reduce the velocity to 25 km/hour. Assuming the deceleration to be uniform, calculate
(i) the deceleration in m/s² of the aircraft
(ii) the velocity in m/s of the aircraft, 30 seconds after landing

Note that all three questions can be solved by graphical methods. The reader may use this method if he wishes, but the solutions below are by calculation.

After reading the following material, the reader shall:

3 Be aware of the relationship between force, mass and acceleration.
3.1 State that acceleration is the result of a net force being applied.
3.2 State that force is proportional to the product of mass and acceleration, i.e. $F \propto ma$.
3.3 Identify the newton as the unit of force which makes the constant of proportionality equal to unity, so that $F = ma$.
3.4 Identify 1 N as equivalent to 1 kg m/s^2.
3.5 State that the unit of force is the newton.
3.6 Describe 'free-fall' as being constant acceleration.
3.7 State that the acceleration due to gravity on earth is approximately 9.81 m/s^2.

In order to make a motor vehicle move, a force must be applied to it. This force may take the form of willing helpers pushing or pulling, or it may originate from the engine, applied to the wheels via the propellor shaft. In either case, the force must be a *net* force in the sense that it must overcome the reluctance to move of sliding and rotating parts.

All readers will be aware that a larger force is necessary to move a stationary lorry than to move a family car: the larger the mass to be moved, then the larger the applied force. Similarly, the more quickly the vehicle is moved from rest, the larger the applied force (the larger the acceleration, the larger the applied force).

Hence there is a connection between the applied force, the mass to be accelerated and the acceleration. Isaac Newton perceived the relationship between these quantities, a relationship which may be expressed as:

Force is proportional to the product of mass and acceleration.
Using F as the symbol for force
m as the symbol for mass
a as the symbol for acceleration
then $F \propto ma$

The proportional sign ($\propto$) can be replaced by an equals sign ($=$) providing a constant of proportionality is introduced, e.g.

$F = ma \times$ a constant

In earlier sections of this book, the newton (symbol N) has been used as the SI unit of force. In SI one newton is defined as *the force required to give a mass of 1 kg an acceleration of 1 m/s^2.*

Substituting $F = 1$ N, $m = 1$ kg and $a = 1$ m/s^2 in the above expression:

$$1\,[\text{N}] = 1\,[\text{kg}] \times 1 \left[\frac{\text{m}}{\text{s}^2}\right] \times \text{a constant}$$

Hence the constant is equal to unity. Thus providing that the force F is in newtons, the mass m is in kg and the acceleration a is in m/s^2, then:

$$F = ma$$

Note that the unit of force can be expressed in terms of the units of mass, length and time, e.g.

1 newton is equivalent to 1 kg $\times$ 1 m/s^2

or N is equivalent to $\dfrac{\text{kgm}}{\text{s}^2}$

Acceleration due to gravity

If an object is dropped from a high building, or an aeroplane, the velocity of the object increases as it falls (until it reaches a certain terminal value, depending upon air resistance). Thus, the object accelerates. It might appear that, not only is the velocity increasing as the object falls, but also that the acceleration is increasing. This is not so; an object in 'free fall' is subjected to constant acceleration throughout the period of free fall.

Because the earth is not a perfect sphere, free fall acceleration varies in magnitude at different points on the earth's surface. However, at

Solutions to self-assessment questions

15

$$v_1 = 0$$
$$v_2 = 20 \text{ m/s}$$
$$t = 60 \text{ s}$$
$$\text{Using } v_2 = v_1 + at$$
$$20 = 0 + a \times 60$$
$$\text{Acceleration, } a = \frac{20}{60}\left[\frac{\text{m/s}}{\text{s}}\right] = 0.333 \text{ m/s}^2$$

16

$$v_1 = 50 \text{ km/hour} = \frac{50 \times 10^3}{3\,600} \text{ m/s}$$
$$= 13.89 \text{ m/s}$$
$$\text{Acceleration, } a = 2 \text{ m/s}^2$$

(i)

$$t = 10 \text{ s}$$
$$\therefore v_2 = 13.89\left[\frac{\text{m}}{\text{s}}\right] + (2 \times 10)\left[\frac{\text{m}}{\text{s}^2}\text{s}\right]$$
$$= (13.89 + 20) \text{ m/s}$$
$$\text{Velocity, } v_2 = 33.89 \text{ m/s}$$

(ii)

$$t = 15 \text{ s}$$
$$v_2 = 13.89\left[\frac{\text{m}}{\text{s}}\right] + (2 \times 15)\left[\frac{\text{m}}{\text{s}^2}\text{s}\right]$$
$$v_2 = (13.89 + 30) \text{ m/s}$$
$$\text{Velocity, } v_2 = 43.89 \text{ m/s}$$

17 (i)

$$v_1 = 250 \text{ km/hour} = \frac{250 \times 10^3}{3\,600} \text{ m/s}$$
$$= 69.44 \text{ m/s}$$
$$v_2 = 25 \text{ km/hour} = 6.944 \text{ m/s}$$
$$t = 2 \text{ minutes} = 120 \text{ s}$$
$$\therefore 6.944 = 69.44 + (a \times 120)$$
$$-62.496 = a \times 120$$
$$\text{Acceleration, } a = -\frac{62.496}{120} = -0.52 \text{ m/s}^2$$

The negative sign indicates a deceleration.

(ii)

$$v_1 = 69.44 \text{ m/s}$$
$$t = 30 \text{ s}$$
$$a = -0.52 \text{ m/s}^2$$
$$\therefore v_2 = 69.44 + (-0.52) \times 30$$
$$= (69.44 - 15.6) \text{ m/s}$$
$$\text{Velocity, } v_2 = 53.84 \text{ m/s}$$

any given point on the surface of the earth, the acceleration in free fall is constant. This acceleration is called *the acceleration due to gravity*; it is denoted by the symbol g, and may be assumed to have a magnitude of 9.81 m/s^2. (In London at sea level, g is almost exactly 9.81 m/s^2; at the equator, g is approximately 9.78 m/s^2; at the poles, g is approximately 9.832 m/s^2.)

Thus an object falling to earth experiences an acceleration, which can be assumed to be 9.81 m/s^2. The object has mass. Hence the object has a force applied to it. This force is different from most other forces in that its direction never changes – it *always* acts towards the centre of the earth. Because it acts in one direction only, it is given a distinguishing name – *weight*. Thus weight is the gravitational force exerted by every object which has mass.

Consider a mass of 1 kg resting on a table; the force exerted by the mass on the table is the gravitational force (i.e. its weight): if the table was removed, the mass would accelerate at approximately 9.81 m/s^2. The magnitude of the acceleration is known, and therefore the magnitude of the gravitational force can be calculated.

Remembering that

$$F\,[\mathrm{N}] - m\,[\mathrm{kg}] \times a\,[\mathrm{m/s^2}]$$

and substituting $m = 1$ kg and $a = 9.81$ m/s^2

$$F\,[\mathrm{N}] = 1 \times 9.81 \left[\frac{\mathrm{kgm}}{\mathrm{s^2}}\right]$$

i.e. $F = 9.81$ N

Hence, the weight of 1 kg is 9.81 N, and in more general terms:

Weight = mass × g.

Self-assessment questions

18 Select the correct alternative, only one of which is correct, from the list below. (The symbol ≡ means 'equivalent to'.)

(i) $N \equiv kg/m/s^2$
(ii) $N \equiv kg/m/s$
(iii) $N \equiv kgm/s^2$
(iv) $N \equiv kgm/s$

19 Calculate the weights of the following quantities, assuming $g = 9.81$ m/s^2:

(i) 10 kg of lead
(ii) 2 kg of polypropylene
(iii) 5 kg of copper
(iv) 10 kg of feathers

20 Statement 1: If a body is in free fall, its acceleration is constant.
Statement 2: Gravitational force always acts towards the centre of the earth.

Underline the correct answer:

(*a*) Only statement 1 is true.
(*b*) Only statement 2 is true.
(*c*) Neither statement is true.
(*d*) Both statements are true.

After reading the following material, the reader shall:

3.8 Solve problems using the equation $v = u+at$, including motion under gravity.

It has been established that

$$s = \left(\frac{v_1+v_2}{2}\right)t \text{ assuming constant acceleration}$$

$$\text{and } v_2 = v_1+at \text{ assuming constant acceleration}$$

where s = displacement (in metres)
v_1 = initial velocity (in m/s)
v_2 = velocity at a later instant in time (in m/s)
a = acceleration (in m/s^2)
t = time taken for the change in velocity (in seconds)

These equations were derived considering the *horizontal* movement of bodies. They can be applied to the *vertical* movement of bodies, providing that the air resistance is considered to be negligible. The force exerted by the air is dependent upon the surface area of the bodies, but for the purpose of the present analysis, the air resistance will be assumed to be zero.

Example 8
A cricket ball is thrown vertically upwards at a velocity of 25 m/s. Find

(i) the time taken to reach maximum height
(ii) the maximum height reached (assume $g = 9.81\ m/s^2$)

As the ball moves upward, its velocity decreases by 9.81 m/s every second. It slows until its speed is zero at maximum height.

(i) Using the equation

$$v_2 = v_1 + at$$
$$v_1 = 25 \text{ m/s}, \ v_2 = 0, \ a = -9.81 \text{ m/s}^2$$
$$0 = 25 + (-9.81) \times t$$
$$\therefore 9.81\, t = 25$$
$$\text{Time, } t = 2.548 \text{ s}$$

(ii) Using the equation

$$s = \left(\frac{v_1 + v_2}{2}\right) t$$
$$s = \left(\frac{25+0}{2}\right) \times 2.548$$
$$= 12.5 \times 2.548 \left[\frac{\text{m}}{\text{s}} \times \text{s}\right]$$
$$\therefore \text{Height, } s = 31.85 \text{ m}$$

Self-assessment questions

In each of the following questions, assume that $g = 9.81$ m/s^2, and that air resistance is negligible.

21 A stone is dropped from a building and hits the ground 3.5 seconds after its release. Find

(i) the velocity of the stone on impact with the ground

(ii) the height from which the stone is dropped

22 A ball is thrown vertically upwards at 30 m/s. Find

(i) the maximum height reached

(ii) its velocity after 2 seconds

Solutions to self-assessment questions

18 The correct alternative is (iii): N ≡ kg m/s^2.

19 Using the relationship:
Weight [N] = mass [kg] × 9.81 [m/s^2]
the weights are

(i) 98.1 N

(ii) 19.62 N

(iii) 49.05 N

(iv) 98.1 N

Solutions to self-assessment questions

20 Both statements are true. Hence (*d*) is the correct assertion, since:

(i) An object in free fall is subjected to constant acceleration throughout the period of its fall.

(ii) Gravitational force is termed 'weight', and therefore always acts towards the earth's centre.

21 (i) Using the equation

$$v_2 = v_1 + at$$
$$v_1 = 0,\ a = 9.81\ \text{m/s}^2,\ t = 3.5\ \text{s}$$
$$v_2 = 0 + (9.81 \times 3.5)\left[\frac{\text{m}}{\text{s}^2}\,\text{s}\right]$$
$$= (9.81 \times 3.5)\,\frac{\text{m}}{\text{s}}$$

$\therefore$ Velocity, $v_2 = 34.34$ m/s

(ii) Using the equation

$$s = \left(\frac{v_1 + v_2}{2}\right)t$$
$$v_1 = 0,\ v_2 = 34.34\ \text{m/s},\ t = 3.5\ \text{s}$$
$$s = \left(\frac{0 + 34.34}{2}\right) \times 3.5\left[\frac{\text{m}}{\text{s}} \cdot \text{s}\right]$$
$$s = (17.17 \times 3.5)\ \text{m}$$

Height, $s = 60.1$ m

22 The maximum height can be found from the equation $s = \left(\frac{v_1 + v_2}{2}\right)t$, but first, the time t must be found from

$$v_2 = v_1 + at$$
$$v_2 = 0,\ v_1 = 30\ \text{m/s},\ a = -9.81\ \text{m/s}^2$$
$$0 = 30 + (-9.81 \times t)$$
$$\therefore 9.81\,t = 30$$

$\therefore$ Time, $t = 3.058$ s

(i)

$$s = \left(\frac{v_1 + v_2}{2}\right)t$$
$$s = \left(\frac{30 + 0}{2}\right) \times 3.058\left[\frac{\text{m}}{\text{s}} \cdot \text{s}\right]$$
$$= (15 \times 3.058)\ \text{m}$$

$\therefore$ Height, $s = 45.87$ m

(ii) The velocity after 2 seconds can be found from

$$v_2 = v_1 + at$$
$$v_1 = 30\ \text{m/s},\ a = -9.81\ \text{m/s}^2,\ t = 2\ \text{s}$$
$$\therefore v_2 = 30 + (-9.81 \times 2)$$
$$= (30 - 19.62)\ \text{m/s}$$

$\therefore$ Velocity, $v_2 = 10.38$ m/s

Section 2

After reading the following material, the reader shall:

4 Be familiar with the concept of static friction between two surfaces in contact.

4.1 Define friction force with the aid of a diagram.

4.2 Compare the direction of friction force and the intended direction of motion.

4.3 State the factors affecting friction force size and direction.

4.4 State that the magnitude of the friction force is independent of the area of contact between the surfaces.

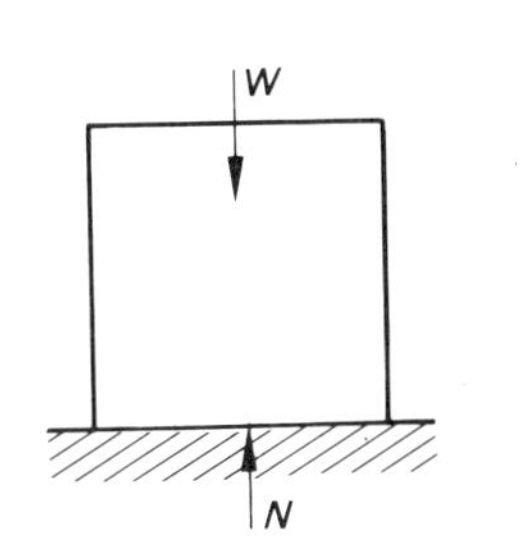

Figure 140 *Weight and normal reaction*

Suppose an object of mass m is resting on a horizontal table as in Figure 140. The object exerts a downward force on the table, the force being the weight of the object, so that

$$W = mg$$

where g = acceleration due to gravity.

The table will exert a force equal and opposite to the weight. This force is denoted in Figure 140 by the symbol N.

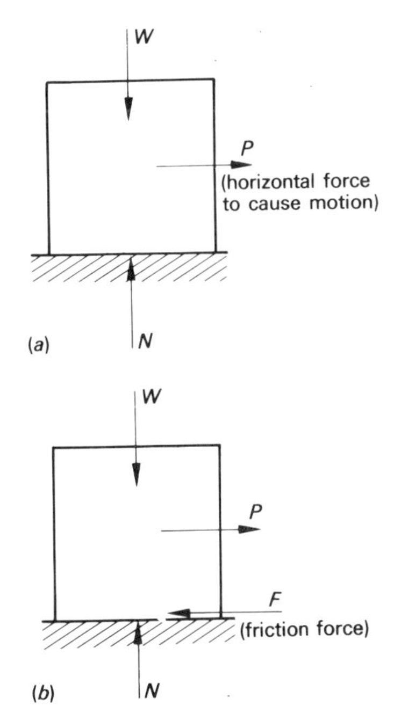

Figure 141 *Force to overcome friction*

Suppose a gradually increasing horizontal force P is applied to the object, intended to cause the object to move horizontally (see Figure 141(a)). If the body is to move, the force P must be large enough to overcome the friction which exists between the two surfaces in contact. Thus the magnitude of P must be increased until it is greater than the *friction force* F in Figure 141(b).

When the object is just about to move, but not moving, then the forces P and F are equal in magnitude, but opposite in direction. Since the weight W and the normal reaction N are also equal in magnitude but opposite in direction, the object is in horizontal and vertical equilibrium. The object does not move until P is greater than the friction force F.

The reader will probably be very familiar with the effects of friction, and how it can be reduced. For instance, any object is easier to pull across a smooth surface than across a rough surface. Similarly, an object with a small mass is easier to pull across a given surface than an object with a larger mass. The placing of a lubricant, such as oil, between the two surfaces in contact reduces the effort required to move an object.

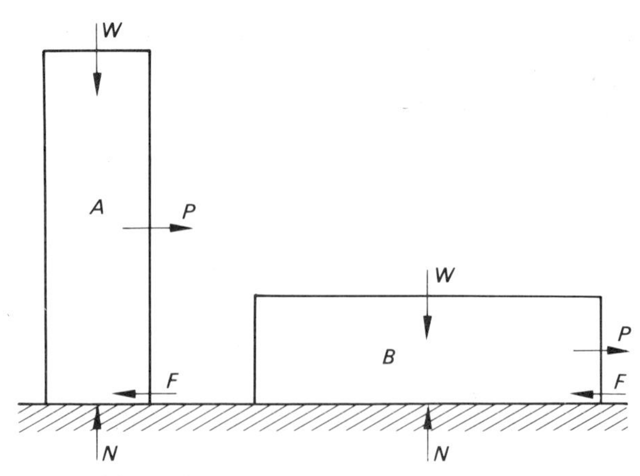

Figure 142 *Effect of area in contact on friction force*

However, consider a situation similar to that shown diagrammatically in Figure 142. Two bodies, of exactly the same mass and made from the same material, rest on the same horizontal surface. Although their masses are the same, body A has a smaller area in contact with the surface than body B. Since the masses of the bodies are the same, then the weight of each body and the normal reaction on each body is the same. It can be shown experimentally that, even though the areas in contact with the horizontal surface are different, the friction force acting on each area is the same. In other words, the friction force is independent of the area of contact between the surfaces.

The discussion so far can be summarized in the following two statements which are sometimes called 'the laws of friction'. Strictly, they are not laws, and are better classified as experimental observations:

(*a*) When the surfaces in contact are clean and dry, the friction force always opposes motion, and is dependent upon the character and surface finish of the materials from which the contact surfaces are made.

(*b*) The friction force is independent of the area of contact between the surfaces.

Self-assessment questions

23 Indicate, by drawing arrows on Figure 143, the directions of
(i) the friction force
(ii) the normal reaction

24 At the instant before the body in Figure 143 begins to move, the force P is:
(i) greater than the friction force
(ii) less than the friction force
(iii) equal to the friction force
(iv) independent of the friction force
Select the correct alternative.

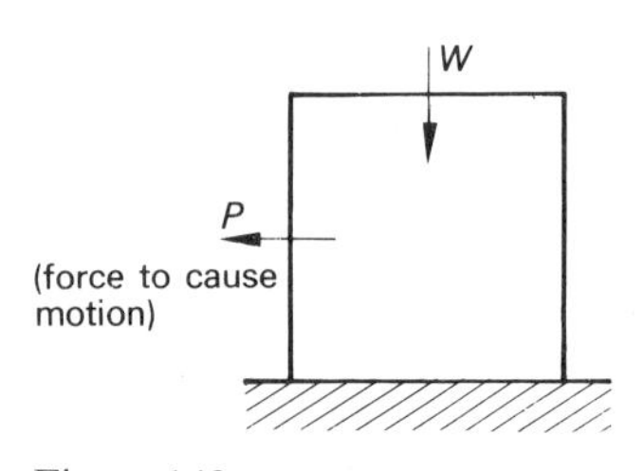

Figure 143

25 Two bodies of equal mass, and made from the same material, rest on a horizontal surface. The area in contact of body X is 100 mm^2, the area in contact of body Y is 1 000 mm^2. The *minimum* force required to move body Y is:

(i) 10 times that required to move body X
(ii) $\frac{1}{10}$ that required to move body X
(iii) equal to that required to move body X
(iv) 100 times that required to move body X

Select the correct alternative.

After reading the following material, the reader shall:

4.5 Identify the relationship between friction force and normal force between two surfaces.

4.6 Solve simple problems involving the use of the equation

$F = \mu N$

It has been suggested that the larger the mass of a body, the larger the force required to overcome friction, and to begin to move the body across a surface. Thus, there is a relationship between friction force and the weight of the body, and since the weight of the body is equal to the normal reaction N, there is a relationship between the friction force and the normal reaction. Expressed in symbolic form,

$F = N \times \text{a constant}$

This relationship has been verified by experimental work on dry surfaces. The constant of proportionality is usually written in symbol form by μ (Greek letter mu), so that

$F = \mu N$
where μ is termed the coefficient of friction,
F has been defined as the friction force,
N is the normal reaction, i.e. it is a force.

Substituting newtons (N) as the unit of force in the above equation:

$F\,[\text{N}] = \mu\,[?] \times N\,[\text{N}]$

Thus μ, the coefficient of friction has no units, and is simply a number.

The value of coefficient of friction depends not only upon the materials in contact, but also upon the nature of the surfaces in contact. For example, consider two steel surfaces in contact: an average value of coefficient of friction is about 0.2. This value increases quite markedly if the surfaces in contact are very rough; conversely, the value is reduced if the surface finish of the materials is smooth.

Typical average values of coefficient of friction are shown in the table.

materials	coefficient of friction
metal on metal	0.2
leather or rubber on metal	0.4
asbestos on cast iron	0.45
rubber on road surface	0.9

Note: The cleanliness and roughness of the surfaces involved has a marked effect upon coefficient of friction. Average values are quoted, but widely different values can be obtained by changing the surface finish of the surfaces.

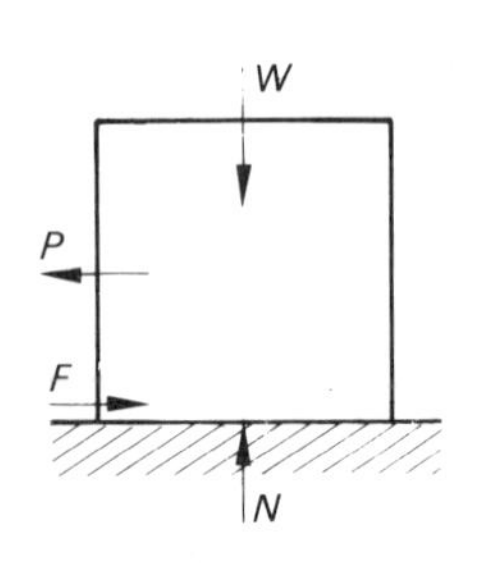

Figure 144

Example 9

A packing case of mass 500 kg is to be pulled along a level floor. If the coefficient of friction between the case and the floor is 0.4 calculate the force required to initiate movement. Assume $g = 9.81$ m/s^2.

Referring to Figure 144,

$$\text{Weight} = W = mg$$

$$\therefore W = (500 \times 9.81)\ \text{kg} \times \frac{\text{m}}{\text{s}^2}$$

$$\text{or } W = 4\,905\ \text{N}$$

$$\therefore \text{Normal reaction} = N = 4\,905\ \text{N}$$

$$\text{Friction force } F = \mu N$$

$$F = (0.4 \times 4\,905)\ \text{N}$$

$$\therefore F = 1\,962\ \text{N}$$

Since the case is about to move, $P = F$

$\therefore$ Force to initiate motion $= 1\,962$ N

Solutions to self-assessment questions

23 The diagram should be completed as in Figure 145, with F, the friction force, opposing the force tending to cause motion, and N, the normal reaction, opposing the weight.

24 The correct alternative is (iii), since the body must be in equilibrium until motion occurs. Thus P equals F.

25 The correct answer is (iii) because,

(*a*) and (*b*) the friction force is independent of the area of contact of the two surfaces, the minimum force P required to move either body X or body Y must equal the friction force.

W
P
m = 500 kg
F
N

Figure 145

The discussion so far has been limited to the instant before an object begins to move. However, the conclusions apply equally to moving objects, *providing that the speed of motion is constant and relatively low*. In other words, there must be no accelerating force acting on the object. Thus the equation connecting F, μ and N is applicable, providing that the motion is uniform, and the magnitude of the speed is relatively small.

Example 10

A mass of 1 tonne (1 000 kg) rests on a horizontal surface. If the coefficient of friction between the mass and the surface is 0.3 calculate the horizontal force required to move the mass at a low uniform speed along the surface. Assume $g = 9.81$ m/s^2.

Gravitational force acting on a mass of 1 tonne is

$$W = (1\,000 \times 9.81)\ \text{kg} \times \frac{\text{m}}{\text{s}^2}$$

$$\therefore W = 9\,810\ \text{N}$$

$$\therefore \text{Normal reaction} = N = 9\,810\ \text{N}$$

$$\text{Friction force } F = \mu N$$

$$= (0.3 \times 9\,810)\ \text{N}$$

$$\therefore F = 2\,943\ \text{N}$$

Since the mass is moving at a low uniform speed,

$$P = F$$

$$\therefore \text{Required horizontal force} = P = 2\,943\ \text{N}$$

$$\underline{\text{Force, } P = 2.943\ \text{kN}}$$

Self-assessment questions

In the following three questions, assume g, the acceleration due to gravity to be 9.81 m/s^2.

26 A casting with a mass of 50 kg is pulled along a horizontal floor at a constant velocity. If the coefficient of friction between the casting and the floor is 0.3, find the horizontal force required.

27 A loaded packing case having a mass of 2 700 kg is to be pulled along a horizontal floor. If the coefficient of friction is 0.25, calculate the horizontal force required to cause motion.

28 A loaded trolley has a mass of 550 kg, and requires a horizontal force of 730 N to maintain a uniform speed over a horizontal surface. Calculate

(i) the normal reaction between the trolley and the surface

(ii) the coefficient of friction between the trolley and the surface

After reading the following material, the reader shall:

4.7 State examples of

(*a*) practical applications of friction forces

(*b*) design implications of friction forces

Readers will be familiar with the operation of a domestic spin drier. The main components are an electric motor driving a barrel at high speed, via leather belts. Friction forces in the electric motor are kept to a minimum, in order to reduce power losses; similarly, the friction forces at the bearing on the shaft which carries the barrel must be reduced to a minimum.

Solutions to self-assessment questions

26

$$\text{Weight of casting} = (50 \times 9.81)\ \text{kg} \times \frac{\text{m}}{\text{s}^2}$$

$$\therefore W = 490.5\ \text{N}$$

$$\therefore \text{Normal reaction} = N = 490.5\ \text{N}$$

$$\text{Friction force} = F = \mu N$$

$$F = (0.3 \times 490.5)\ \text{N}$$

$$\therefore F = 147.15\ \text{N}$$

Since the casting is moving with constant velocity, $P = F$

$\therefore$ Force to move casting $= P = 147.15$ N

27

$$\text{Weight of packing case} = W = (2\,700 \times 9.81)\ \text{kg} \times \frac{\text{m}}{\text{s}^2}$$

$$\therefore W = 26\,490\ \text{N}$$

$$\therefore \text{Normal reaction} = N = 26.49\ \text{kN}$$

$$\text{Friction force} = F = \mu N$$

$$\therefore F = (0.25 \times 26.49 \times 10^3)\ \text{N}$$

$$\therefore F = 6.662\ \text{kN}$$

Since the case is just about to move, $P = F$

$\therefore$ Force to move packing case $= 6.662$ kN

28 (i)

$$\text{Weight of loaded trolley} = W = (550 \times 9.81)\ \text{kg} \times \frac{\text{m}}{\text{s}^2}$$

$$W = 5\,395.5\ \text{N}$$

$$\therefore W = 5.40\ \text{kN approximately}$$

$$\therefore \text{Normal reaction} = N = 5.40\ \text{kN}$$

(ii) Since the trolley is moving at uniform speed,

$$P = F$$

$$\therefore F = 730\ \text{N}$$

$$\text{But } F = \mu N$$

$$\therefore 730 = \mu N$$

$$\text{Also, } N = 5.4\ \text{kN} = 5\,400\ \text{N}$$

$$\therefore 730 = \mu \times 5\,400$$

$\therefore$ Coefficient of friction, $\mu = 0.135$

However, the belt connecting the electric motor to the drive shaft depends upon high friction forces, since the higher the friction force between the belt and the pulleys, the higher the efficiency of power transmission.

When high friction forces are necessary, engineers choose the most suitable material for the particular application, e.g. leather, rubber or asbestos, which all have a high coefficient of friction when in contact with other materials.

When high friction forces are undesirable, materials with a low coefficient of friction may be used, e.g. nylon or polytetrafluorethylene (PTFE). Nylon is used as a material for gears and bearings, but where high forces are involved, engineers select a metal such as cast steel for gears or a carbon steel for bearings. A good surface finish of the contact faces is very important where metals are used, as is the lubrication of the moving surfaces. The best lubricants (i.e. those which reduce the friction force to a minimum) for metals are oil and grease. It is essential that the lubricant is not squeezed out from between the surfaces in contact. One advantage of using a polymer instead of a metal for either gears or bearings is that polymers do not in general require lubrication.

Self-assessment question

29 The basic propulsion principle of a motor vehicle can be summarized as follows:
Power is generated in the engine, and transmitted through the clutch to the gearbox. The drive then passes to the rear axle (when the back wheels are driven) or the drive shafts (in the case of a front wheel-drive vehicle), and then to the wheels.

List the components, and suggest materials for those components in which:

(*a*) Frictional effects should be reduced to a minimum.
(*b*) High friction forces are necessary.

Further self-assessment questions

30 Complete the following statements by inserting the correct word in each of the statements.

(i) Speed is the rate of change of ______ with respect to time.
(ii) Velocity is the rate of change of ______ with respect to time.
(iii) Acceleration is the rate of change of ______ with respect to time.
(iv) Force is proportional to the product of mass and ______.

31 Complete the following table by selecting the appropriate SI units from the list on the right of the table. Note that some units may be used more than once. If it is thought that a quantity has no units, write 'none' in the right hand column.

quantity	units	*list of units*
speed		N
displacement		kg
time		m/s
force		s
acceleration		m/s^2
velocity		m
distance		
weight		
friction force		
coefficient of friction		

32 Write down the SI units of force, mass and acceleration which enable the equation

Force = mass × acceleration

to be true.

33 A motor car, travelling at constant speed, takes 9 hours to cover a distance of 540 km. Find

(i) the distance covered in 3.5 hours

(ii) the time taken to travel 480 km

Solution to self-assessment question

29 (*a*) Components in which frictional effects should be reduced to a minimum include:

(i) The engine – pistons, cylinders, crankshaft, camshaft, etc.; typical materials are steel, cast iron and aluminium alloy.

(ii) The gearbox, propellor shaft and drive shafts; typical materials are steel and cast iron.

(*b*) Components where high friction forces are necessary include:

(i) The clutch, since the clutch plates depend upon friction during operation; a typical material for the plates is asbestos.

(ii) The tyres and brakes. A high friction force is necessary to maintain effective tyre adhesion with the road surface. (Compare the coefficient of friction for rubber on a road surface in the table on page 186, with other values in the table.) A high friction force is also necessary for the brakes to fulfil their essential function. A typical material for the brake linings or pads is asbestos. Note that the presence of a lubricant between the contact surfaces (e.g. water between tyres and road surface, or between brake linings and drum) drastically reduces the friction force.

34 A car travels 15 km northwards, and then travels 20 km in a westerly direction. What is its displacement from the original position?

35 A cyclist covers the first 20 km of a journey in 70 minutes. He rests for a period of 20 minutes, then resumes his journey, covering the next 31 km in 90 minutes. Calculate his average speed over the whole 3 hours in km/hour and m/s.

36 An aeroplane travels in a due westerly direction. It covers the first 1 000 km of its journey at an average velocity of 400 km/hour. What must be its average velocity over the remaining 600 km, in order that the average velocity over the whole 1 600 km may be 500 km/hour?

37 The displacement of a cyclist is estimated every second, the results being tabulated in the table below:

time (s)	0	1	2	3	4	5	6	7	8
displacement (m)	0	5.9	11.6	17.5	29.2	40.8	52.5	68.8	85.0

Plot a graph of displacement against time, joining the points by straight lines. Hence, find the three velocities of the cyclist during the measured time.

38 The table below shows how the velocity of a car varies with time.

velocity (m/s)	0	1.45	3.00	4.50	4.87	5.25	5.62	6.00	6.37
time (s)	0	0.5	1.0	1.5	2.0	2.5	3.0	3.5	4.0

velocity (m/s)	6.75	7.13	7.50	6.50	4.95	3.42	1.90	0
time (s)	4.5	5.0	5.5	5.7	6.0	6.3	6.6	7.0

Plot a graph of velocity against time, assuming the points are joined by straight lines. Determine

(i) the acceleration of the car between the time intervals 0 to 1.5 s and 2.0 to 5.0 s

(ii) the retardation of the car during the last 1.5 s

39 The velocity of a car falls from 75 km/hour to zero in a uniform manner. If the time taken is 5 seconds, calculate the retardation.

40 A motorcycle brakes to a halt from a velocity of 60 km/hour on a level surface without skidding. The time from the instant the brakes are applied until the motorcycle is stationary is measured as 4 seconds. Assuming the braking is uniform, calculate

(i) the retardation in m/s^2

(ii) the distance covered during braking

41 An object is dropped from the top of a building, and hits the ground with a velocity of 40 m/s. Assuming the air resistance to be negligible and taking g as 9.81 m/s^2, calculate

(i) the time of descent

(ii) the height of the building

42 A train accelerates uniformly from rest along a straight level track at a rate of 0.25 m/s^2. Find

(i) the time to reach a velocity of 70 km/hour

(ii) the displacement when the velocity is 70 km/hour

(iii) the velocity 20 s after the start

(iv) the time for the velocity to increase from 40 km/hour to 60 km/hour

43 Select from the list below those factors which affect the magnitude of the friction force between two surfaces in contact:

(*a*) the materials of the surfaces

(*b*) the area of materials in contact

(*c*) the surface finish of the surfaces in contact

(*d*) the force between the surfaces

44 The table of a planing machine has a mass of 600 kg, and moves horizontally on its bed. If the coefficient of friction between the table and the bed is 0.07, find the force required to overcome friction when a casting having a mass of 800 kg is mounted on the machine table. Assume $g = 9.81$ m/s^2.

45 A trolley having a mass of 450 kg is pulled along a level surface at a steady speed by a force of 1.1 kN, applied horizontally. Calculate the horizontal force required if a mass of 570 kg is added to the trolley. Assume $g = 9.81$ m/s^2.

Answers to self-assessment questions

30 (i) Speed is the rate of change of *distance* with respect to time.
(ii) Velocity is the rate of change of *displacement* with respect to time.
(iii) Acceleration is the rate of change of *velocity* with respect to time.
(iv) Force is proportional to the product of mass and *acceleration*.

quantity	units
speed	m/s
displacement	m
time	s
force	N
acceleration	m/s^2
velocity	m/s
distance	m
weight	N
friction force	N
coefficient of friction	none

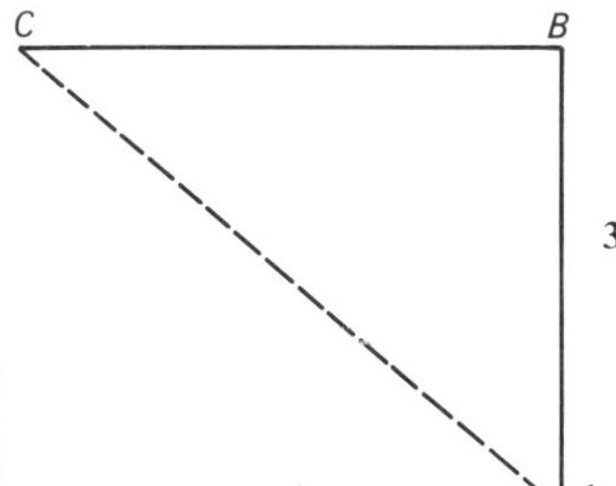

Figure 146

32 Force: newtons (N)
Mass: kilograms (kg)
Acceleration: metre/second^2 (m/s^2)

33 (i) Distance covered in 3.5 hours is 210 km.
(ii) Time taken to travel 480 km is 8 hours.

34 The travel path of the car is shown in Figure 146, A being the initial position and C the final position. Displacement is 25 km from A at an angle of 53°6′ from the north direction.

35 Total distance covered is 51 km in a total time of 3 hours.
Average speed = 17 km/hour = 4.72 m/s.

36 Total flight time of aircraft is 3.2 hours.
Flight time for first 1 000 km is 2.5 hours.

$$\text{Average velocity} = \frac{600}{0.7}\ \frac{\text{km}}{\text{hours}} = 857 \text{ km/hour}$$

37 The graph is made up of three straight lines representing velocity:
Velocity between 0 and 3 s is 5.8 m/s.
Velocity between 3 and 6 s is 11.7 m/s.
Velocity between 6 and 8 s is 16.3 m/s.

38 The graph is made up of three straight lines, the slope of the lines representing acceleration or deceleration:
Acceleration between 0 and 1.5 s is 3 m/s^2.
Acceleration between 2 and 5 s is 0.75 m/s^2.
Deceleration during final 1.5 s is −5 m/s^2.

39 $v_1 = 20.83$ m/s, $v_2 = 0$
Retardation = −4.167 m/s^2

40 $v_1 = 16.67$ m/s, $v_2 = 0$
(i) Retardation = −4.167 m/s^2
(ii) Distance covered = 33.34 m

41 $v_1 = 0$, $v_2 = 40$ m/s, $a = -9.81$ m/s^2
(i) Time of descent = 4.078 s
(ii) Height of building = 81.55 m

42 (i) $v_1 = 0$, $v_2 = 19.44$ m/s, $a = 0.25$ m/s^2
Time to reach a velocity of 70 km/hour is 77.76 s.

(ii) $v_1 = 0$, $v_2 = 19.44$ m/s, $t = 77.76$ s
Displacement = 756 m

(iii) $v_1 = 0$, $t = 20$ s, $a = 0.25$ m/s^2
Velocity 20 s after start = 5 m/s

(iv) $v_1 = 11.11$ m/s, $v_2 = 16.67$ m/s, $a = 0.25$ m/s^2
Time = 22.24 s

43 The magnitude of the friction force between two surfaces in contact is affected by the materials from which the surfaces are made, the force between the surfaces and the surface finish of the surfaces; the area of the materials in contact has no effect upon the magnitude of the friction force.

44 Weight of table and casting (equal to the normal reaction N) is $(1\,400 \times 9.81)$ N = 13.73 kN
Force to overcome friction = 961 N

45 Weight of unloaded trolley is (450×9.81) N = 4.41 kN
Coefficient of friction = 0.249
Weight of loaded trolley is $(1\,020 \times 9.81)$ N = 10 kN
Force to move loaded trolley at steady speed is 2.49 kN.

Topic area: Statics

Section 1

After reading the following material, the reader shall:

1 Be aware of the distinction between scalar and vector quantities.
1.1 Define a scalar quantity.
1.2 Define a vector quantity.
1.3 State that force is a vector quantity.
1.4 Identify vector and scalar quantities from a given list.

Certain physical quantities can be specified by quoting just the magnitude of the quantity, for example a mass of 5 kg, or a speed of 50 km/hour. Other quantities are not completely specified by their magnitude alone, because a direction also needs to be specified; for example a wind velocity of 15 km/hour in a south-easterly direction. Other quantities for which both magnitude and direction need to be specified are acceleration and force. At first glance it might appear that only the magnitude of a force needs to be specified, but if the reader refers to the topic area concerned with friction (pages 00), he will appreciate that the direction of a force also needs to be specified.

A quantity which has both magnitude and direction is a *vector* quantity; a quantity which possesses magnitude only, is a *scalar* quantity. Thus mass and speed are scalar quantities, whereas force and velocity are vector quantities.

Self-assessment question

1 The following table lists a number of physical quantities with which the reader will be familiar. Complete the right hand side of the table by writing the letter 'S' for those quantities which are scalars, and the letter 'V' for those quantities which are vectors.

quantity	identification S or V	quantity	identification S or V
acceleration		time	
energy		density	
displacement		distance	
coefficient of friction		work done	
electric current		volume	
weight			

After reading the following material, the reader shall:

1.5 Identify a vector quantity as a quantity which can be represented graphically.

1.6 Recognize the main advantage and the main disadvantage of vector diagrams.

1.7 Determine graphically the resultant of two co-planar forces acting at a point.

Figure 147 *Gravitational force as a vector*

A vector quantity has been defined as a quantity which possesses both magnitude and direction. It can therefore be represented by a straight line; the length of the line represents to scale the magnitude of the quantity, and the direction of the line is parallel to the line of action of the quantity. Such a line is termed a *vector*. Thus the gravitational force (i.e. weight) of a body can be represented as in Figure 147: the direction is vertical, and the length *ab* represents to scale the magnitude of the force. The arrow on the vector indicates that the force is acting downwards from *a* to *b*.

In order to illustrate the use of vectors and vector diagrams, self-assessment question 3 in Topic area: Dynamics (page 165) is reproduced here as Example 1.

Solution to self-assessment question

1 The table should be completed as shown below:

quantity	identification S or V
acceleration	V
energy	S
displacement	V
coefficient of friction	S
electric current	V
weight	V
time	S
density	S
distance	S
work done	V
volume	S

Acceleration, displacement and electric current are vector quantities, since both magnitude and direction need to be specified. Weight is a force in a particular direction. Work done is defined as the product of a force and the distance moved in the direction of the force.

Coefficient of friction, time, density, distance and volume can be completely specified by quoting an appropriate magnitude alone. Energy is also a scalar quantity since it does not act in a particular direction (consider radiant heat energy for an example).

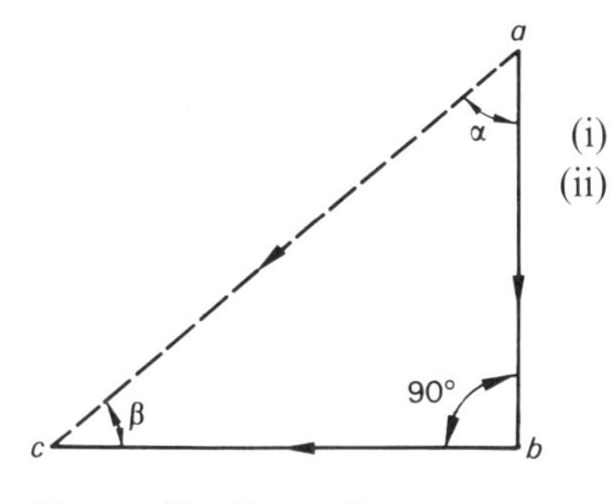

Figure 148 *Vector diagram – displacement*

Example 1

A car travels 3 km southwards and then travels 4 km westwards.

(i) What distance has the car travelled?

(ii) What is the displacement of the car from the initial position?

This problem was solved by calculation using the theorem of Pythagoras and trigonometry. It could equally well be solved by drawing a vector diagram as illustrated in Figure 148.

ab represents the initial 3 km of travel in magnitude and direction; *bc* represents the second 4 km of travel in magnitude and direction; *ac* represents the *resultant* of the two vectors *ab* and *bc*, and if the vector *ac* is measured, it will be found to represent 5 km. The vector *ac* is the displacement of the car from its initial position *a* to its final position *c*. Thus the arrow on the vector points from a to c.

Since Figure 148 is drawn to scale, the angle α (or the angle β) can simply be measured, rather than calculated, when it will be found that $\alpha = 53°$ and $\beta = 37°$.

If the answers obtained from the vector diagram are compared with those obtained by calculation (page 168) a close correlation can be seen. However, unless very great care is taken, the graphical method yields approximate answers. The converse is that in many cases the graphical method is very much simpler, and results in fewer opportunities for error than the analytical method. For example, consider the following:

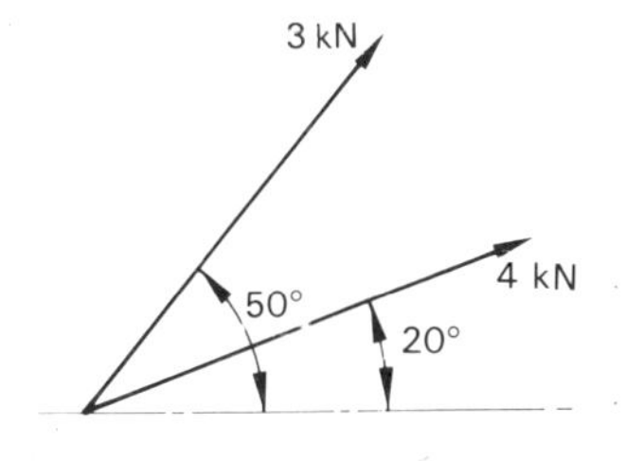

Figure 149 *Two forces acting at a point*

Example 2

Two forces act at a point as shown in Figure 149. Determine the resultant of the forces.

The vector diagram is drawn to scale in Figure 150. *ab* represents the 4 kN force in magnitude and direction; *bc* represents the 3 kN force in magnitude and direction; *ac* represents the resultant of the two forces in magnitude and direction. Thus,

Resultant force = 6.85 kN

at an angle of 33.5° to the datum. Note that the vector representing the 3 kN force (*bc*) is not drawn from the same point as the vector representing the 4 kN force (*ab*), even though the forces act at the same point. In order to add vectorially (i.e. in order to determine the resultant of a number of vectors) the vectors must be drawn in the manner a → b, b → c, c → d etc. The sum of the vectors, or the resultant, is then found by joining the initial and final points.

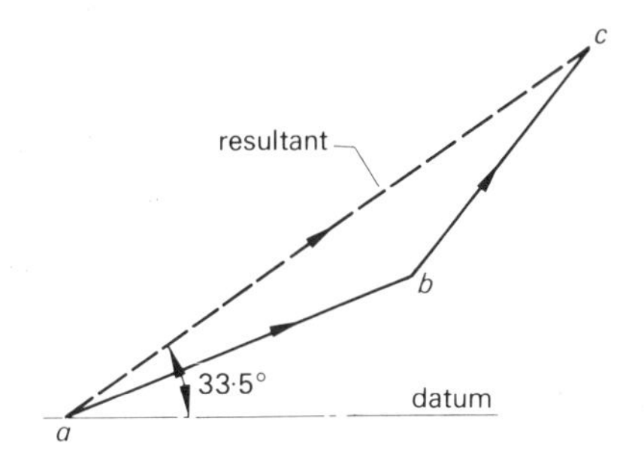

Figure 150 *Two co-planar force vector diagrams*

Self-assessment questions

2 Two equal and opposite forces (each of magnitude *F*) act upon the same point in a body. No other forces act on the body. The resultant force on the body is equal to:
(i) $2F$
(ii) F
(iii) zero
(iv) $F/2$
Select the correct alternative.

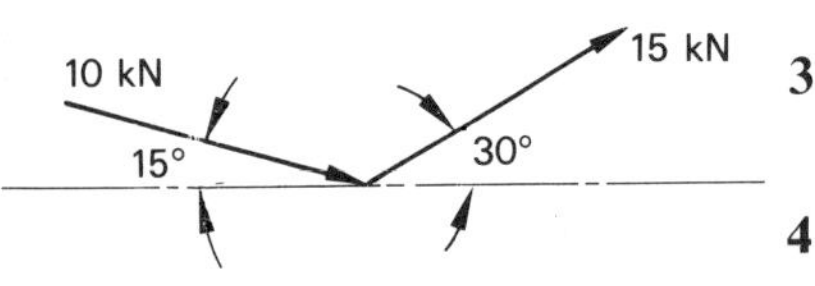

Figure 151 *Two co-planar forces*

3 Two forces act at a point as shown in Figure 151. Find graphically, the resultant force.

4 A ship is being towed at constant speed by two tugs *A* and *B*. The angle made by the tow rope of tug *A* with the direction of motion of the ship is 50°, and the force in the rope is 40 kN; the angle made by the tow rope of tug *B* with the direction of motion of the ship is 35°, and the force in the rope is 65 kN. Determine graphically, the resultant of the forces in the tow ropes acting on the ship.

5 A person walks 1 km in a north-easterly direction, followed by 2 km in a due east direction and then 1.5 km in a northerly direction. Find graphically, the person's displacement from his original position.

After reading the following material, the reader shall:

2 Be familiar with the concept of static equilibrium.
2.1 State Newton's third law of motion.
2.2 Describe stable, unstable and neutral equilibrium.

The reader will probably be familiar with the words 'static' and 'equilibrium', but not with the scientific meaning of the words. The simplest case of a body in equilibrium is one in which two exactly equal forces in magnitude (but opposite in direction) act upon it. In essence, this statement reflects Newton's third law of motion which can be written as: *If a state of equilibrium exists, then to every force there is an equal and opposite force.*

Newton's law explains equilibrium, but no mention has yet been made of the term 'static'. The reader could be forgiven for thinking that 'static' means 'stationary' or 'not moving', but this is only half true in a scientific sense. A more complete definition of static refers not only to stationary bodies, but also to bodies *moving at constant velocity*. The emphasis on constant velocity is important; a body which is accelerating (i.e. a body whose velocity is changing) is subjected to an accelerating force, and is referred to as being in a *dynamic* state.

Thus static equilibrium refers to a body which is stationary or moving with constant velocity. The forces acting on the body are equal and opposite; in other words, the resultant force on the body is zero. This is almost, but not quite, a complete definition of static equilibrium. There is one further qualification to add, which will be left until later in this section.

Stable, unstable and neutral equilibrium

The reader may be familiar with a relatively simple toy designed for young children. The toy frequently takes the form of a teddy bear designed to stand on the floor. No matter how much the child pushes the toy – even if it lies horizontal on the floor – when the child releases the applied force, the bear returns to the upright position. The toy is described as being in *stable equilibrium*, since it tends to return to its original position. It is as though the toy wants to return to its original 'happy' state of static equilibrium.

Alternatively, the toy can be modified so that the deflection of the toy tends to increase from the upright position, even though the child ceases to push it; this condition is called *unstable equilibrium* (as though the toy has a 'death wish' compared with the original version).

The toy can be modified a second time so that, when deflected from the upright position, it remains in this new deflected position when the applied force is removed. This condition is called *neutral equilibrium*, and the 'will' of the toy could now be described as 'indifferent'.

Very occasionally, new transport vehicles, new structures or mechanisms have been found to be unstable or neutrally stable when tested; for example, there are recorded instances of ships capsizing as they are launched. It is essential that modifications to produce stable equilibrium are incorporated before the vehicle enters service, or before full production is begun.

Self-assessment questions

6 Newton's third law of motion is concerned with forces in static equilibrium.

(i) Does the term *static* refer only to stationary bodies?

(ii) Complete the following statement:
If static equilibrium exists, to every force there is an equal and ______ ______.

7 A ship is steering a straight course when a heavy sea causes it to steer to starboard (to the right, looking forward). Describe the response of the ship, assuming no action by the helmsman if it is in a state of

(i) stable

(ii) unstable

(iii) neutral equilibrium

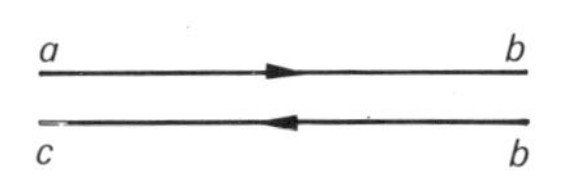

Figure 152 *Vector diagram: equal and opposite forces*

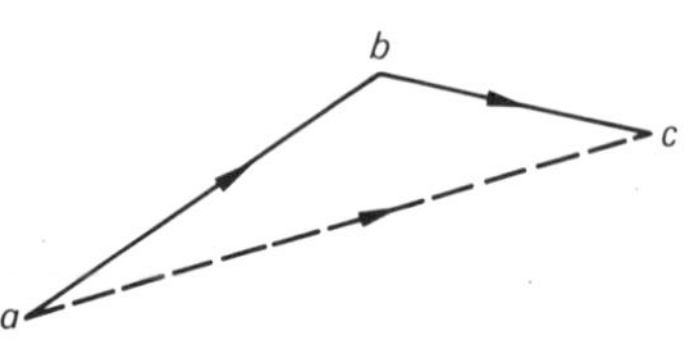

Figure 153 *Resultant of the two co-planar forces*

Solutions to self-assessment questions

2 The correct alternative is (iii), since if the vector diagram for the system is drawn, it would appear as in Figure 152. The two vectors *ab* and *bc* are drawn slightly apart for clarity. In reality, the two vectors should appear as one line so that *c* coincides with *a*. Hence, the resultant force is zero.

3 The vector diagram is shown in Figure 153, with *ab* representing the 10 kN force, and *bc* representing the 15 kN force. The resultant is *ac* in the direction *a* to *c*. Its magnitude is 23.9 kN at an angle of 12° to the horizontal.

Note that the vector diagram could be constructed by drawing the vector representing the 15 kN force first, followed by the vector representing the 10 kN force. The diagram then differs from that shown in Figure 153, but yields exactly the same answer.

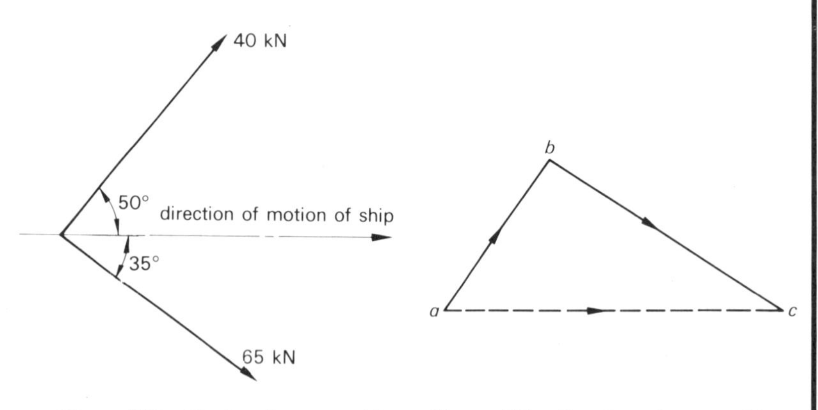

Figure 154 *Towing forces on ship*

Figure 155 *Resultant force on ship*

4 The tow ropes are drawn diagrammatically in Figure 154, and the corresponding vector diagram in Figure 155. *ab* represents the 40 kN force in magnitude and direction. *bc* represents the 65 kN force in magnitude and direction. Thus the resultant is *ac* in the direction *a* to *c*. Its magnitude is 75 kN at an angle of 3° to the direction of motion of the ship.

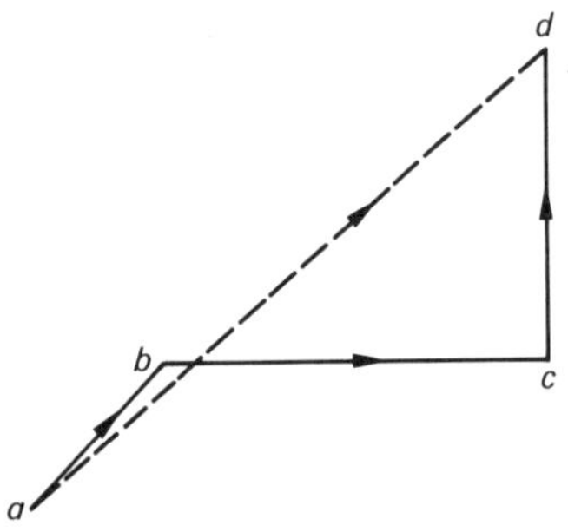

5 The vector diagram is drawn in Figure 156. *ab* represents the 1 km displacement, *bc* the 2 km displacement and *cd* the 1.5 km displacement. Thus the displacement of the person from his original position is represented by the vector *ad*, in the direction *a* to *d*. Its magnitude is 3.5 km at an angle of 50° to the north direction.

Figure 156 *Resultant displacement*

After reading the following material, the reader shall:

2.3 Define the moment of a force about a point.
2.4 Identify the units of the moment of a force about an axis.
2.5 Calculate the magnitude of the moment of a force about an axis, given simple practical examples.

When a door is pushed, it rotates about the vertical axis through the hinges. In technical language, the force applied to the door produces a *moment* about the hinge axis. The moment of a force about an axis is dependent upon the magnitude of the applied force, and the perpendicular distance of the line of action of the force from the axis, i.e.

Moment of force about an axis = Force × perpendicular distance between the line of action of the force and the axis

Thus the magnitude of the moment can be varied (i) by changing the magnitude of the force or (ii) by varying the distance between the line of action of the force and the axis. This distance is given the name *moment arm*.

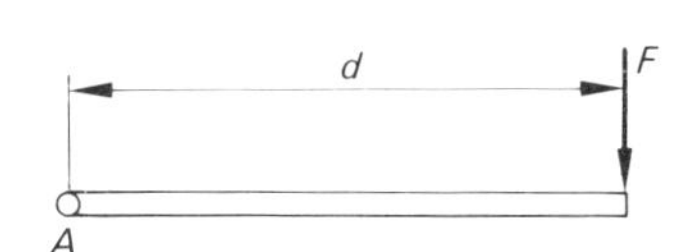

Figure 157 *Moment of force about an axis*

Note that the moment of a force refers specifically to an *axis*. However, it has become accepted custom to refer to the moment of a force about a *point*, for example, in Figure 157, the moment of F about the hinge point A is $F \times d$. What is actually meant is the moment of F about an axis through A perpendicular to the plane of the paper.

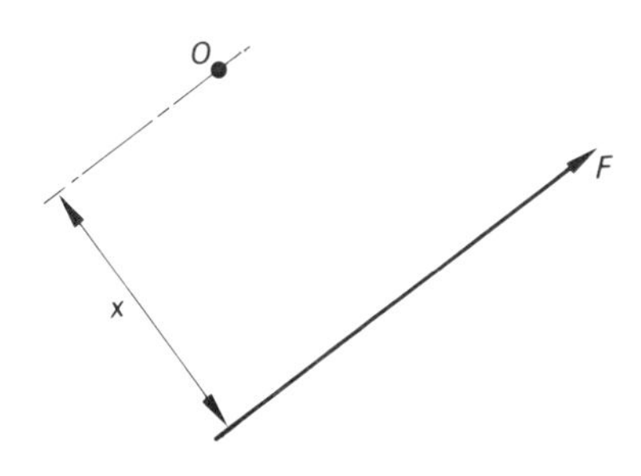

Figure 158

Self-assessment questions

8 Referring to Figure 158, write down the moment of the force F about an axis through 0 perpendicular to the paper.

9 The SI units of the moment of a force about an axis are:
(i) N
(ii) Nm
(iii) m
(iv) N/m
Select the correct alternative.

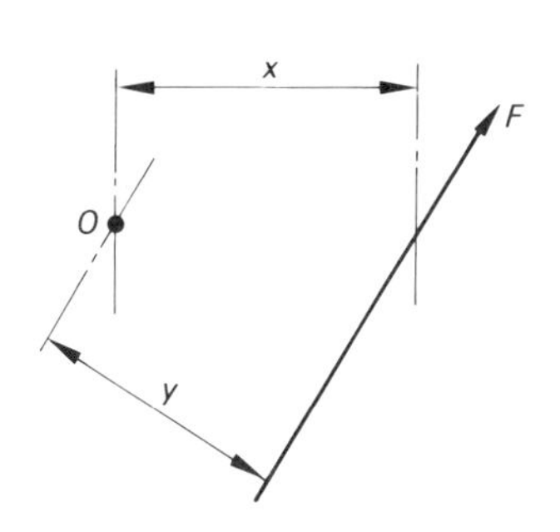

Figure 159

10 Referring to Figure 159, the moment of the force F about an axis through 0, perpendicular to the paper is either $F \times x$ or $F \times y$. Select the correct alternative.

11 The effective length of a spanner is 100 mm, and a force of 100 N is applied at one end. Find the moment of the force about an axis through the bolt at the opposite end of the spanner from the applied force. See Figure 160.

If the effective length of the spanner is increased to 200 mm, what force must be applied in order that the same moment is applied to the bolt?

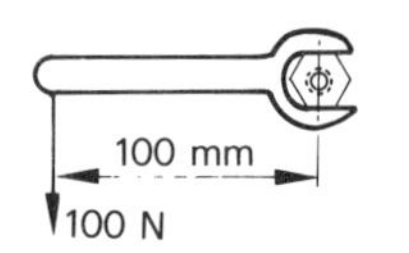

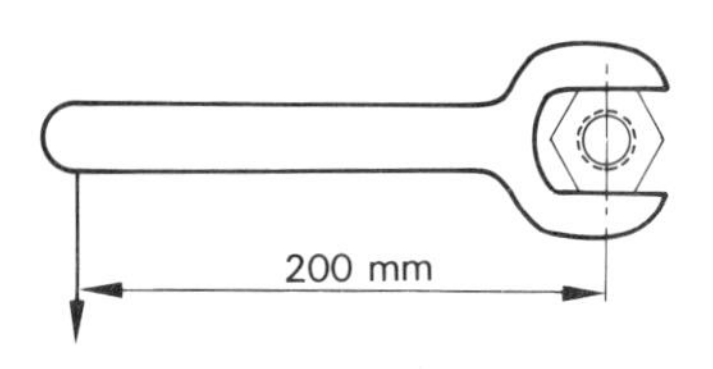

Figure 160

After reading the following material, the reader shall:

2.6 State the principle of moments.
2.7 Solve simple beam problems.

Earlier in this section, reference was made to Newton's third law of motion: 'If a state of equilibrium exists, then to every force there is an equal and opposite force.'

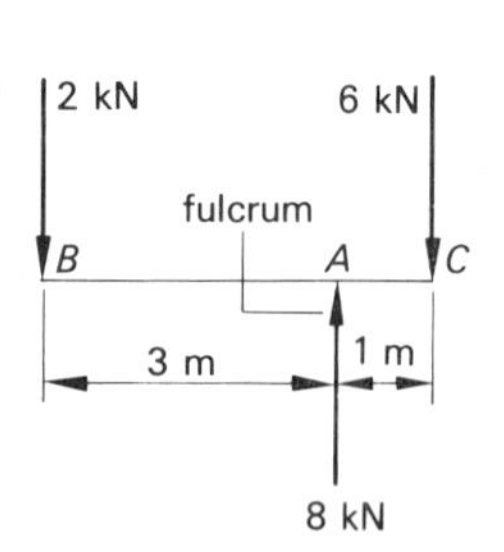

Figure 161 *Principle of moments*

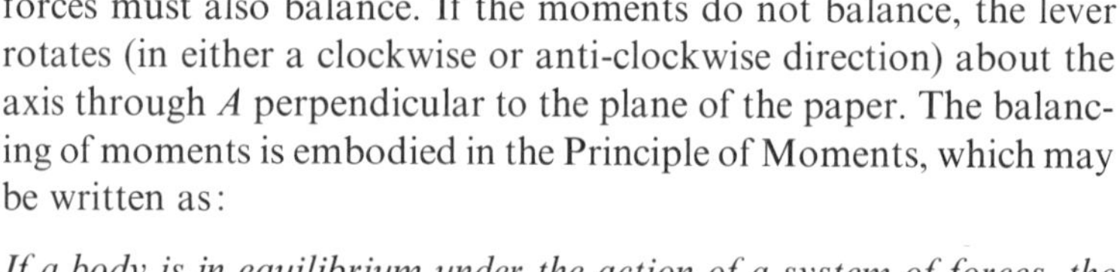

Suppose this law is applied to the lever shown in Figure 161. The applied forces of 2 kN and 6 kN are balanced by the equal and opposite reaction of 8 kN at the fulcrum *A*. This however fulfils only one of the conditions for equilibrium, since the moments of the forces must also balance. If the moments do not balance, the lever rotates (in either a clockwise or anti-clockwise direction) about the axis through *A* perpendicular to the plane of the paper. The balancing of moments is embodied in the Principle of Moments, which may be written as:

If a body is in equilibrium under the action of a system of forces, the resultant moment about any axis is zero.

Thus, the sum of the clockwise moments about any axis is equal to the sum of the anti-clockwise moments about the same axis.

Solutions to self-assessment questions

6 (i) The term static does not refer only to stationary bodies, but also to bodies moving with constant velocity.
(ii) If static equilibrium exists, to every force there is an equal and *opposite force*.

7 (i) The ship tends to return to its original course if it is stable.
(ii) If the ship is unstable, the angle between the new course of the ship and its original course tends to increase.
(iii) If the ship is neutrally stable, it tends to steer the new course.

The system of forces in Figure 161 will now be analysed to check whether it obeys the principle of moments.

Take moments about an axis through the fulcrum at point A (note that the 8 kN force produces no moment about an axis through A, because its line of action passes through A).

$$\text{Anti-clockwise moment} = 2\,[\text{kN}] \times 3\,[\text{m}] = 6\text{ kNm}$$
$$\text{Clockwise moment} = 6\,[\text{kN}] \times 1\,[\text{m}] = 6\text{ kNm}$$

Thus the lever in Figure 161 fulfils both of the conditions for static equilibrium to exist:

(i) The resultant force in any direction acting on the body is zero.
(ii) The resultant moment about any axis acting on the body is zero.

Self-assessment question

12 The principle of moments states that the sum of the clockwise moments about *any* axis is equal to the sum of the anti-clockwise moments *about the same axis*. The reader is asked to confirm this statement by taking moments of the forces in Figure 161 about two other axes – those through points B and C perpendicular to the paper.

The two conditions of static equilibrium – those concerning forces and moments – are applied in the analysis of many technological problems. One of the most common problems is the analysis of forces applied to beams.

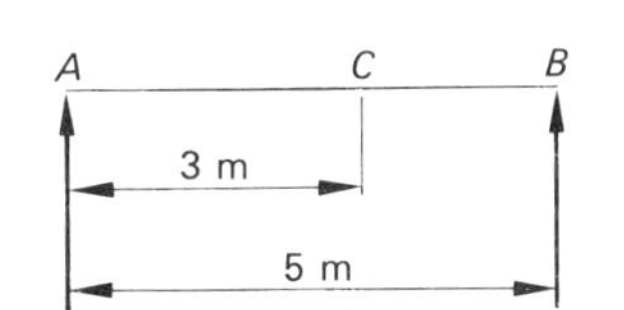

Figure 162 *Simply supported beam*

Example 3
A beam of negligible mass, compared with the loading, rests on supports A and B, 5 m apart, as shown in Figure 162. If a mass of 10 kg is hung on the beam at point C, 3 m from A, calculate the reaction forces at A and B. Assume $g = 9.81\text{ m/s}^2$.

$$\text{Gravitational force at } C = 5\text{ kg} \times 9.81\text{ m/s}^2$$
$$= 49.05\text{ N}$$

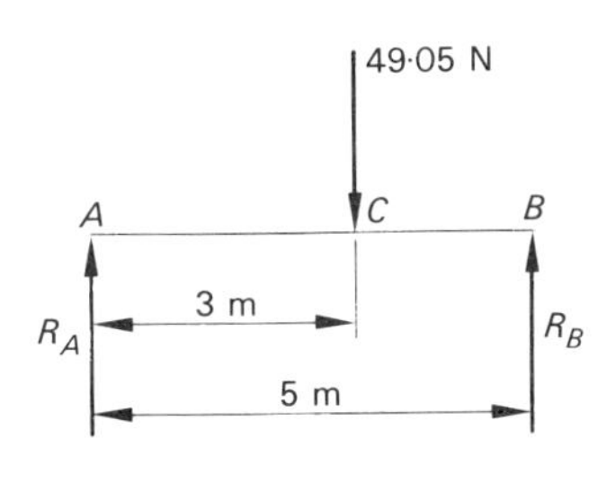

Figure 163 *Forces acting on a beam*

Let the reaction force at A be R_A, and the reaction force at B be R_B. The system of forces acting on the beam is then as shown in Figure 163.

Taking moments about the axis through B, when the force at B will produce no moment:

$$\text{Anti-clockwise moment} = 49.05\,[\text{N}] \times 2\,[\text{m}] = 98.10\text{ Nm}$$
$$\text{Clockwise moment} = R_A\,[\text{N}] \times 5\,[\text{m}] = 5\,R_A\text{ Nm}$$

For equilibrium,

$$\text{clockwise moments} = \text{anti-clockwise moments}$$
$$\therefore 5\ R_A = 98.10$$
$$\text{Reaction force at } A,\ R_A = \frac{98.10}{5} \frac{[\text{Nm}]}{[\text{m}]} = 19.62\ \text{N}$$

Taking moments about the axis through A, when the force at A will produce no moment:

$$\text{Clockwise moment} = 49.05\ [\text{N}] \times 3\ [\text{m}] = 147.15\ \text{Nm}$$
$$\text{Anti-clockwise moment} = R_B\ [\text{N}] \times 5\ [\text{m}] = 5\ R_B\ \text{Nm}$$

Equating clockwise and anti-clockwise moments,

$$5\ R_B = 147.15$$
$$\text{Reaction force at } B,\ R_B = \frac{147.15}{5} \frac{[\text{Nm}]}{[\text{m}]} = 29.43\ \text{N}$$

Solutions to self-assessment questions

8 The moment of the force F about an axis through 0 is $F \times x$.

9 The correct alternative is (ii), i.e. Nm. A moment is the product of a force and a distance; the SI unit of force is the newton, and the SI unit of length is the metre.

10 The correct alternative is $F \times y$, since a moment about an axis is the product of a force and the perpendicular distance between the line of action of the force and the axis.

11 When the effective length of the spanner is 100 mm:

$$\text{Moment about an axis through the bolt} = 100 \times 100\ \text{Nmm}$$
$$= 100 \times 0.1\ \text{Nm}$$
$$= 10\ \text{Nm}$$

If the effective length of the spanner is 200 mm, then:

$$\text{Moment} = F \times d$$
$$10\ [\text{Nm}] = F \times 0.2$$
$$\therefore \text{Force, } F = \frac{10}{0.2}\left[\frac{\text{Nm}}{\text{m}}\right] = 50\ \text{N}$$

Solution to self-assessment question

12 If moments are taken about an axis through B, the 2 kN force produces no moment:

$$\text{Clockwise moment} = 6\ [\text{kN}] \times 4\ [\text{m}] = 24\ \text{kNm}$$
$$\text{Anti-clockwise moment} = 8\ [\text{kN}] \times 3\ [\text{m}] = 24\ \text{kNm}$$

If moments are taken about an axis through C the 6 kN force produces no moment:

$$\text{Clockwise moment} = 8\ [\text{kN}] \times 1\ [\text{m}] = 8\ \text{kNm}$$
$$\text{Anti-clockwise moment} = 2\ [\text{kN}] \times 4\ [\text{m}] = 8\text{kNm}$$

Thus the principle of moments applies about any axis of the lever.

To confirm the arithmetic and method, it is advisable to check that the beam fulfils the condition regarding forces for static equilibrium:

Total downward force $= 49.05$ N
Total upward force $= (19.62+29.43)$ N
$= 49.05$ N

Hence the resultant force on the beam is zero.

The reader could have noticed that, once R_A has been determined by taking moments, the reaction at B could be found by considering the vertical equilibrium of the beam, instead of taking moments for a second time; the amount of calculation involved is then reduced. However, until the reader is sufficiently well practised, he is recommended to use the method suggested above, and consider the equilibrium of vertical forces as a check on his work. One other point needs to be noted concerning the method of solution:

If moments are taken about an axis through C, then

Anti-clockwise moment $= 2 \times R_B$
Clockwise moment $= 3 \times R_A$

Equating clockwise and anti-clockwise moments,

$$2\,R_B = 3\,R_A$$

This equation cannot be solved unless either R_A or R_B is already known. Thus the first step in the solution is to eliminate one of the unknowns, either R_A or R_B, by taking moments about an axis through either A or B.

Consider a rather more complex example.

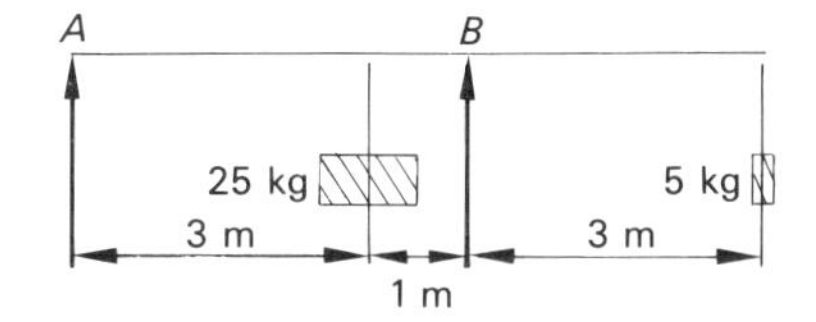

Figure 164 *Beam with overhang*

Example 4
A beam of negligible mass is 7 m long, and rests on two supports A and B, 4 m apart, as shown in Figure 164. The beam carries two masses of 25 kg and 5 kg as shown. If $g = 9.81$ m/s^2

(i) Determine the reaction forces at A and B.
(ii) If the 5 kg mass is increased to 10 kg, what then are the reactions at A and B?

(i) Let R_A and R_B be the reaction forces at A and B respectively.

Gravitational force on 25 kg mass $= (25 \times 9.81)$ N
$= 245.3$ N
Gravitational force on 5 kg mass $= (5 \times 9.81)$ N
$= 49.05$ N

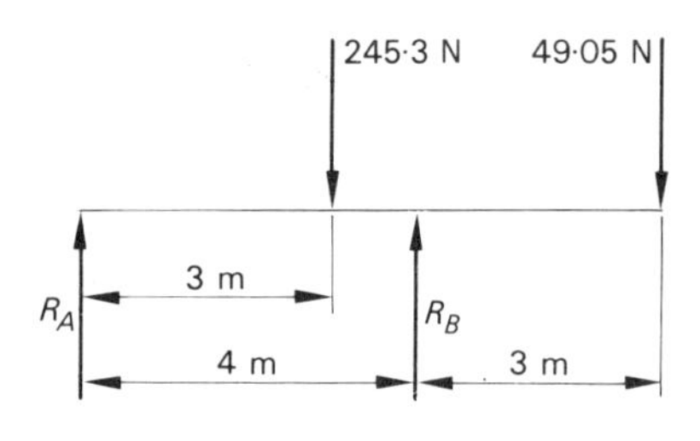

Figure 165 *Reaction forces on beam with overhang*

The system of forces acting on the beam is now as shown in Figure 165:

Take moments about an axis through B to eliminate one of the unknowns,

$$\text{Clockwise moments} = [(4 \times R_A) + (49.05 \times 3)] \text{ Nm}$$
$$\therefore \text{Clockwise moments} = [4\,R_A + 147.2] \text{ Nm}$$
$$\text{Anti-clockwise moments} = [245.3 \times 1] \text{ Nm}$$
$$= 245.3 \text{ Nm}$$

Equating clockwise and anti-clockwise moments,

$$4\,R_A + 147.2 = 245.3 \text{ Nm}$$
$$4\,R_A = 98.1 \text{ Nm}$$
$$\text{Reaction force at } A,\ R_A = \frac{98.1\ [\text{Nm}]}{4\ \ [\text{m}]} = 24.53 \text{ N}$$

Take moments about an axis through A:

$$\text{Clockwise moments} = [(245.3 \times 3) + (49.05 \times 7)] \text{ Nm}$$
$$= [735.9 + 343.4] \text{ Nm}$$
$$= 1\,079.3 \text{ Nm}$$
$$\text{Anti-clockwise moments} = 4 \times R_B$$

Equating clockwise and anti-clockwise moments

$$4\,R_B = 1\,079.3 \text{ Nm}$$
$$\text{Reaction force at } B,\ R_B = \frac{1\,079.3\ [\text{Nm}]}{4\ \ [\text{m}]} = 269.82 \text{ N}$$

As a check on the answers, consider the vertical equilibrium of the beam:

$$\text{Upward forces} = R_A + R_B = (24.53 + 269.82) \text{ N} = 294.35 \text{ N}$$
$$\text{Downward forces} = (245.3 + 49.05) \text{ N} = 294.35 \text{ N}$$

Thus the answers for R_A and R_B are correct.

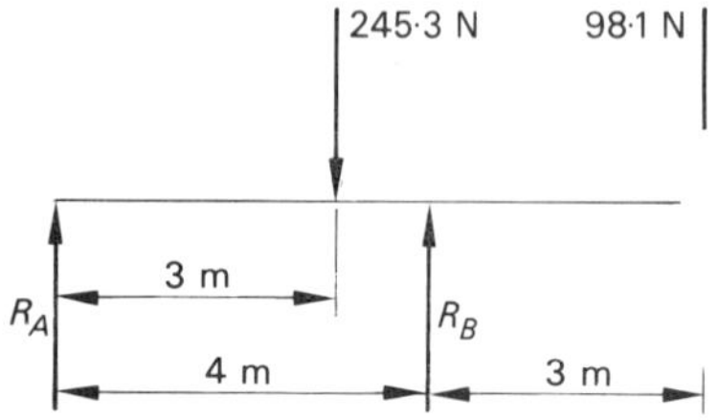

Figure 166 *Effect of increasing overhang force*

(ii) If the 5 kg mass is increased to 10 kg, the gravitational force on the mass is 2×49.05 N, i.e. 98.1 N; the system of forces acting on the beam is now as shown in Figure 166. Take moments about an axis through A:

$$\text{Clockwise moments} = [(245.3 \times 3) + (98.1 \times 7)] \text{ Nm}$$
$$= [735.9 + 686.7] \text{ Nm}$$
$$= 1\,422.6 \text{ Nm}$$
$$\text{Anti-clockwise moments} = 4 \times R_B$$

Equating clockwise and anti-clockwise moments:

$$4\ R_B = 1\,422.6 \text{ Nm}$$

$$\text{Reaction force at } B,\ R_B = \frac{1\,422.6}{4}\frac{[\text{Nm}]}{[\text{m}]} = 355.65 \text{ N}$$

Take moments about an axis through B,

$$\begin{aligned} \text{Clockwise moments} &= [(4 \times R_A) + (98.1 \times 3)] \text{ Nm} \\ &= [4\ R_A + 294.3] \text{ Nm} \\ \text{Anti-clockwise moments} &= [245.3 \times 1] \text{ Nm} \\ &= 245.3 \text{ Nm} \end{aligned}$$

Equating clockwise and anti-clockwise moments

$$\begin{aligned} 4\ R_A + 294.3 &= 245.3 \text{ Nm} \\ 4\ R_A &= -49 \text{ Nm} \\ \text{Reaction force at } A,\ R_A &= -\frac{49}{4}\frac{[\text{Nm}]}{[\text{m}]} = -12.25 \text{ N} \end{aligned}$$

The negative sign indicates that the direction assumed for R_A in Figure 166 is incorrect, i.e. the reaction force at A acts downwards, and not upwards.

Checking the equilibrium of the vertical forces:

$$\begin{aligned} \text{Downward forces} &= (245.3 + 98.1 + 12.25) \text{ N} = 355.65 \text{ N} \\ \text{Upward force} &= 355.65 \text{ N} \end{aligned}$$

Thus the magnitudes and directions of R_A and R_B are correct.

Self-assessment questions

13 A straight lever is 1.5 m long. If a downward force of 300 N is necessary at one end of the lever to balance a mass of 90 kg at the other end, find the distance of the fulcrum from the mass. Assume $g = 9.81 \text{ m/s}^2$.

14 A beam of negligible mass is 5 m long, and is supported at its two ends A and B. Two forces are applied to the beam; one of 300 N at a distance of 1.5 m from A, and the other of 225 N at a distance of 3.5 m from A. Calculate the reaction forces at A and B.

15 A beam of negligible mass is 3 m long, and rests on two supports, one at the end of the beam, the second 0.5 m from the opposite end of the beam.

(i) Calculate the upward reactions at the supports when a mass of 100 kg is hung midway between the supports.

(ii) If a 50 kg mass is added at the tip of the overhanging section of the beam, determine the reactions at the supports. Take g as 9.81 m/s^2.

Solutions to self-assessment questions

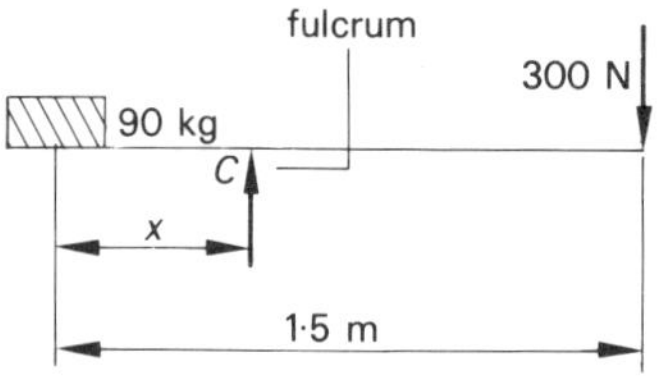

Figure 167

13 The problem is resolved into the calculation of the distance x in Figure 167.
Gravitational force on 90 kg mass $= 90\ [\text{kg}] \times 9.81\ [\text{m/s}^2] = 882.9$ N

Take moments about an axis through C:

$$\text{Clockwise moments} = [300 \times (1.5 - x)]\ \text{Nm}$$
$$\text{Anti-clockwise moments} = 882.9 \times x\ \text{Nm}$$

Equating clockwise and anti-clockwise moments

$$300\,(1.5 - x) = 882.9x$$
$$300 \times 1.5 - 300x = 882.9x$$
$$300 \times 1.5 = 882.9x + 300x$$
$$1\,182.9x = 450$$
$$\text{Distance, } x = \frac{450\ [\text{Nm}]}{1\,182.9\ [\text{m}]} = 0.38\ \text{m}$$

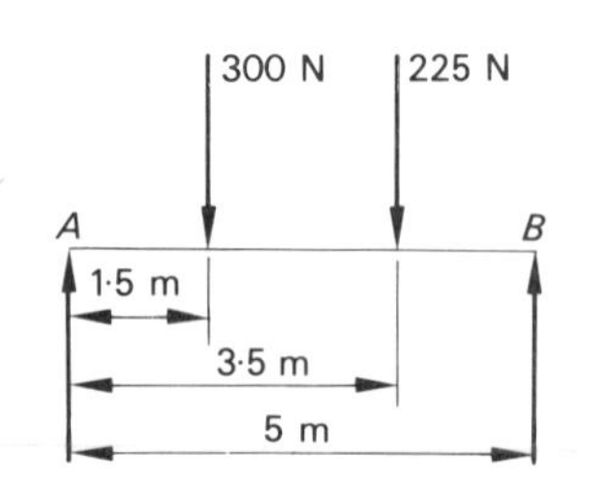

Figure 168

14 Referring to Figure 168, let the reaction forces at A and B be R_A and R_B respectively.
Take moments about an axis through B:

$$\text{Clockwise moments} = (R_A \times 5)\ \text{Nm}$$
$$\text{Anti-clockwise moments} = [(300 \times 3.5) + (225 \times 1.5)]\ \text{Nm}$$
$$= [1\,050 + 337.5]\ \text{Nm}$$
$$= 1\,387.5\ \text{Nm}$$

Equating clockwise and anti-clockwise moments:

$$5\,R_A = 1\,387.5\ \text{Nm}$$
$$\text{Reaction force at } A,\ R_A = \frac{1\,387.5\ [\text{Nm}]}{5\ [\text{m}]} = 277.5\ \text{N}$$

Take moments about an axis through A:

$$\text{Clockwise moments} = [(300 \times 1.5) + (225 \times 3.5)]\ \text{Nm}$$
$$= [450 + 787.5]\ \text{Nm}$$
$$= 1\,237.5\ \text{Nm}$$
$$\text{Anti-clockwise moments} = (R_B \times 5)\ \text{Nm}$$

Equating clockwise and anti-clockwise moments

$$5R_B = 1\,237.5\ \text{Nm}$$
$$\text{Reaction force at } B,\ R_B = \frac{1\,237.5\ [\text{Nm}]}{5\ [\text{m}]} = 247.5\ \text{N}$$

Checking the equilibrium of the vertical forces:

$$\text{Upward forces} = (277.5 + 247.5)\ \text{N} = 525\ \text{N}$$
$$\text{Downward forces} = (300 + 225)\ \text{N} = 525\ \text{N}$$

Thus the magnitude and directions of the reaction forces are correct.

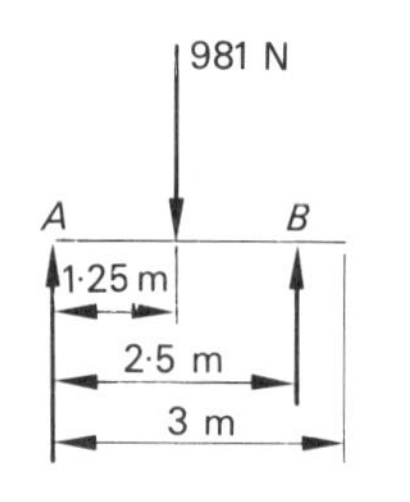

Figure 169

15 (i) Referring to Figure 169:
Gravitational force on 100 kg mass $= 100\ [\text{kg}] \times 9.81\ [\text{m/s}^2] = 981$ N

Let the reactions at A and B be R_A and R_B respectively.
Take moments about an axis through B:

$$\text{Clockwise moments} = (R_A \times 2.5)\ \text{Nm}$$
$$\text{Anti-clockwise moments} = (981 \times 1.25)\ \text{Nm}$$
$$= 1\,226.25\ \text{Nm}$$

Equating clockwise and anti-clockwise moments:

$$2.5\,R_A = 1\,226.25\ \text{Nm}$$
$$\text{Reaction force at } A,\ R_A = \frac{1\,226.25\ [\text{Nm}]}{2.5\ [\text{m}]} = 490.5\ \text{N}$$

Take moments about an axis through A:

$$\text{Clockwise moments} = (981 \times 1.25)\ \text{Nm}$$

$$\text{Anti-clockwise moments} = 2.5\ R_B\ \text{Nm}$$

$$\therefore 2.5\ R_B = 1\,226.25\ \text{Nm}$$

$$\text{Reaction force at } B,\ R_B = \frac{1\,226.25\ [\text{Nm}]}{2.5\ \ [\text{m}]} = 490.5\ \text{N}$$

Checking the equilibrium of vertical forces:

$$\text{Upward forces} = 2 \times 490.5 = 981\ \text{N}$$

This is equal to the downward force, and hence the magnitude and direction of the reaction forces are correct.

Note that for this part of the problem (because the beam has negligible mass), the overhang to the right of the support at B is irrelevant to the solution.

In essence, the beam is loaded symmetrically, with the two reactions supporting an equal share of the applied force.

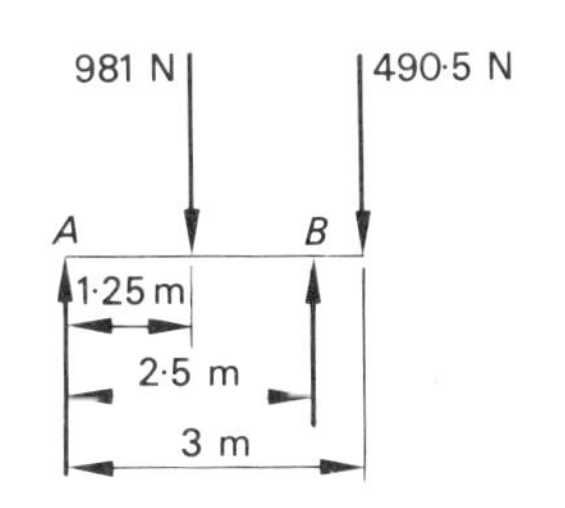

Figure 170

(ii) Referring to Figure 170:

$$\text{Gravitational force on 100 kg mass} = 981\ \text{N}$$

$$\text{Gravitational force on 50 kg mass} = 490.5\ \text{N}$$

Let the reaction forces at A and B be R_A and R_B respectively. Take moments about an axis through B:

$$\text{Clockwise moments} = [2.5\ R_A + (490.5 \times 0.5)]\ \text{Nm}$$

$$= [2.5\ R_A + 245.25]\ \text{Nm}$$

$$\text{Anti-clockwise moments} = (981 \times 1.25)\ \text{Nm}$$

$$= 1\,226.25\ \text{Nm}$$

Equating clockwise and anti-clockwise moments:

$$2.5\ R_A + 245.25 = 1\,226.25\ \text{Nm}$$

$$2.5\ R_A = 981\ \text{Nm}$$

$$\text{Reaction force at } A,\ R_A = \frac{981\ [\text{Nm}]}{2.5\ \ [\text{m}]} = 392.4\ \text{N}$$

Take moments about an axis through A:

$$\text{Clockwise moments} = [(981 \times 1.25) + (490.5 \times 3)]\ \text{Nm}$$

$$= [1\,226.25 + 1\,471.5]\ \text{Nm}$$

$$= 2\,697.75\ \text{Nm}$$

$$\text{Anti-clockwise moments} = 2.5\ R_B$$

Equating clockwise and anti-clockwise moments:

$$2.5\ R_B = 2\,697.75\ \text{Nm}$$

$$\text{Reaction force at } B,\ R_B = \frac{2\,697.75\ [\text{Nm}]}{2.5\ \ [\text{m}]} = 1\,079.1\ \text{N}$$

Checking the equilibrium of vertical forces:

$$\text{Upward forces} = (392.4 + 1\,079.1)\ \text{N} = 1\,471.5\ \text{N}$$

$$\text{Downward forces} = (981 + 490.5)\ \text{N} = 1\,471.5\ \text{N}$$

Thus the magnitude and direction of the reactions are correct.

After reading the following material, the reader shall:

3 Know the meaning of the phrase 'centre of gravity'.
3.1 Define centre of gravity.
3.2 Show with the aid of sketches, the position of the centre of gravity of
(*a*) a thin uniform rod
(*b*) a rectangular lamina
(*c*) a circular lamina

Any object or body consists of a large number of particles; the gravitational forces acting on the particles can be considered as a series of virtually parallel forces, because the physical size of the object is so very much smaller than that of the earth. All of these forces can be replaced by a single resultant force, i.e. the weight of the object. The line of action of the weight passes through a point defined as the *centre of gravity*: all of the mass of the object can be assumed to be concentrated at this point. The technically correct term for centre of gravity is centre of mass, but it has become customary to call it *centre of gravity*, and abbreviate the words to *c.g.*

It is often possible to discover the position of the centre of gravity by experiment. For example, Figure 171(*a*) shows a lamina (i.e. a thin plate) of irregular shape suspended by means of a thin cord passed through a very small hole drilled at point *X*. The force in the suspension cord is equal to the weight of the lamina, and hence the centre of gravity must lie somewhere on the dotted line of Figure 171(*a*).

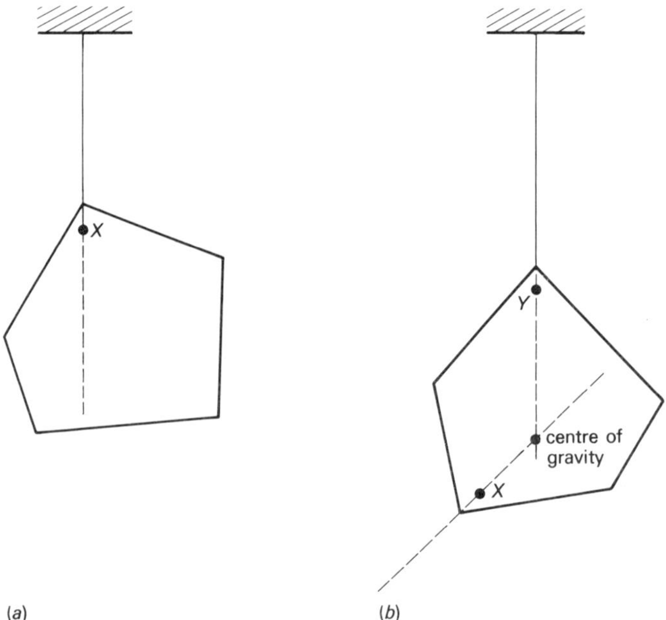

Figure 171 *Determination of centre of gravity of an irregular laminar*

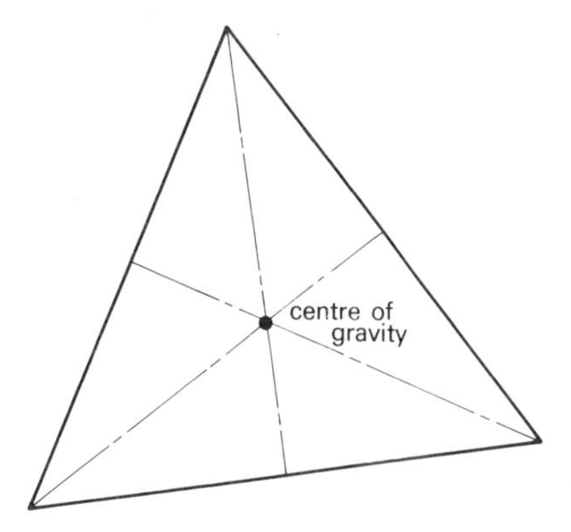

Figure 172 *Centre of gravity of a triangle*

If the lamina is now suspended from another point such as Y in Figure 171(b), then the centre of gravity must lie on the dotted line projected from Y. Thus the centre of gravity must be at the intersection of the two lines in Figure 171(b).

If the lamina is of regular shape, then the centre of gravity can often be determined graphically as in Figure 172, which illustrates the centre of gravity of a triangle at the intersection of lines drawn from each vertex to the mid-point of the opposite side. Thus the centre of gravity of a regular lamina lies at the geometrical centre, and at mid-thickness of the lamina. Similarly, the centre of gravity of a straight uniform rod is on the axis of the rod at mid-length.

Self-assessment questions

16 The line of action of the gravitational force acting on a body passes through the centre of gravity.

TRUE/FALSE

17 Show with the aid of sketches, the position of the centre of gravity of
(i) a rectangular lamina
(ii) a circular lamina

18 A rod XY is 2 m long, and is of uniform section. The rod, which has a mass of 6 kg is placed on a fulcrum positioned 0.7 m from X. Determine the vertical force required at X to prevent rotation of the rod. Take g as 9.81 m/s^2.

Further self-assessment questions

19 Complete the following statements by inserting the missing word(s):
(i) A vector quantity has both ______ and ______.
(ii) A scalar quantity has ______ only.

20 A barge is towed along a canal by a rope inclined at 30° to the direction of the canal. The force in the rope is 3 kN. Find the components of the force in the tow rope, one parallel to the direction of the canal, the other perpendicular to the direction of the canal. (Hint: treat the force in the tow rope as a ıltant; draw a vector diagram beginning with the resultant.)

21 Complete the following statements by inserting the missing phrase or word:
(i) The moment of a force about an axis is equal to the product of the force and ______________________________

____
(ii) If a body is in static equilibrium, the resultant force acting on the body is ______.
(iii) If a body is in static equilibrium, the resultant moment acting on the body is ______.

22 A body is in a state of static equilibrium when an external force is applied to it, and then removed. Identify the following statements concerning the equilibrium of the body as either true or false.

(i) If the body is in stable equilibrium, the displacement caused by the force tends to reduce to zero.

TRUE/FALSE

(ii) If the body is in unstable equilibrium, the displacement caused by the force tends to increase.

TRUE/FALSE

(iii) If the body is in neutral equilibrium, the displacement caused by the force remains essentially constant.

TRUE/FALSE

23 A lever AB, of negligible mass and 2 m long, rests on a fulcrum at C, which is 0.5 m from A. A downward force of 10 kN is applied at A. Calculate

(i) the force required at B to maintain equilibrium
(ii) the reaction force at the fulcrum

24 A beam is 10 m long, and is supported at its ends A and B. Forces of 50 kN, 100 kN and 70 kN are applied to the beam at distances of 2 m, 5 m and x metre respectively from A. Neglecting the mass of the beam, determine the distance x which makes the reaction forces at the supports equal.

25 The mass of a uniform steel beam is 750 kg. The beam is 3 m long, is supported at one end, A, and at a point 2 m from A, so that there is an overhang of 1 m. Calculate the reaction forces at the supports. Take g as 9.81 m/s^2.

26 A connecting rod of an engine has a mass of 20 kg. To find the position of the centre of gravity of the rod, it is placed on two knife-edged supports 1.5 m apart. The reaction forces at the supports are measured and found to be 130.8 N and 65.4 N. Find the position of the centre of gravity from the most highly loaded support. Take g as 9.81 m/s^2.

27 Two men carry a scaffolding pole, each supporting one end. The pole is 8 m long, has a mass of 35 kg and the centre of gravity is 3 m from one end.

(i) How much of the mass of the pole does each man support?
(ii) If a bucket of cement having a mass of 10 kg is carried by hanging it from the pole, find the distance from either end of the pole that it should be hung, in order that the total mass is shared equally by the two men. Take g as 9.81 m/s^2.

Solutions to self-assessment questions

16 The statement is *true*, since centre of gravity is defined as the point at which all the mass of a body can be considered to act. Thus the line of action of the gravitational force acting on the mass passes through this point.

17 The positions of the centre of gravity are shown in Figure 173(*a*) and (*b*). Note that for a regular lamina, the centre of gravity coincides with the centre of area.

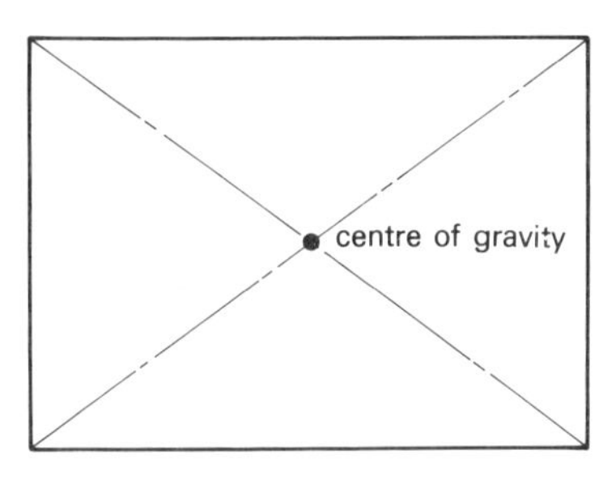

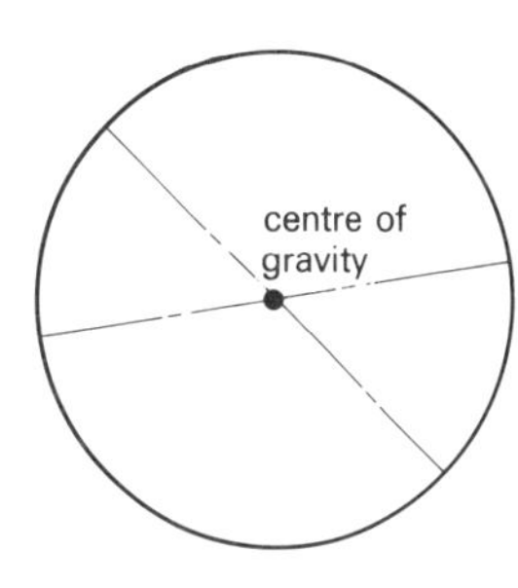

Figure 173 (*a*) (*b*)

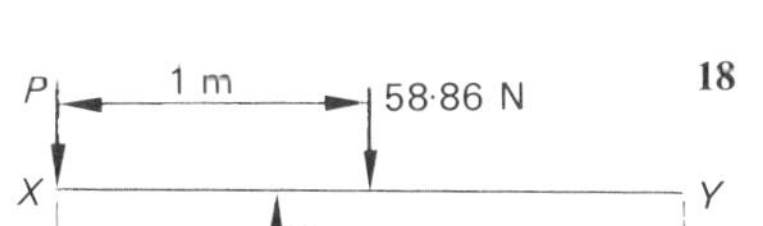

Figure 174

18 Gravitational force = 6 [kg] × 9.81 [m/s²] = 58.86 N.
The centre of gravity is 1 m from either X or Y, so that the arrangement is as shown in Figure 174.
Let the force at X be P, acting vertically down, and let the reaction force at the fulcrum be R_F, acting vertically up. Take moments about an axis through the fulcrum:

$$\text{Clockwise moment} = (58.86 \times 0.3)\ \text{Nm} = 17.66\ \text{Nm}$$

$$\text{Anti-clockwise moment} = P \times 0.7\ \text{Nm}$$

Equating clockwise and anti-clockwise moments

$$0.7\,P = 17.66\ \text{Nm}$$

$$\text{Force, } P = \frac{17.66\ [\text{Nm}]}{0.7\ \ [\text{m}]} = 25.23\ \text{N}$$

If necessary, this answer could be checked by (*a*) taking moments about an axis through X and finding the reaction force R_F, and (*b*) checking the vertical equilibrium of the rod.

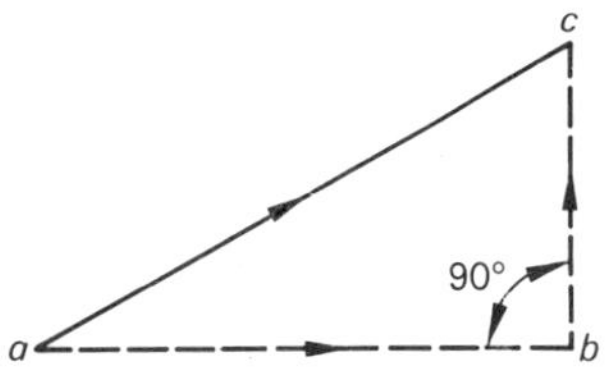

Answers to self-assessment questions

19 (i) A vector quantity has both *magnitude* and *direction*.
(ii) A scalar quantity has *magnitude* only.

20 The vector diagram is shown in Figure 175, with *ac* representing the 3 kN force in the tow rope. Hence the vectors *ab* and *bc* represent the components of the force parallel to and normal to the direction of the canal, respectively. Component parallel to direction of canal is 2.6 kN. Component normal to direction of canal is 1.5 kN.

Figure 175 *Components of the resultant force*

21 The statements should be completed as follows:
(i) The moment of a force about an axis is equal to the product of the force and *the perpendicular distance between the line of action of the force and the axis.*
(ii) If a body is in static equilibrium, the resultant force acting on the body is *zero*.
(iii) If a body is in static equilibrium, the resultant moment acting on the body is *zero*.

22 All the statements are *true*.

23 (i) Force required at B = 3.33 kN downward.
(ii) Reaction force at the fulcrum = 13.33 kN upward.

24 $R_A = R_B = 110$ kN
Distance x = 7.14 m from A.

25 Reaction force at A = 1.84 kN
Reaction force at other support = 5.52 kN.

26 Distance of c.g. from most highly loaded support is 0.5 m.

27 (i) 13.12 kg and 21.88 kg
(ii) 7.5 m or 0.5 m

Section 2

After reading the following material, the reader shall:

4 Be familiar with the concept of pressure.
4.1 Define pressure.
4.2 State the units of pressure to be N/m^2 or pascals.
4.3 Calculate pressure, given force and area.

When a motor car or bicycle tyre is inflated, the pressure inside the tyre is increased. The pressure is caused by the compressed air exerting forces on the inner surfaces. There is a relationship between pressure, force and the area upon which the force acts, which may be expressed as

$$\text{Pressure} = \frac{\text{force}}{\text{area upon which the force acts}}$$

The SI unit of force is the newton (N), and one of the SI units of area is the square metre (m^2). Hence the unit of pressure is N/m^2. In some text books, the unit of pressure is termed the pascal (abbreviated to Pa) such that

1 Pa = 1 N/m^2

However, N/m^2 (or multiples of this unit) are used in the main, in the following text.

Self-assessment question

28 Which one of the four alternatives which follow yields the largest pressure?
(i) A force of 1 N acting on an area of 1 m^2.
(ii) A force of 1 N acting on an area of 2 m^2.
(iii) A force of 1 N acting on an area of 0.5 m^2.
(iv) A force of 6 N acting on an area of 4 m^2.

For practical use 1 N/m^2 is a very small pressure indeed – in essence it is a very small force (1 N) acting on a relatively large area (1 m^2); more practical units of pressure are the multiples kN/m^2 and MN/m^2.

Self-assessment questions

29 A force of 7.5 kN is applied to a rectangular surface measuring 1.5 m by 3.2 m. Find the pressure acting on the surface.

30 Steam under pressure is applied to a piston having a diameter of 0.25 m. If the force on the piston is estimated at 65 kN, calculate the steam pressure.

31 A pneumatic cylinder contains air at a pressure of 2.5 MN/m^2. If the piston has a diameter of 150 mm, determine the force applied to it.

After reading the following material, the reader shall:

4.4 State that there is a pressure due to the atmosphere.
4.5 Name the value of standard atmospheric pressure in N/m^2.
4.6 Identify the thermodynamic unit of pressure as the bar.
4.7 Define the bar as 10^5 N/m^2.

Every person, indeed every object on earth, is subjected to atmospheric pressure. As is well known, this pressure varies with altitude from sea level – the pressure at the top of Mount Everest (8 840 m above sea level) is approximately 31% of that at sea level. However, air pressure at sea level, or at any given altitude also varies, since it is subject to the prevailing climatic conditions; readers will have heard or read weather forecasts with a phrase such as 'An area of high [or low] pressure is approaching.' Such a phrase prompts the question, 'High [or low] compared with what standard?' Just as there is a standard kilogram mass or a standard metre length, so there is a standard atmospheric pressure which is defined as 101 325 N/m^2 (101.325 kN/m^2).

The units of pressure are sometimes expressed in *bars*, particularly when considering the thermodynamic properties of vapours. A *bar* is defined as 10^5 N/m^2. This unfortunately does not conform with the rule 10^3, 10^6, 10^9, etc. recommended in SI. Some of the reasons why the bar is used as a unit of pressure are rather complex, but one of the reasons is that standard atmospheric pressure $= 101\,325$ N/m^2 $= 1.013\,25 \times 10^5$ N/m^2, which is very nearly 1 bar. In other words, 1 bar corresponds very closely to standard atmospheric pressure, which is convenient in the solution of some thermodynamic problems. It may also be of interest that the bar, or a derivative of the bar (i.e. the millibar) is the unit of pressure favoured by meteorologists.

Solution to self-assessment question

28 The pressures are:
(i) 1 N/m^2
(ii) 0.5 N/m^2
(iii) 2 N/m^2
(iv) 1.5 N/m^2
Hence alternative (iii) results in the largest pressure.

Self-assessment questions

32 Convert the units of the following pressures to bars:

(i) 356 kN/m^2
(ii) 0.65 kN/m^2
(iii) 5.2 MN/m^2

33 Convert the units of the following pressures to kN/m^2:

(i) 1 bar
(ii) 3.46 bar
(iii) 0.024 bar

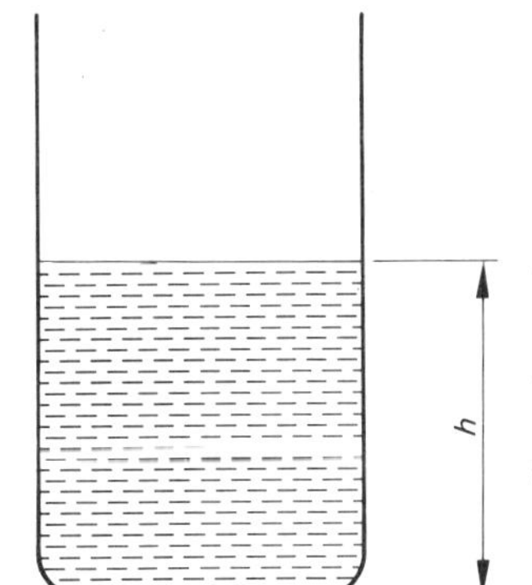

(b)

Figure 176 *Factors affecting pressure in a fluid*

After reading the following material, the reader shall:

5 Know the effects of pressure on fluids.
5.1 State the factors which determine the pressure at any point in a fluid; fluid density, depth, g, surface pressure.
5.2 State that the pressure at any given point in a liquid is equal in all directions.
5.3 State that the pressure in a liquid is independent of the shape of the vessel.
5.4 State that the pressure in a liquid acts in a direction normal to its containing surface.

Suppose two containers are filled with the same fluid as shown in Figure 176. The following questions relate to the magnitude of the pressure exerted by the fluid upon the walls of the containers.

Self-assessment question

34 Answer each of the following questions by either Yes or No:

(i) Does the magnitude of the pressure on the base increase as the height h of the fluid (see Figure 176) is increased?
(ii) Does the magnitude of the pressure at any given height increase, if the fluid is changed for one of greater density?
(iii) Does the magnitude of the pressure at any given height change if the pressure acting on the surface of the fluid changes?
(iv) Does the magnitude of the pressure at any given height change, if the shape of the container is changed from the shape of Figure 176(*a*) to the shape of Figure 176(*b*)? Note that the height h is the same in both containers, as is the quantity of fluid.

If the reader has answered the questions, and has read the solutions overleaf, he should appreciate that the pressure at any point in a fluid depends upon the height of the fluid, the fluid density and the surface pressure, but is independent of the shape of the container.

There is however one other factor, which determines the pressure in the fluid. It has been established that the pressure depends upon the mass of fluid, because gravitational force acts on the mass. Gravitational force depends upon the acceleration due to gravity. Hence the other factor which determines the pressure in the fluid is the acceleration due to gravity.

Thus the magnitude of the pressure at any point in a fluid depends upon the height of the fluid, the fluid density, the acceleration due to gravity and the surface pressure. Before discussing the relationship between pressure and these factors, some consideration should be given to the direction of the pressure acting in a fluid.

The pressure in (say) a bicycle tyre acts in every direction, and at any point is the same in all directions. At a point on the surface of the tyre, the pressure must act at right angles to the surface. If these

Solutions to self-assessment questions

29

$$\text{Applied force} = 7.5\ \text{kN}$$

$$\text{Area upon which force is acting} = (1.5 \times 3.2)\ \text{m}^2 = 4.8\ \text{m}^2$$

$$\text{Pressure} = \frac{\text{force}}{\text{area upon which force is acting}}$$

$$\text{Pressure} = \frac{7.5\ [\text{kN}]}{4.8\ [\text{m}^2]} = 1.56\ \text{kN/m}^2$$

30

$$\text{Applied force} = 65\ \text{kN}$$

$$\text{Area of piston} = \frac{\pi d^2}{4} = \left(\frac{\pi \times 0.25^2}{4}\right)\text{m}^2 = 0.049\,1\ \text{m}^2$$

$$\text{Pressure} = \frac{\text{force}}{\text{area of piston}} = \frac{65\ [\text{kN}]}{0.049\,1\ [\text{m}^2]}$$

$$\text{Pressure} = 1\,324\ \text{kN/m}^2 = 1.324\ \text{MN/m}^2$$

31

$$\text{Area of piston} = \frac{\pi d^2}{4} = \left(\frac{\pi}{4} \times 100^2\right)\text{mm}^2 = \left(\frac{\pi}{4} \times \frac{100^2}{10^6}\right)\text{m}^2$$

$$= 0.785 \times 10^{-2}\ \text{m}^2$$

$$\text{Pressure} = \frac{\text{force}}{\text{area of piston}}$$

$$\therefore \text{Force} = \text{pressure} \times \text{area of piston}$$

$$= 2.5 \times 10^6 \left[\frac{\text{N}}{\text{m}^2}\right] \times 0.785 \times 10^{-2}\ [\text{m}^2]$$

$$= (2.5 \times 10^6 \times 0.785 \times 10^{-2})\ \text{N}$$

$$= 1.963 \times 10^4\ \text{N}$$

$$\text{Force} = 19.63\ \text{kN}$$

32 (i) $356\ \text{kN/m}^2 = 356\,000\ \text{N/m}^2 = 3.56\ \text{bar}$

(ii) $0.65\ \text{kN/m}^2 = 650\ \text{N/m}^2 = 0.006\,5\ \text{bar}$

(iii) $5.2\ \text{MN/m}^2 = 5\,200\,000\ \text{N/m}^2 = 52\ \text{bar}$

33 (i) $1\ \text{bar} = 100\,000\ \text{N/m}^2 = 100\ \text{kN/m}^2$

(ii) $3.46\ \text{bar} = 346\,000\ \text{N/m}^2 = 346\ \text{kN/m}^2$

(iii) $0.024\ \text{bar} = 2\,400\ \text{N/m}^2 = 2.4\ \text{kN/m}^2$

statements were not true, then one wall of the tyre would be inflated more than another. The same arguments also apply to the pressure in a liquid.

Self-assessment questions

35 Which one of the following factors does *not* affect the magnitude of the pressure in a liquid?
(i) density
(ii) shape of vessel
(iii) quantity of fluid
(iv) acceleration due to gravity

36 (i) The direction of pressure in a liquid is dependent upon the shape of the containing surface.
TRUE/FALSE
(ii) Pressure at a given level in a liquid is equal in all directions.
TRUE/FALSE

After reading the following material, the reader shall:

5.5 State that the pressure due to a column of liquid depends upon the density of a liquid and the height of the column.
5.6 Solve simple problems using $p = \rho g h$.

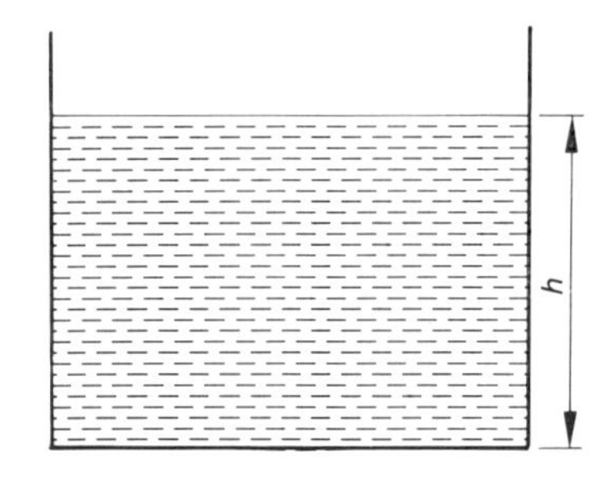

Figure 177 *Pressure due to a column of liquid*

Suppose a liquid of density ρ is poured into a container until the height of the liquid is h, as in Figure 177. Let the acceleration due to gravity acting on the liquid be g, and let the pressure at any point in the liquid be p. If the cross-sectional area of the container base is A, then

$$\begin{aligned} \text{Volume of liquid} &= A\,[\text{m}^2] \times h\,[\text{m}] \\ &= Ah\,[\text{m}^3] \\ \text{Mass of liquid} &= \text{density} \times \text{volume} \\ &= \rho \left[\frac{\text{kg}}{\text{m}^3}\right] \times Ah\,[\text{m}^3] \\ &= \rho Ah\,[\text{kg}] \end{aligned}$$

Solution to self-assessment question

34 The answers to the four questions are:
(i) Yes. If the height h is increased, there is a larger mass of fluid, which exerts a greater gravitational force on the same area.
(ii) Yes. A denser fluid produces a greater gravitational force on the same area.
(iii) Yes. The pressure in the fluid is related to the surface pressure.
(iv) No. The magnitude of the pressure at any given height is independent of the shape of the container.

Let the gravitational force acting on the mass of liquid be F. Thus

$$F = \text{mass of liquid} \times \text{acceleration due to gravity}$$

$$= \rho Ah\,[\text{kg}] \times g \left[\frac{\text{m}}{\text{s}^2}\right]$$

or $F = \rho Ahg\,[\text{N}]$

The area upon which the force F acts is the area of the container base, A. Dividing both sides of the equation by A

$$\frac{F}{A} = \rho hg \left[\frac{\text{N}}{\text{m}^2}\right]$$

or, Pressure $= \rho gh$

The inference that can be drawn is that as the density of the liquid, or the height of the liquid are varied, then the pressure in the liquid varies. Thus the equation confirms the statements made earlier regarding the factors affecting the pressure in a liquid.

Example 5

A rectangular storage tank has a base of length 3 m and width 2 m, and its sides are perpendicular to the base. The tank contains oil of density 750 kg/m^3 to a depth of 2 m. Calculate

(i) the pressure at a height of 1 m from the base

(ii) the pressure acting on the base of the tank

(iii) the total force acting on the base of the tank

Assume the acceleration due to gravity is 9.81 m/s^2.

(i) $\rho = 750$ kg/m^2, $g = 9.81$ m/s^2, $h = 1$ m

$\therefore$ Pressure at a height of 1 m is $750 \times 9.81 \times 1 \left[\frac{\text{N}}{\text{m}^2}\right]$

Pressure $= 7\,357.5$ N/m^2 $= 7.36$ kN/m^2

(ii) ρ and g remain as in part (i), $h = 2$ m

Pressure acting on the base $= 750 \times 9.81 \times 2 \left[\frac{\text{N}}{\text{m}^2}\right]$

Pressure $= 14.72$ kN/m^2

(iii)

$$\text{Pressure} = \frac{\text{force}}{\text{area acted upon by the force}}$$

$\therefore$ Force $=$ pressure $\times$ area acted upon by the force

Area of base $= 3\text{ m} \times 2\text{ m} = 6\text{ m}^2$

$\therefore$ Force acting on base $= 14.72 \times 10^3 \left[\frac{\text{N}}{\text{m}^2}\right] \times 6\,[\text{m}^2]$

Force $= 88.32 \times 10^3$ N $= 88.32$ kN

Self-assessment questions

37 A cylindrical water storage tank of diameter 0.5 m, contains water to a depth of 1.2 m. If the density of water is 1 000 kg/m^3, and the acceleration due to gravity is 9.81 m/s^2, determine:

(i) the pressure exerted on the base of the tank

(ii) the total force acting on the base

38 Figure 178 shows two containers each filled with water to the same height of 0.6 m. If the density of water is 1 000 kg/m^3 and g is 9.81 m/s^2, find the pressure acting on the base of each container.

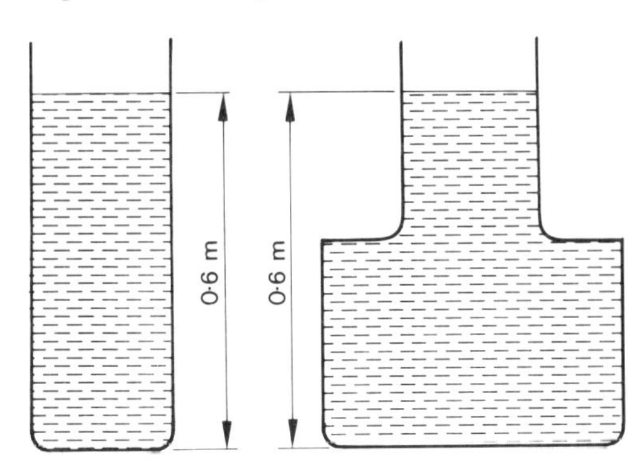

Figure 178

39 The pressure on the base of a cylindrical tank containing oil cannot exceed 25 kN/m^2. If the density of the oil is 800 kg/m^3 and g is 9.81 m/s^2, calculate the maximum height to which the tank can be filled.

After reading the following material, the reader shall:

6 Be aware of some of the methods of measuring pressure.

6.1 Distinguish between absolute pressure, atmospheric pressure and gauge pressure.

6.2 Measure gas pressure using

(*a*) a U-tube manometer

(*b*) a pressure gauge

6.3 Label the connections of a U-tube manometer as 'vessel pressure' and 'atmospheric pressure'.

6.4 Calculate gas pressures given U-tube manometer readings, fluid density and acceleration due to gravity.

Solutions to self-assessment questions

35 The correct alternative is (ii), since the vessel can be any shape and not affect the magnitude of pressure.

36 Both statements are *true*, since

(i) the pressure in a liquid acts at right angles to the containing surface

and (ii) if the pressure were not equal in all directions, the fluid would move in the direction of the larger pressure.

Liquid column barometer

One of the simplest methods of measuring atmospheric pressure is the liquid column barometer. If a glass tube, sealed at one end is placed in a liquid as shown in Figure 179(*a*), the air pressure in the tube is the same as the atmospheric pressure outside the tube. Hence the liquid level in the tube is the same as that in the vessel.

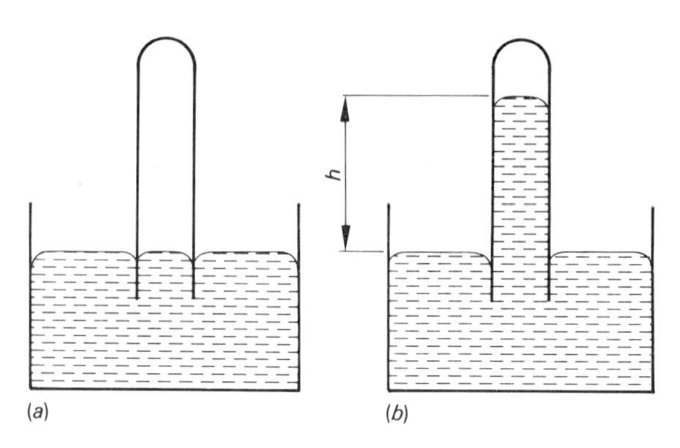

Figure 179 *Liquid column barometer*

If all the air in the tube is now removed, creating a vacuum, the liquid level rises to some height h, as shown in Figure 179(*b*). The height is such that the pressure on the liquid in the vessel (due to the mass of liquid in the tube) equals the atmospheric pressure outside the tube. Providing the density of the liquid is known, then the atmospheric pressure can be found using the formula:

Pressure $= \rho g h$

If the atmospheric pressure falls, then the height of the liquid column decreases; if the atmospheric pressure rises, then the height of the column increases.

If water is used as the liquid (density 1 Mg/m^3) in the vessel and tube, then it can be shown that at standard atmospheric pressure of 101.325 kN/m^2, the height of water in the tube is 10.33 m. A column of water 10.33 m high is in normal circumstances impractical. To reduce the length of tube required, mercury is usually chosen as the liquid. Assuming the density of mercury as 13.6 Mg/m^3, the height of the column at standard atmospheric pressure is 0.76 m (760 mm). This particular number may be familiar to readers.

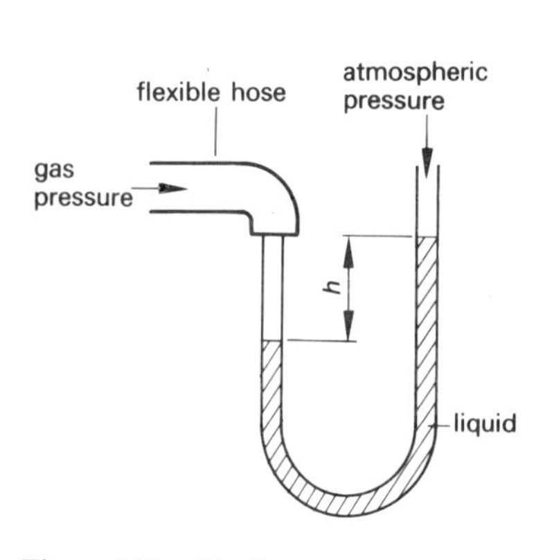

Figure 180 *U-tube manometer*

The manometer

A manometer is used to measure the difference between the pressure of a gas or vapour and atmospheric pressure. Figure 180 illustrates

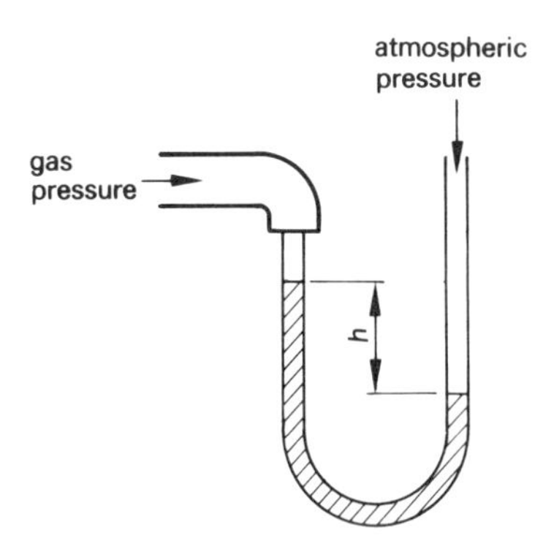

Figure 181 *U-tube manometer: gas pressure less than atmospheric pressure*

the basic principle of a U-tube manometer, which consists of a U-tube containing liquid (usually mercury or water) and a measuring scale. One end of the U-tube is open to atmosphere, the other is connected to a vessel containing gas at unknown pressure.

If the gas pressure is greater than atmospheric pressure, the additional pressure of the gas causes the liquid levels to move as shown in Figure 180, so that their difference in height is h. Then:

Pressure of gas = atmospheric pressure
+pressure due to liquid column of height h

If the pressure of the gas is less than atmospheric pressure, then the liquid in the connected column of the U-tube is higher than in the open column as shown in Figure 181. Under these circumstances:

Pressure of gas = atmospheric pressure
−pressure due to liquid column of height h

Solutions to self-assessment questions

37 (i) $\rho = 1\,000 \text{ kg/m}^3$, $g = 9.81 \text{ m/s}^2$, $h = 1.2$ m

$$\therefore \text{Pressure on the base} = 1\,000 \times 9.81 \times 1.2 \left[\frac{\text{N}}{\text{m}^2}\right]$$

$$\underline{\text{Pressure} = 19\,620 \text{ N/m}^2 = 19.62 \text{ kN/m}^2}$$

(ii)

$$\text{Area of base} = \frac{\pi d^2}{4} = \frac{\pi \times 0.5^2}{4} [\text{m}^2]$$

$$= 0.196 \text{ m}^2$$

$$\text{Force on base} = \text{pressure} \times \text{area of base}$$

$$= 19.62 \times 10^3 \left[\frac{\text{N}}{\text{m}^2}\right] \times 0.196 [\text{m}^2]$$

$$\underline{\text{Force} = 3\,850 \text{ N} = 3.85 \text{ kN}}$$

38 The height of the liquid in each container is the same; the density of the liquids is the same; so is the acceleration due to gravity. Thus the pressure acting on the base of each container must be the same. Another way to arrive at the same conclusion is to remember that the pressure in a liquid is independent of the shape of the vessel.

$$\text{Pressure on either base} = 1\,000 \times 9.81 \times 0.6 \left[\frac{\text{N}}{\text{m}^2}\right]$$

$$\underline{\text{Pressure} = 5\,886 \text{ N/m}^2 = 5.89 \text{ kN/m}^2}$$

39

$$\text{Pressure} = \rho g h$$

$$\therefore \text{Height of liquid} = \frac{\text{pressure}}{\rho g}$$

$$= \frac{25 \times 10^3}{800 \times 9.81} \frac{\left[\frac{\text{N}}{\text{m}^2}\right]}{\left[\frac{\text{kgm}}{\text{s}^2} \times \frac{1}{\text{m}^3}\right]}$$

$$\underline{\text{Height} = 3.186 \text{ m}}$$

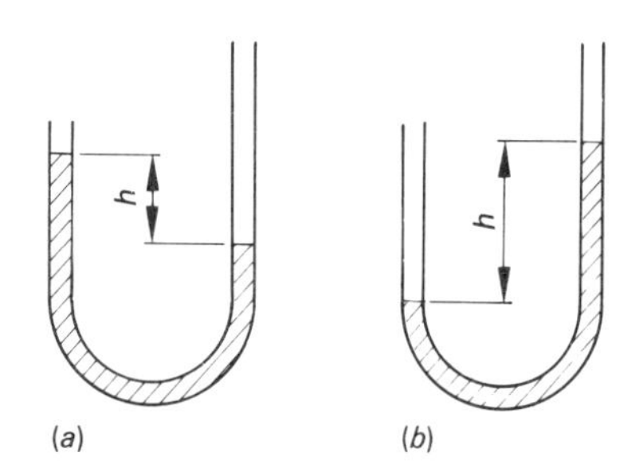

Figure 182 *Pressure readings on U-tube manometer*

The pressure measured by a manometer, or any other pressure measuring instrument, is termed the *gauge pressure*; the actual pressure of the gas is termed the *absolute pressure*. Thus, if the manometer reading is as shown in Figure 180, then

Absolute pressure = atmospheric pressure + gauge pressure

However, if the manometer reading is as shown in Figure 181, then

Absolute pressure = atmospheric pressure − gauge pressure

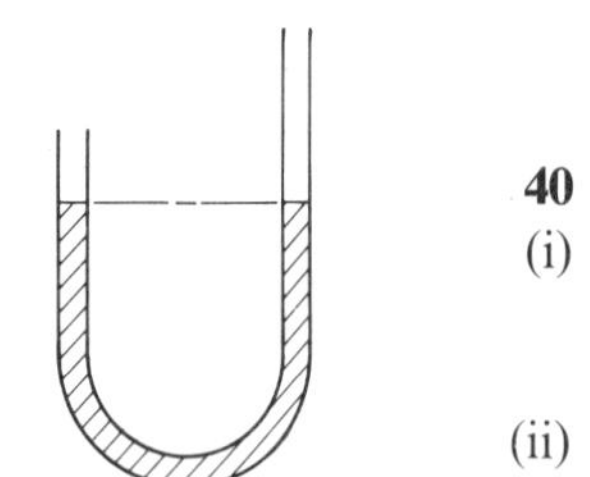

Figure 183 *U-tube manometer reading*

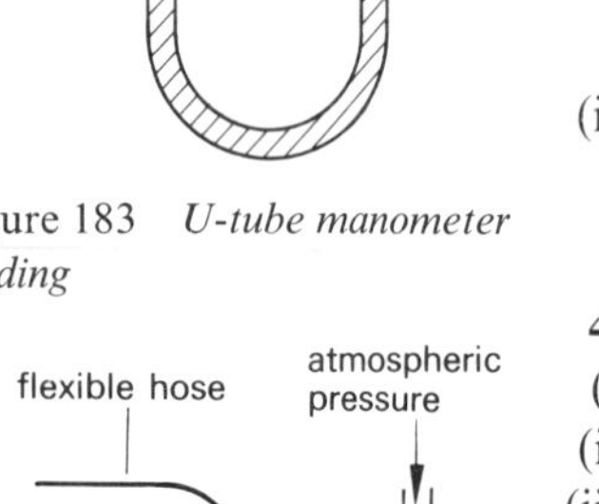

Figure 184

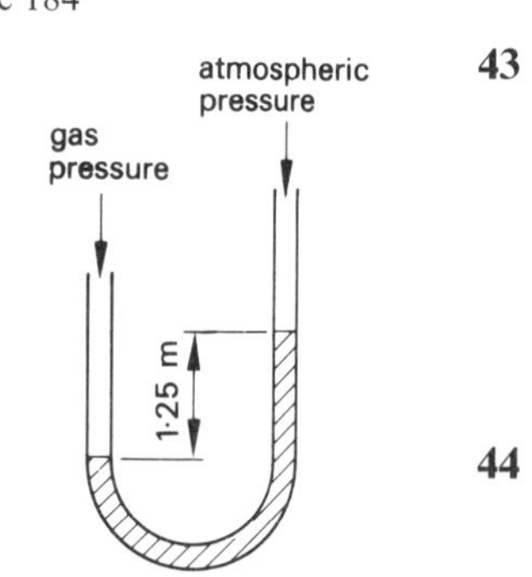

Figure 185 *U-tube manometer: differential pressure*

Self-assessment questions

40 Figure 182 illustrates two pressure readings on a U-tube manometer.
(i) On Figure 182(*a*) label the left and right hand tubes as either vessel pressure or atmospheric pressure, if the vessel pressure being measured is *less* than atmospheric pressure.
(ii) On Figure 182(*b*) label the left and right hand tubes as either vessel pressure or atmospheric pressure, if the vessel pressure being measured is *greater* than atmospheric pressure.

41 Figure 183 indicates that the vessel pressure being measured is
(i) greater than atmospheric pressure
(ii) less than atmospheric pressure
(iii) zero
(iv) equal to atmospheric pressure
Select the correct alternative.

42 The height h in Figure 184 represents
(i) absolute pressure of the gas
(ii) gauge pressure of the gas
(iii) atmospheric pressure of the gas
(iv) actual pressure of the gas
Select the correct alternative.

43 Earlier in this section during the discussion of measurement of atmospheric pressure, it was stated that, at standard atmospheric pressure, the liquid column barometer would read either 10.33 m of water or 0.76 m of mercury. Confirm these readings, given that the acceleration due to gravity is 9.81 m/s^2, standard atmospheric pressure is 101.325 kN/m^2, density of water is 1 Mg/m^3 and density of mercury is 13.6 Mg/m^3.

44 A U-tube manometer containing mercury is connected to a cylinder of oxygen. If the reading on the manometer is 1.25 m, as shown in Figure 185, find the absolute pressure of the oxygen. Assume acceleration due to gravity is 9.81 m/s^2, density of mercury is 13.6 Mg/m^3 and standard atmospheric pressure is 101.325 kN/m^2.

After reading the following material, the reader shall:

6.5 Label a simple sketch of a Bourdon pressure gauge.

6.6 Explain why a Bourdon pressure gauge is used for measuring large gauge pressures, in preference to a U-tube manometer.

Suppose a U-tube manometer containing mercury is used to measure a gauge pressure of (say) 800 kN/m^2. The difference in the heights of the liquid is then very nearly 6 m. A glass tube of this length is rather impractical; hence to measure high pressures, a Bourdon pressure gauge is often used.

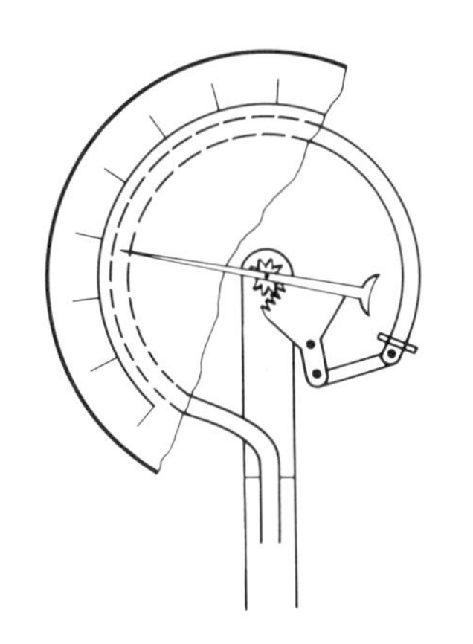

Figure 186 *Bourdon pressure gauge*

The principle of operation of a Bourdon pressure gauge is illustrated in Figure 186. It consists of a hollow tube bent into an arc. One end of the tube is connected to the source of the pressure to be measured, the other end is closed. If pressure is released into the tube, it tends to straighten out; this motion is transmitted, via a rack and pinion, to a pointer moving over a calibrated scale.

This type of gauge, generally used for measuring very high pressures, can be used, with a slight modification, for measuring pressures below atmospheric pressures. In this role it is known as a vacuum gauge.

Self-assessment questions

45 State the main disadvantage of the U-tube manometer if it is used for measuring high pressures.

46 Figure 186 illustrates an outline sketch of a Bourdon pressure gauge. Using the figure, label the following: calibrated scale, hollow tube, connection to vessel pressure, pointer, rack, pinion.

Further self-assessment questions

47 Write down the value of standard atmospheric pressure in N/m^2. What is its value in bars?

48 Convert:

(i) 30 N/m^2 into bars

(ii) 500 kN/m^2 into bars

(iii) 3 bars into kN/m^2

49 A force of 40 kN is distributed over a surface. Calculate the area of the surface in mm^2 in order that the pressure acting on the surface is 6.5 MN/m^2.

50 Steam at a pressure of 54 bars is applied to a piston having a diameter of 150 mm. Calculate the force in kN on the piston.

Solutions to self-assessment questions

40 (i) Since the vessel pressure is less than atmospheric pressure, the left hand tube must be connected to the vessel, and the right hand tube open to atmosphere as depicted in Figure 187(*a*).

(ii) Since the vessel pressure is greater than atmospheric pressure, the left hand tube must be connected to the vessel, and the right hand tube open to atmosphere as depicted in Figure 187(*b*).

41 One of the tubes is connected to the vessel and the other tube is open to atmosphere; so if the liquid levels are equal, the pressures must be equal. Hence the correct alternative is (iv).

42 The correct alternative is (ii), since a manometer measures the difference between the actual pressure of the gas and atmospheric pressure.

43 $p = 101.325\ \text{kN/m}^2$, $g = 9.81\ \text{m/s}^2$
If water is the liquid in the barometer,
$\rho = 1\ \text{Mg/m}^3$. Using $p = \rho g h$

$$h = \frac{p}{\rho g} = \frac{101.325 \times 10^3}{1 \times 10^3 \times 9.81} \frac{\left[\dfrac{\text{N}}{\text{m}^2}\right]}{\left[\dfrac{\text{kg}}{\text{m}^3} \cdot \dfrac{\text{m}}{\text{s}^2}\right]}$$

$$= \frac{101.325 \left[\dfrac{\text{kgm}}{\text{s}^2} \cdot \dfrac{1}{\text{m}^2}\right]}{9.81 \left[\dfrac{\text{kgm}}{\text{s}^2} \cdot \dfrac{1}{\text{m}^3}\right]}$$

$$= 10.329 \left[\frac{\text{m}^3}{\text{m}^2}\right]$$

Height, $h = 10.33$ m approximately

If mercury is the liquid in the barometer,
$\rho = 13.6\ \text{Mg/m}^3$. Using $p = \rho g h$

$$h = \frac{p}{\rho g} = \frac{101.325 \times 10^3}{13.6 \times 10^3 \times 9.81}\ \text{m} = 0.759\ \text{m}$$

Height, $h = 0.76$ m approximately

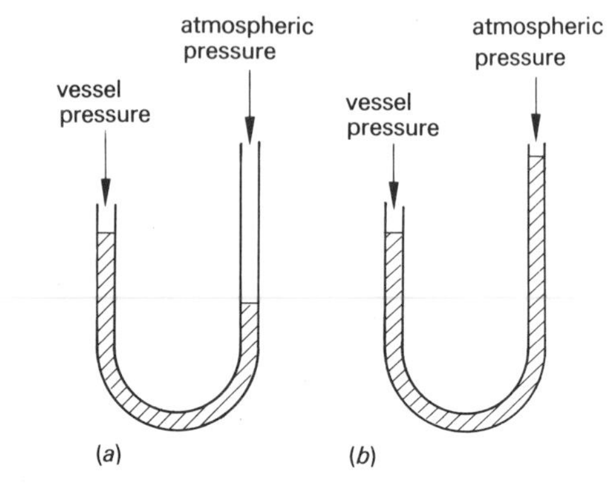

Figure 187 *Solution to self-assessment question 40*

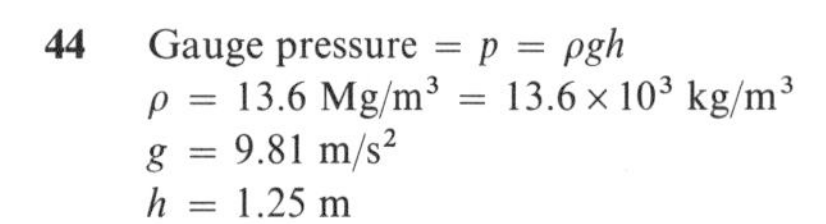

44 Gauge pressure $= p = \rho g h$
$\rho = 13.6\ \text{Mg/m}^3 = 13.6 \times 10^3\ \text{kg/m}^3$
$g = 9.81\ \text{m/s}^2$
$h = 1.25$ m

$$\therefore \text{Gauge pressure} = 13.6 \times 10^3 \left[\frac{\text{kg}}{\text{m}^3}\right] \times 9.81 \left[\frac{\text{m}}{\text{s}^2}\right] \times 1.25\,[\text{m}]$$

$$= 13.6 \times 10^3 \times 9.81 \times 1.25 \left[\frac{\text{kg}}{\text{ms}^2}\right]$$

$$= 166.7 \times 10^3\ \text{N/m}^2$$

$$= 166.7\ \text{kN/m}^2$$

Absolute pressure = atmospheric pressure + gauge pressure
$= (101.325 + 166.7)\ \text{kN/m}^2$

Absolute pressure $= 268\ \text{kN/m}^2$ approximately

51 A cylindrical concrete vertical pillar has a diameter of 300 mm. If it supports a force of 25 kN, determine the pressure on the cross section.

52 The most extreme pressure occurring naturally at the bottom of the sea is thought to be at the bottom of the Mariana trench in the Pacific Ocean, where the depth is 10.9 km. Assuming the density of sea water is 1 030 kg/m^3 and g as 9.81 m/s^2, estimate the pressure at this depth.

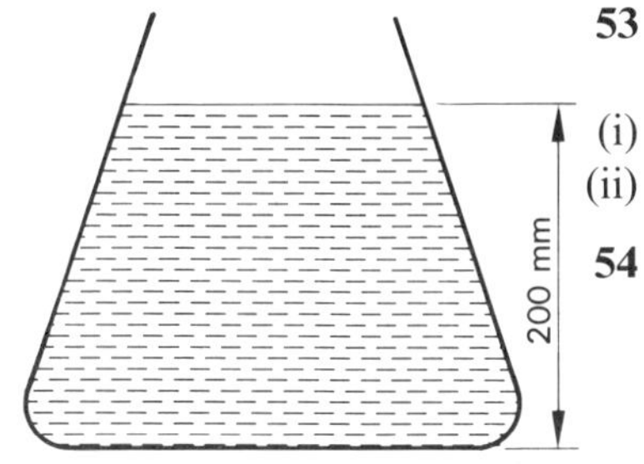

Figure 188

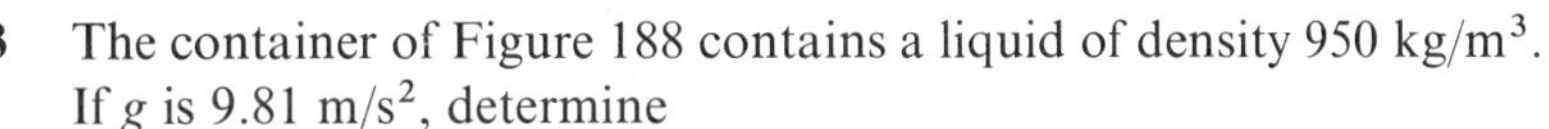

53 The container of Figure 188 contains a liquid of density 950 kg/m^3. If g is 9.81 m/s^2, determine

(i) the pressure acting on the base of the container

(ii) the total force acting on the base, if the base diameter is 250 mm

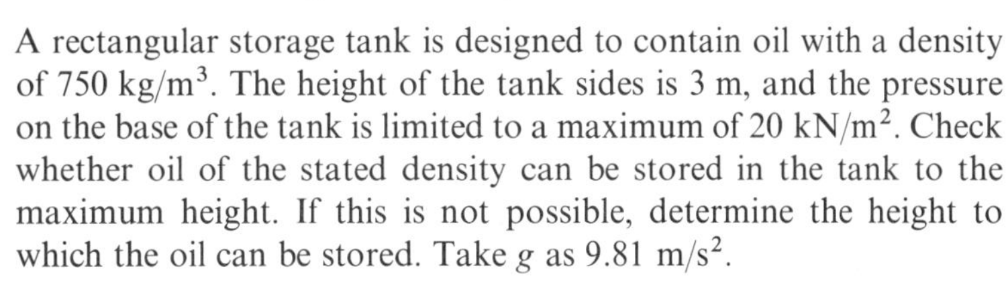

54 A rectangular storage tank is designed to contain oil with a density of 750 kg/m^3. The height of the tank sides is 3 m, and the pressure on the base of the tank is limited to a maximum of 20 kN/m^2. Check whether oil of the stated density can be stored in the tank to the maximum height. If this is not possible, determine the height to which the oil can be stored. Take g as 9.81 m/s^2.

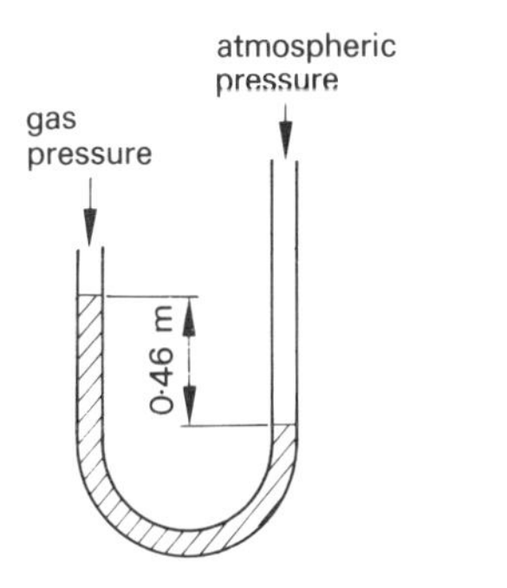

Figure 189

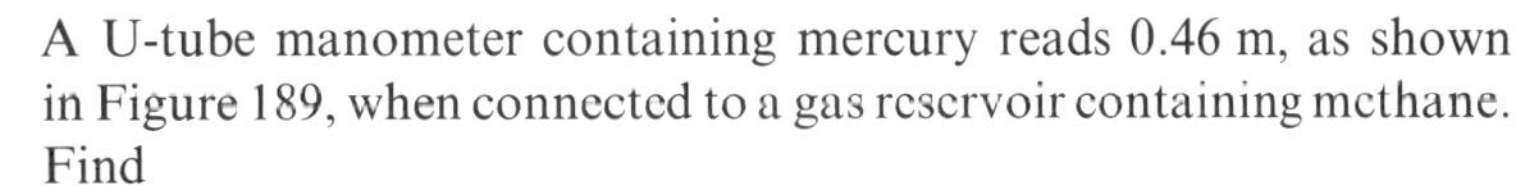

55 A U-tube manometer containing mercury reads 0.46 m, as shown in Figure 189, when connected to a gas reservoir containing methane. Find

(i) the gauge pressure

(ii) the absolute pressure of the methane

Take g as 9.81 m/s^2 and the density of mercury as 13.6 Mg/m^3. Assume standard atmospheric pressure where appropriate.

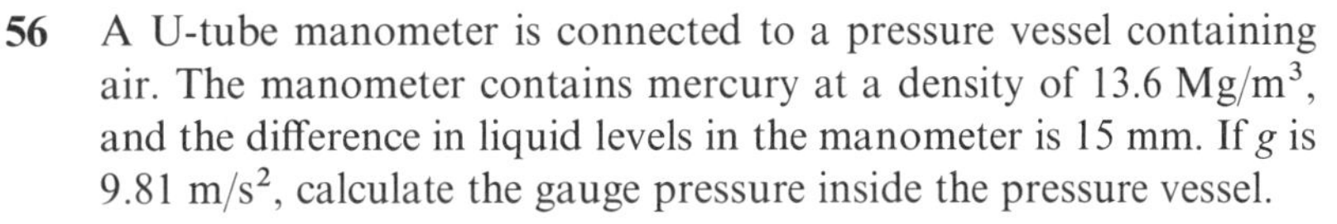

56 A U-tube manometer is connected to a pressure vessel containing air. The manometer contains mercury at a density of 13.6 Mg/m^3, and the difference in liquid levels in the manometer is 15 mm. If g is 9.81 m/s^2, calculate the gauge pressure inside the pressure vessel.

What is the difference in the levels of mercury, if the pressure inside the pressure vessel is 3 kN/m^2?

Solutions to self-assessment questions

45 A U-tube manometer, used to measure very high pressures, necessitates an excessively long tube.

46 A labelled sketch of a Bourdon pressure gauge is shown in Figure 190.

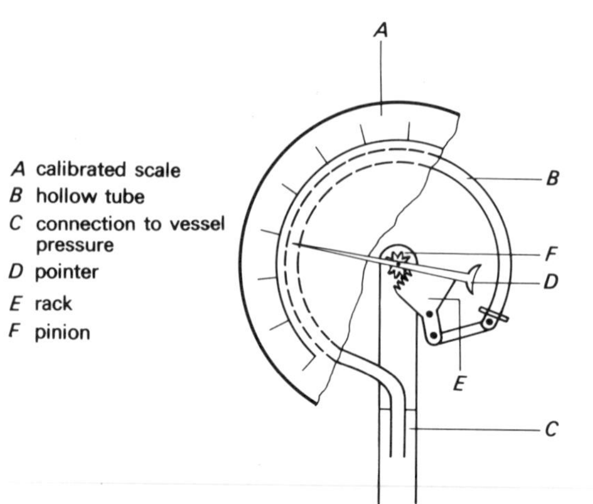

Figure 190

47 101.325×10^3 N/m^3; 1.013 25 bars.

48 (i) 0.000 3 bars; (ii) 5 bars; (iii) 300 kN/m^2.

49 6 150 mm^2.

50 95.4 kN.

51 353.7 kN/m^2.

52 110 MN/m^2 approximately (i.e. about 1 000 times atmospheric pressure).

53 (i) 1.864 kN/m^2; (ii) 91.5 kN.

54 The tank cannot be filled to a height of 3 m. Maximum storage height is 2.72 m.

55 (i) Gauge pressure = 61.37 kN/m^2.
(ii) Absolute pressure = 39.96 kN/m^2.

56 Gauge pressure = 2 kN/m^2.
Difference in mercury levels = 22.49 mm.

Topic area: Chemical reactions

After reading the following material, the reader shall:

1 Understand the principles of oxidation.
1.1 State that air is a mixture mainly of oxygen and nitrogen.
1.2 Describe how a substance (such as copper) gains mass when heated in air.
1.3 State that oxygen is taken from the air when substances like copper are heated.
1.4 Describe how substances burning in air combine with oxygen.
1.5 Describe an oxide as a compound of an element and oxygen.
1.6 Describe how oxygen and water are involved in rusting.
1.7 Give examples of damage done by rusting.
1.8 Name processes which will help restrict rusting of iron.

The separation of physical science into different sections, such as biology, physics, astronomy and chemistry, is in many ways artificial. It is, however, convenient to obtain a number of views of nature. The universe seems to present a complex and confusing aspect; before it is possible to understand the whole pattern it is necessary to study the important components.

So far, the work in this book has been mainly concerned with what is generally accepted as physics. This part deals with topics closer to chemistry. Physics deals with the general properties of matter, with mass, motion, force, change of state; chemistry deals with the particular properties of different kinds of matter, and changes from one kind of matter to another. Physics is involved at all times with energy and energy conversions; chemistry deals only with energy associated with chemical changes.

The work that follows describes a few of the chemical changes which involve the vital element oxygen. Oxygen, under ordinary conditions, is a tasteless, odourless, colourless gas. It freezes to a solid at about 55 K and boils to a gas at about 90 K. When it is a liquid, it is a clear blue colour, and, rather unexpectedly, is strongly attracted by a magnet. Gaseous or liquid oxygen is chemically very active. Even at the very low temperature of liquid oxygen, iron burns vigorously in the liquid (i.e. at about −183 °C).

When oxygen combines chemically with another material the process is called *oxidation*. The material is said to have been oxidized.

Slow oxidation is involved in many well-known processes. All mammals, and many other animals, sustain life by the slow oxidation of food in their bodies. The oxidation can only be brought about by the animal breathing in air which contains oxygen. The rusting of iron and the decay of dead wood is brought about by slow oxidation. When rapid oxidation of a material occurs, the process is called burning or *combustion*. The combustion of petrol in air is extremely rapid; in common with all burning processes, it is accompanied by the liberation of heat energy and light energy.

Air is a mixture of gases and water vapour. Nitrogen and oxygen form the great majority of the air, but also present are small amounts of other gases. The volumetric composition of air is approximately: oxygen 21%, nitrogen 78%, argon 0.9%, carbon dioxide 0.03%, water vapour 0.5% to 3.0%. Traces of other gases are also present. Note that the values do not add up to 100%, as the percentage of water vapour varies from place to place, and over time.

The atmosphere – the envelope of air round the earth – is vital to life as it is known on earth. The oxygen in the upper layers of the atmosphere shields the earth from the fierce ultra-violet radiations coming from the sun. The air blanket ensures that the earth does not suffer the enormous temperature changes between night and day that take place on planets like Mercury.

Oxygen has a great many uses. Firefighters, divers, and mountaineers on very high peaks carry supplies of oxygen which are suitably diluted with other gases.

Nitrogen, the other major constituent of air, is a colourless, odourless, not very reactive gas. It is not poisonous but can cause death by suffocation. Nitrogen is an important factor in the growth of plants. It is also an important raw material in industry. Large quantities are used in the manufacture of nitric acid. In the electronics industry nitrogen is used to create the inert atmosphere needed in the manufacture of parts such as transistors.

The presence of water vapour in the atmosphere is shown by the formation of mists, dew, frosts and clouds. The amount of water vapour present in a given volume of air varies considerably. This can have a noticeable effect on bodily comfort; if the air is 'saturated' with water vapour, the body cannot control its temperature by evaporating water from the skin, and conditions feel 'sticky' and uncomfortable.

Carbon dioxide is always present in the atmosphere, although only in small amounts. The increased burning of fuels like oil are increasing the concentration of carbon dioxide in the air; this, some scientists speculate, may result in an increase in the air temperatures at the surface of the earth. This increase might have serious consequences if the polar ice caps began to melt.

Gases like neon, krypton, argon and helium – sometimes called the inert gases because they are not very active chemically – occur in the atmosphere. Neon lights are familiar in advertising; krypton is used for safety in the cap lamps of miners; argon is used in special welding processes, and is also put into most electric light bulbs. Helium is used in the breathing mixture used by deep sea divers.

Air, then, is a mixture of many substances. Oxygen and nitrogen, however, account for nearly all the volume; the other constituents can be thought of as minor additions.

One final comment about air. It dissolves, although not very readily, in water. Oxygen is essential for fishes to survive, and they breathe the oxygen dissolved in the water. Metals such as iron rust when below water because the dissolved oxygen combines with the iron, causing oxidation – commonly called rusting.

Oxygen, a strongly active chemical element, combines with many substances when these are heated. For instance, sulphur, carbon and phosphorus burn readily in oxygen. A few metals also burn in oxygen, including sodium, calcium, and magnesium.

Most metals combine with oxygen, but do so too slowly to be described as 'burning'. The process of oxidation is much slower.

For instance, if a known mass of copper powder is heated strongly for a few minutes, allowed to cool, and then remeasured, an increase in mass is noted. If the heating and measuring is repeated until no further increase in mass can be detected then it can be assumed that all the copper has been turned into copper oxide. Clearly, the oxygen has been extracted from the air near the copper.

The events can be summarized in a chemical equation:

Copper + oxygen → copper oxide
(Chemists use their own type of symbols, and the equation can be written:
$2\,Cu + O_2 \rightarrow 2\,CuO$)

Other metals, such as iron and lead, behave in a similar fashion.

All instances of the chemical joining of oxygen and an element are classed as oxidation. Burning or combustion is the particular case of oxidation when heat and light are released. Chemists prefer to use the word combustion; 'burning' is a less precise word.

An element, such as sodium, burning in air produces its oxide. In general when an element burns in air or in oxygen, a new substance made up of atoms of the element and of atoms of oxygen are formed. This compound is called the oxide of the element. More complicated substances – compounds – produce more than one product. Most fuels (e.g. coal) are mixtures of several compounds. These contain chemical energy, a portion of which is released during the time that the burning takes place.

Different substances start to burn at different temperatures. Vapours generally ignite much more readily than solids or liquids. Over the surface of a substance like oil there is a cloud of oil vapour. This can easily be ignited. Yet if a hot wire is plunged into the liquid nothing very much happens, for the heat from the wire is rapidly conducted away, and the wire is not enough to ignite the liquid oil.

Combustion is a rapid process; oxidation of metals may take a very long time. Iron, and its derivative steel, are the most commonly used metals in the world. If they are exposed to a damp atmosphere they corrode, forming a reddish-brown powdery surface. This readily flakes off, and more iron or steel is exposed, and corrodes in its turn.

This corrosion of iron and steel is called *rusting*. Rusting is a complex process, and experiments show that

(i) If the air or oxygen in contact with the iron is absolutely dry, then no rusting takes place.

(ii) If iron is immersed in water from which all dissolved oxygen has been removed, then no rusting occurs.

Air and moisture are needed together to cause rusting. To prevent rusting, the iron or steel must be protected. The metal can be coated with protective coverings. These can be of paint, oil, or grease, or of a metal which has properties different from iron and does not readily corrode. Nickel and chromium do not normally rust; by a process called electro-plating a thin layer of either of these metals can be placed over the surface of the iron or steel. Zinc also does not react readily with the atmosphere; the process of protecting the iron or steel by using a layer of zinc is called galvanizing.

Self-assessment questions

1 (*a*) What is the approximate percentage (by volume) of oxygen in the atmosphere?

(*b*) Name two of the inert gases in the atmosphere.

(*c*) Is the percentage of water vapour present in the atmosphere always the same?

2 Explain what is meant by ‘combustion’.

3 (*a*) When iron rusts, a chemical reaction takes place between the iron and a supply of oxygen. If a piece of iron lies on the bed of a stream it still rusts. Explain this.

(*b*) If iron bolts are to be stored in a sealed jar, suggest how rusting could be prevented.

4 If iron materials are left in moist air in a sealed vessel, how are (i) the iron and (ii) the air affected?

5 If a flaming piece of paper is plunged into an open vessel containing methylated spirits, the vapour over the liquid catches fire. The paper, on entering the liquid, ceases to burn. Explain what happens.

6 When lighting a fire in a grate, why is it essential that the materials (coal, kindling, paper) be loosely packed?

7 What is the main reason for placing 'Fire Doors' in many public buildings?

8 Name three ways in which society makes use of manufactured oxygen.

9 In what ways are combustion and rusting similar? In what ways are they different?

10 Describe ways in which the rusting of iron can be minimized.

11 Give three examples of iron or steel rusting in everyday circumstances.

After reading the following material, the reader shall:

2 Know the effects of electricity on substances.
2.1 Describe metals (and carbon) as good conductors, in the solid state, of electricity.
2.2 Explain conduction in liquids as due to ions, charged parts of a molecule.
2.3 Describe electro-plating.

It has been shown that a drift of electrons through a substance is an electric current. To establish a drift of electrons through a solid substance there must be (i) free electrons in random motion between the atoms in the substance, and (ii) a difference in the state of charge between the ends of the substance. Substances which contain an abundance of free electrons have low resistivity and are classified as 'good' conductors. Metals are examples of 'good' conductors as they have low resistivity when compared with other solid non-metals such as glass, PVC, rubber and wood. A substance with an abundance of free electrons but which is not a metal is carbon.

Solutions to self-assessment questions

1 (*a*) Oxygen forms about 21% of air by volume.

(*b*) Inert gases mentioned in the text include neon, krypton, argon and helium. Others that readers may have read of include xenon and radon.

(*c*) No. The percentage can vary considerably. In hot desert regions the air may be almost completely without water vapour. On a misty day in temperate climates the air may be completely saturated with water vapour.

2 Combustion is the particular case of oxidation where heat and light are released.

3 (*a*) Iron only rusts in the presence of both water and oxygen (or air). Air dissolves to some extent in water. Thus the conditions on the bed of the stream ensure that the iron is in the presence of the two substances necessary to cause rusting.

(*b*) It is unlikely that the air will be evacuated from a jar storing iron bolts. Oxygen is thus present. To prevent rusting, the water vapour in the air must be removed. This can be accomplished by placing a drying agent (e.g. anhydrous calcium chloride) in the jar.

4 (i) The iron slowly oxidizes (rusts) and increases in mass.

(ii) The oxygen is slowly absorbed by the iron, and eventually the air in the jar contains a very much reduced percentage of oxygen. The proportions of the mixture is then very different from the air outside the vessel.

5 The vapour over the methylated spirits is intimately mixed with air. The burning paper is above the ignition temperature needed to cause combustion of the vapour, so the vapour catches fire.

On entering the liquid, little oxygen is present; also the heat from the paper is rapidly conducted away from the point of entry. In these circumstances there is little chance of the liquid beginning to burn; the paper cannot burn itself if the liquid surrounding it is not freely supplying oxygen.

6 This is so that the oxygen in the air can be rapidly drawn in to those parts where the materials are burning. Combustion is essentially a rapid process.

7 A ready supply of oxygen is needed to maintain combustion. Fire doors 'baffle' the air in a building and restrict the flow of air. Fire doors do not prevent fires, but they can slow down the spread of a fire.

8 Oxygen is used by firemen when dealing with fires where smoke or other choking atmospheres are met, by divers, and by mountaineers at great altitudes.
Oxygen is used in metal cutting, and also in welding processes.
Oxygen is supplied to hospital patients suffering from respiratory diseases (i.e. when the blood is not absorbing enough oxygen in the lungs).

9 Combustion and rusting are similar in that both are oxidation processes. They are different in that combustion is rapid, whilst rusting is slow.

10 Rusting can be reduced by covering the surface of iron or steel with a protective coating of grease, oil, plastic, paint or some metal which is corrosion resistant (e.g. tin, zinc).

11 Examples of rusting include: rusting of iron guttering; rusting of the steel parts of motor cars; rusting of the hulls of ships; rusting of bridges.

When carbon is used as a conductor it is in a crystalline form known as graphite. Graphite can be made into a variety of shapes, and by the addition of copper, both the resistivity and the hardness of the material can be controlled. It is used as a conductor in the form of blocks, granules or powder. The resistivity of graphite decreases with increase in temperature.

Graphite is used as a conductor in electric furnaces, in arc lamps, in batteries, in machines as carbon brushes and in variable resistors.

Investigations of the effects of a 'continuous' flow of current through a solid material followed the invention of the electric battery in 1799. At the same time the effect of a 'continuous' flow of current through liquids were investigated. The effect of a flow of current through water was discovered in 1802. The electric current decomposes the water into its constituent elements, oxygen and hydrogen. It was later found that many substances, in solution or the molten state, conduct electricity and are decomposed by the flow of the current. These substances are called *electrolytes*. The process of passing of a current through a liquid and thereby decomposing it into constituent elements is called *electrolysis*.

Figure 191 shows a *voltameter*. A voltameter is a device in which electrolysis takes place. In this case the diagram shows a copper voltameter, as the electrolyte is copper sulphate solution and the conductors are made of copper. The conductors immersed in the liquid are called *electrodes*. The electrode connected to the negative terminal of the battery is called the *cathode*. The electrode connected to the positive terminal of the battery is called the *anode*.

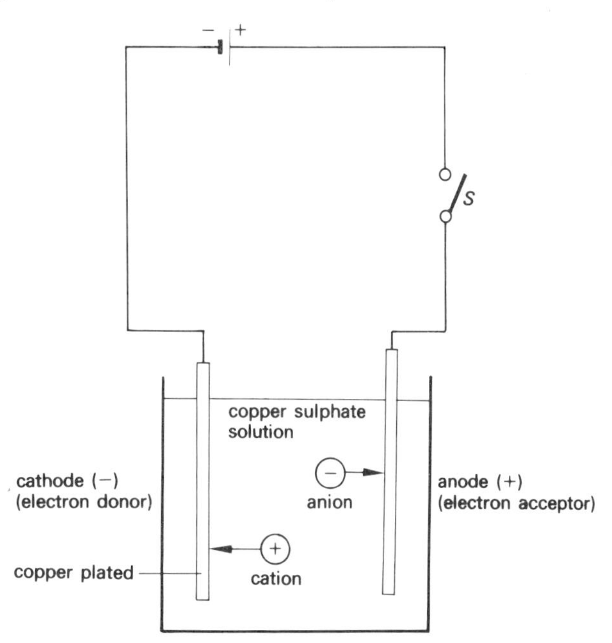

Figure 191 *A copper voltameter (electroplating)*

When the switch, *S*, is closed, the cathode is negatively charged, and exerts a force of attraction on positively charged particles in the electrolyte. At the same time the anode is positively charged and attracts negatively charged particles in the electrolyte. The molecules of copper sulphate which form the electrolyte decompose into positive copper particles and into negative sulphate particles at or near the charged electrodes. These particles are called *ions* (from the Greek 'ion', meaning a wanderer) and the process is called *ionization*. The negative ions, referred to as *anions*, are attracted to the positive anode. The positive ions, referred to as *cations*, are attracted to the negative cathode. At the cathode the positive charge on a cation is neutralized by electrons from the battery, and a neutral copper atom is deposited on the cathode. The negative sulphate anions give up their charge at the positive anode which donates copper cations to the electrolytes, where they combine with sulphate ions to form copper sulphate in the electrolyte. The movement of anions towards the anode and of cations towards the cathode, which is the result of the difference in the state of charge between the anode and the cathode, constitutes a flow of current.

A liquid which produces an abundance of cations and anions at or near differently charged electrodes is called a *strong electrolyte*. Examples of strong electrolytes are solutions of common salt, of sulphuric acid and of copper sulphate. Because strong electrolytes ionize readily, they are 'good' conductors when compared with liquids which do not readily ionize. Examples of liquids which do not readily ionize are formic acid, acetic acid, ammonium hydroxide and water. These are called *weak electrolytes* and, compared with strong electrolytes, are 'poor' conductors.

In the voltameter in Figure 191, electrolysis is used to produce a coat of copper on the copper cathode, the process being called *electroplating*. The copper comes from the anode, which dissolves into the electrolyte. The process can be used to purify copper. In this case the copper anode contains impurities which are deposited on the bottom of the voltameter, whilst pure copper is deposited on the

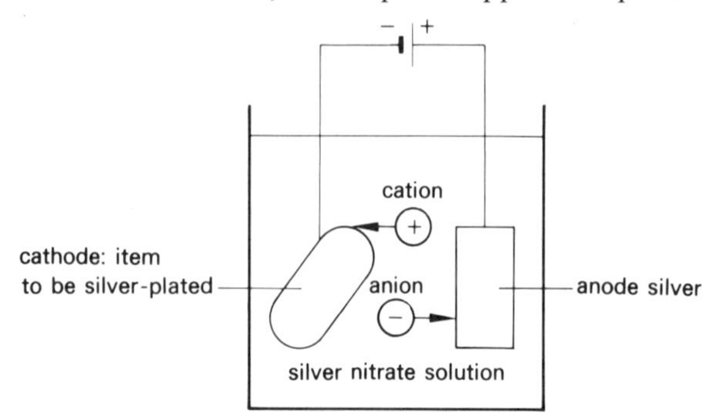

Figure 192 *Silver voltameter*

cathode, which is originally a small mass of pure copper. A similar process is used in gold, silver, nickel and chrome plating. In all of these cases the electrolyte is a solution of the ions of the anode metal. Figure 192 shows a silver voltameter. In each case the anode dissolves into the electrolyte, and is transferred to the cathode. The cathode may be of any conducting material. Thus copper, brass or iron may be plated with metal from the anode. The process is used to enhance the appearance of metals and to protect metals from corrosion.

Self-assessment questions

12 Which of the following statements identify a solid material as a 'good' conductor?
(*a*) an abundance of electrons in random motion
(*b*) high resistivity
(*c*) an abundance of cations
(*d*) low resistivity

13 Complete each of the following statements:
(*a*) The term electrolysis means ______.
(*b*) Conduction in liquids is a drift of cations to the (i) ______ and a drift of anions to the (ii) ______.
(*c*) Charged electrodes in an electrolyte cause the electrolyte at or near the electrodes to ______ or ______.
(*d*) The charged particles in an electrolyte at or near charged electrodes are called (i) ______; (ii) ______.
(*e*) The charged particles in an electrolyte obey the law of charges, which states that ______.
(*f*) An electrolyte which ionizes readily at or near charged electrodes is classed as a (i) ______ electrolyte and is a (ii) ______ conductor.

14 (*a*) The cathode acts as an electron donor in a voltameter.
TRUE/FALSE
(*b*) Cations are attracted to the cathode in a voltameter.
TRUE/FALSE
(*c*) In electro-plating the cathode donates metal to the electrolyte.
TRUE/FALSE
(*d*) The anions have a negative charge.
TRUE/FALSE
(*e*) The anode in a voltameter is connected to the negative plate of the battery.
TRUE/FALSE
(*f*) The anode supplies cations to the electrolyte.
TRUE/FALSE
(*g*) In a voltameter used for electro-plating the metal to be plated is called the cathode.
TRUE/FALSE

After reading the following material, the reader shall:

2.4 Describe a simple cell.

2.5 Use the electrochemical series to make qualitative predictions about pairs of metals in cells.

2.6 Describe corrosion as 'simple cell' action.

2.7 Discuss the damage done by corrosion.

The electric battery as originally invented consisted of two dissimilar metals (zinc and copper), separated by a cloth soaked in a strong electrolyte made from common salt dissolved in water. Later the dissimilar metal electrodes were immersed in an electrolyte. The arrangement is called a *simple cell* or a *voltaic cell.* When the electrodes are connected to an external circuit, current flows. The arrangement and the direction of conventional current and electron flow are shown in Figure 193. In this case the electrodes are copper and zinc; the electrolyte, dilute sulphuric acid, which when ionized forms hydrogen cations and sulphate anions.

When electrons flow in the external circuit through the resistor R, the electrons which constitute the current flow are repelled by excess negative charge in the zinc electrode. The excess negative charge is the result of the zinc ionizing and releasing zinc cations into the electrolyte. There they combine with sulphate anions from the electrolyte to form neutral molecules of zinc sulphate. This electrode is called the negative *plate* or *pole* of the cell.

Solutions to self-assessment questions

12 (*a*) an abundance of electrons in random motion
(*b*) low resistivity

13 (*a*) The term electrolysis means *to split up by means of electricity*.
(*b*) Conduction in liquids is a drift of cations to the (i) *cathode* and a drift of anions to the (ii) *anode*.
(*c*) Charged electrodes in an electrolyte cause the electrolyte at or near the electrodes to *ionize*.
(*d*) The charged particles in an electrolyte at or near charged electrodes are called (i) *negative ions (anions)*; (ii) *positive ions (cations)*.
(*e*) The charged particles in an electrolyte obey the law of charges, which states that *like charges repel, unlike charges attract*.
(*f*) An electrolyte which ionizes readily at or near charged electrodes is classed as a (i) *strong* electrolyte and is a (ii) *good* conductor.

14 (*a*) True.
(*b*) True.
(*c*) False. The anode donates metal.
(*d*) True.
(*e*) False. The anode is connected to the positive plate.
(*f*) True.
(*g*) True.

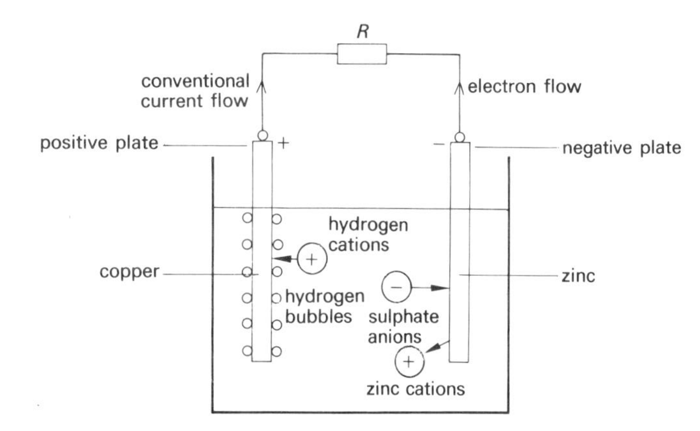

Figure 193 *Simple cell*

The electrons which constitute the current flow transfer the energy acquired in the negative plate to the external circuit, but maintain their negative charge until they arrive at the second plate in the cell. This second plate is called the positive plate or pole. In this plate the electrons join with the hydrogen cations at or near the surface of the plate, and the negative charge on the electron balances the positive charge on the cation to produce hydrogen gas. The hydrogen gas forms a layer of bubbles on the positive plate. This produces an effect called *polarization*.

When hydrogen bubbles have covered the whole of the positive plate the cell is polarized, and the electron flow in the external circuit stops. When no current is flowing in the external circuit, the negative plate continues to ionize into the electrolyte, and also commences to produce hydrogen gas. This gas leaves the surface of the plate, rises through the electrolyte and escapes into the atmosphere. The chemical reactions which take place when a simple cell is not supplying current to an external circuit are called *local action*.

The fact that polarization takes place when current is flowing, and local action occurs when no current is flowing, limits the uses that can be made of simple cells.

Investigations into local action and polarization in simple cells show that when a single electrode is immersed in a solution of its own ions there is a potential difference (p.d.) between the electrode and the solution. In order to measure this p.d. a second electrode (standard electrode) is introduced into the solution. The p.d. between the standard electrode and the solution is regarded as zero volts. When a cell is formed consisting of a standard electrode, the electrode to be tested and a suitable solution, the p.d. measured between the standard electrode and the test electrode is called the *electrode potential*.

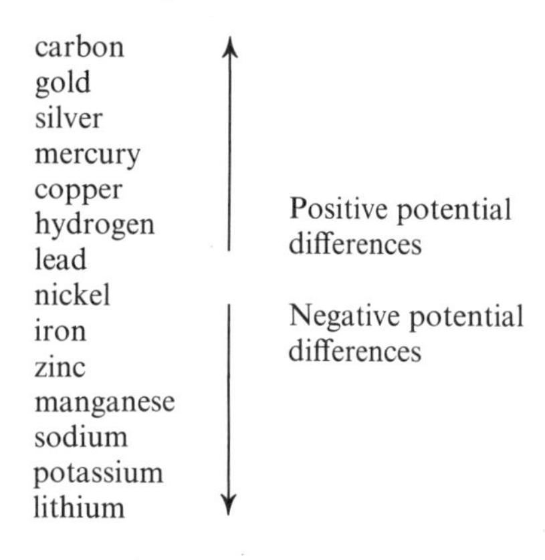

Figure 194 *Electrochemical series*

Electrode potentials range from a small positive p.d. above the standard electrode to a small negative p.d. below the standard electrode. When the electrode potentials of a range of substances are arranged in order of their magnitudes above and below that of the standard electrode, the list is called the *electrochemical series* of the substances.

The list in Figure 194 shows substances arranged into an electrochemical series. Hydrogen is used as the standard for comparison.

When any two of these substances are used as electrodes in a suitable electrolyte, a p.d. exists between them, and a simple cell is formed. Consider copper and carbon: the copper is less positive (i.e. more negative) than carbon. Thus electrons flow from the copper electrode towards the carbon in an external circuit connecting the plates. The p.d. is small when compared to that which occurs when copper and zinc are immersed in a suitable electrolyte. Zinc is much more

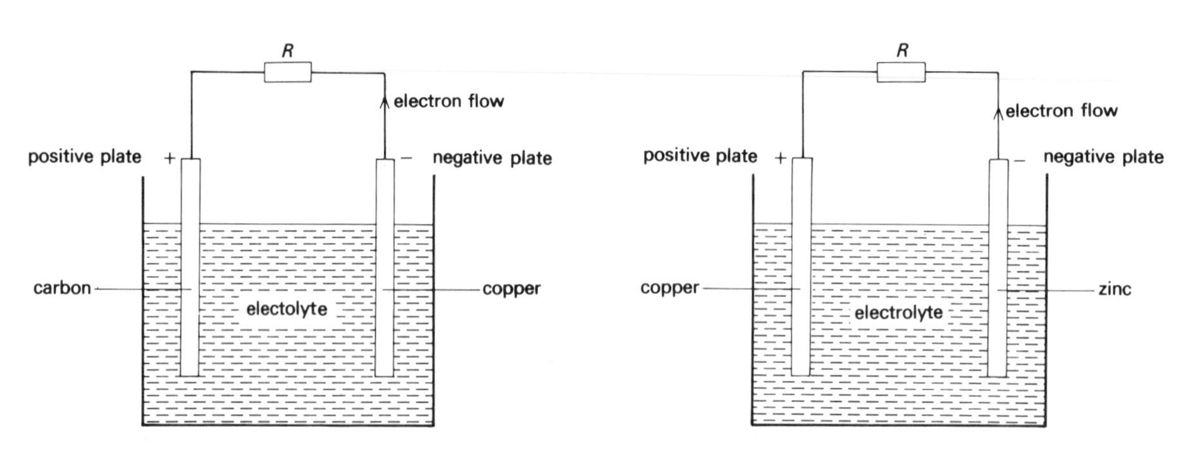

Figure 195 *Carbon–copper simple cell*

Figure 196 *Copper–zinc simple cell*

negative than copper, so that when these electrodes are immersed in an electrolyte, electrons from the zinc flow through an external circuit to the copper electrode.

When metals with different electrode potentials are in contact with an electrolyte, a simple cell is formed. If the electrodes are externally connected, current flows in the external circuit, positive ions are released from the more negative plate into the electrolyte, and this plate slowly dissolves into the electrolyte. In many instances this effect is not advantageous. In these instances, the loss of material from the negative substance is called *electrochemical corrosion*.

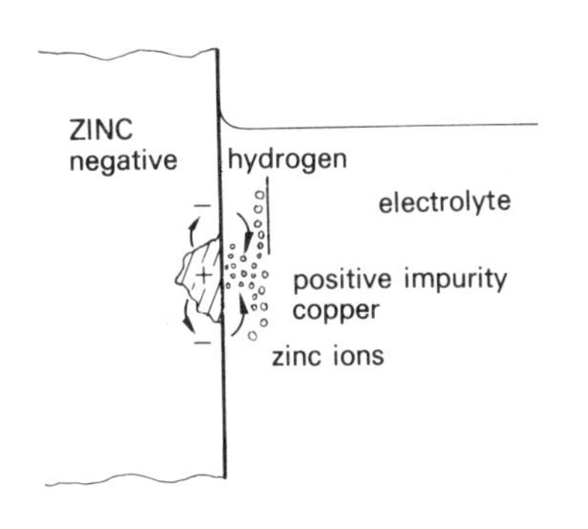

Figure 197 *Local action*

Figure 197 shows a zinc electrode in contact with an electrolyte of dilute sulphuric acid. Pure zinc does not react with sulphuric acid, but most zinc includes impurities such as copper which is more positive than zinc. In this case a simple cell is formed with copper as one plate, zinc as the second plate and sulphuric acid as the electrolyte. The plates of the cell are joined by the bulk of the zinc which forms the external circuit. The current which circulates causes hydrogen and zinc ions to go into solution, hence causing the zinc electrode near the impurity to dissolve.

Electrochemical corrosion of ferrous materials takes place in a similar manner. The iron and the impurities in iron form the plates of a simple cell. The electrolyte may be impure water, salt water or moist air. The bulk of the ferrous material joins together the two plates to form an external circuit. When current flows in the external circuit, iron cations are released, and corrosion takes place on the surface of the ferrous material. This form of corrosion of ferrous materials is *rust*. It varies in colour from bright yellow to deep red-brown. The cost of replacing ferrous material damaged by corrosion is enormous, and great effort is put into devising methods of protecting ferrous materials.

A common method of impeding the formation of rust is to paint the surface with a film of oil. This method gives limited protection, and is designed to limit the contact between the metal and liquids.

A more successful method of limiting the contact between the ferrous material and the liquid is to electro-plate the ferrous material.

Probably the most successful method is to provide a *sacrificial metal*. A sacrificial metal is one with a more negative electrode potential than iron. Zinc is more negative than iron, and is often used to protect iron from corrosion. An example is galvanized iron. Galvanized iron is iron coated with a layer of treated zinc; the zinc prevents moisture coming into contact with the iron, and so prevents the formation of simple cells and the resulting corrosion. Should the zinc coat crack and moisture enter the crack, a simple cell is formed with iron and zinc as the plates. The plates are connected together by the bulk of the iron, and a current flows. As zinc is more negative than iron, cations of zinc are released into the electrolyte. As long as some zinc remains, rusting of the iron is impeded.

Electrochemical corrosion occurs in all instances when two metals with different electrode potentials are in contact with an electrolyte. It can also occur when one metal containing impurities is in contact with an electrolyte. These conditions are present in many cases, and the damage done by corrosion is very widespread and expensive.

Self-assessment questions

15 Complete the following statements:

(*a*) The negative plate in a simple cell which is supplying current supplies ______ to the electrolyte.

(*b*) The positive plate in a simple cell which is supplying current donates electrons which ______ cations.

(*c*) When a cell, made up of a copper plate and a zinc plate immersed in an electrolyte, is not supplying current, the reaction at the negative plate is called ______.

(*d*) The effect of the layer of gas bubbles which forms on the positive plate of a simple cell is called ______.

(*e*) When a simple cell is polarized the current flow in the external circuit ______.

16 What name is given to a list of electrode potentials?

17 By referring to Figure 194, identify the plate which will donate cations to an electrolyte in the following pairs of plates:

(*a*) carbon–lithium
(*b*) zinc–nickel
(*c*) silver–sodium
(*d*) copper–gold

18 Explain how a sacrificial metal impedes corrosion.

19 What is the cause of electrochemical corrosion?

After reading the following material, the reader shall:

3 Be aware of the nature of chemical reactions.
3.1 Describe chemical reactions as interactions between substances which involve a re-arrangement of atoms.
3.2 Distinguish between physical and chemical changes.
3.3 Identify a chemical equation.

A physicist would consider the extremely rapid burning of a petrol–air mixture in a motor car cylinder in terms of the pressures caused, the thickness of metal needed in the cylinder walls to resist the quickly changing forces, the amount of heat conducted away from the cylinder, and the lubrication problems that arise when temperatures are very high. A chemist studying the same situation would consider the elements, mixtures, and compounds involved in the burning of the petrol and the air, the source of the heat liberated during the combustion, and possible changes that might take place in the properties of the materials from which the cylinders and pistons are made.

Chemical changes involve the production of new substances, in this case, exhaust gases.

Most familiar chemical changes do not have such explosive effects as the noisy combustion taking place inside a car engine. In fact, most chemical changes take place slowly. Rust on steel girders forms over a long period of time; silver teapots tarnish by chemical actions that may take months; baking a cake involves chemical change not over months, but preferably not so fast that it is explosive!

Of the elements known to date, ten are gases in their natural state; these include oxygen, nitrogen, chlorine and argon. Two elements occur naturally as liquids – mercury and bromine. The other elements – more than ninety of them – are solid at ordinary temperatures and pressures; these solids include many metals, and materials such as carbon and sulphur. Elements join together to form other materials. Some materials contain two or more elements united in a chemical compound. Carbon dioxide, for instance, is formed by the chemical combination of the elements carbon and oxygen. Sulphuric acid is a compound containing atoms of hydrogen, sulphur and oxygen. Other materials consist of mixtures of elements or mixtures of compounds. Air is a well-known example of a mixture of elements. It is important to distinguish between compounds and mixtures since chemical reactions always involve the building up or breaking down of compounds.

In a chemical reaction there is no change in the total mass of the substances involved. If all the reactants (i.e. those materials present when the chemical reaction begins) and all the products of the reaction are weighed, then it is always found that:

Total mass before chemical reaction = total mass after chemical reaction

No atoms are destroyed or created: they are simply present in new patterns. This is known as the law of conservation of matter.

Readers will be familiar with many cases of physical change. Changes of state (e.g. the freezing of water into ice) and simple cases of dissolving (e.g. sugar in water) are physical changes. Physical changes can in most cases be readily reversed; chemical changes or reactions frequently are very difficult to reverse. In most cases chemical reactions are irreversible. The table below lists the main features of chemical and physical changes.

chemical change	physical change
considerable energy frequently liberated (as in petrol engine cylinder). Sometimes considerable energy absorbed.	little energy change involved (e.g. mixing of gases involves no energy change).
chemical changes always produce at least one new substance.	new substances are never formed (ice may change into water, but no new material is produced).
the masses of the individual new compounds formed are normally different from the masses of the original chemicals (reactants).	each substance retains its mass throughout.
Very difficult to reverse.	Normally simple to reverse.

Every element has been given a symbol by which it can be identified. The first use of these symbols is to represents a single atom of the element. Some of the symbols are a single capital letter (C represents carbon), whilst other symbols are a capital letter followed by a small letter (Cu represents copper).

Solutions to self-assessment questions

15 (*a*) The negative plate in a simple cell which is supplying current supplies *cations* to the electrolyte.

(*b*) The positive plate in a simple cell which is supplying current donates electrons which *neutralize* cations.

(*c*) When a cell, made up of a copper plate and a zinc plate immersed in an electrolyte, is not supplying current, the reaction at the negative plate is called *local action*.

(*d*) The effect of the layer of gas bubbles which forms on the positive plate of a simple cell is called *polarization*.

(*e*) When a simple cell is polarized the current flow in the external circuit *ceases*.

16 The list is called an electrochemical series.

17 (*a*) lithium

(*b*) zinc

(*c*) sodium

(*d*) copper

18 A sacrificial metal must be more negative than the metal it protects. The sacrificial metal forms the negative plate of a simple cell, and the other metal the positive plate. When the plates are immersed in an electrolyte and an external circuit is completed between the plates, cations are given off by the sacrificial metal into the electrolyte. This corrodes the sacrificial metal, and as the positive metal is donating only electrons to the cations, the metal does not corrode. The process is the action of a simple cell.

19 Electrochemical corrosion is caused when two substances with different electrode potentials are in contact with an electrolyte.

The molecules of elements and of compounds are each represented by a chemical formula. For instance O_2 is the formula for the element oxygen. O_2 represents a molecule which contains two atoms of oxygen. NaCl represents a molecule of the compound sodium chloride (common salt). NaCl represents a molecule containing one atom of sodium (Na) and one atom of chlorine (Cl). SO_2 represents a molecule of the compound sulphur dioxide. The molecule contains one atom of sulphur (S) and two atoms of oxygen (O_2). CH_4 represents a molecule of the gas methane (a compound). The molecule contains one atom of carbon (C) and four atoms of hydrogen (H_4).

When chemical reactions take place, new substances are formed. The changes in the materials that have taken place can be represented by a chemical equation which makes use of the symbols representing molecules of the elements and of compounds.

The chemical equation shows the reactants and the products.

A chemical equation is a representation of a chemical reaction using formulae and symbols.

Consider the case of sulphur burning in air and producing the gas sulphur dioxide.

This can be written:

Sulphur + oxygen → sulphur dioxide

reactants product

This can be written as symbols and formulae:

$$S + O_2 \rightarrow SO_2$$

The number of atoms of individual elements on each side of the equation must be the same. In this case there are three atoms on each side of the arrow.

Much more complicated equations can be worked out. For instance:

Zinc + Sulphuric acid → Zinc sulphate + Hydrogen

$$Zn + H_2SO_4 \rightarrow ZnSO_4 + H_2$$

When the equation is balanced (i.e. when the number of atoms on each side is the same), then chemists can use the equation to calculate reacting quantities. This is necessary in all manufacturing processes.

The equations do not give any indication of the conditions under which the chemical reactions take place. For instance, $S + O_2 \rightarrow SO_2$ does not give any idea of how the sulphur is induced to react with the oxygen.

When a chemist has to predict the reaction of a substance in a novel set of circumstances, he can draw on his knowledge of the substance, and use his analytical skill with chemical equations to estimate what are the likely products of the new chemical reaction he has to set up.

After reading the following material, the reader shall:

3.4 Describe the characteristics of acids.

3.5 Describe a base as a substance that removes the acidic properties of acids.

3.6 State that an alkali is a soluble base.

3.7 Describe acidity/alkalinity by means of indicators.

Acids and bases are among the most important industrial chemicals. Widely used acids include sulphuric, hydrochloric, nitric, acetic and oxalic. Bases used in many chemical reactions include ammonia, sodium hydroxide, sodium carbonate and pyridine.

Acids and bases are closely interrelated classes of chemical compounds. They are complicated substances, and it is not surprising that the definitions of them have altered considerably during the evolution of the history of chemistry. Even after extensive studies of chemistry by large numbers of scientists in recent times, some of the problems of defining acids and bases still have not been resolved.

The ideas of acidity and alkalinity have been familiar for many centuries. The word acid is derived from a Latin word meaning 'sour', whilst the word 'alkaline' comes from the Arabic word for 'ashes'.

There are many common acids which are very weak. These occur naturally in vegetables and fruits. However, even weak acids can be dangerous; the oxalic acid found in rhubarb leaves can cause serious poisoning of animals, even though its concentration is low.

Acids can be taken to be substances that:

(i) React with bases (e.g. sodium hydroxide) to form a salt and water only. (The bases are the 'base' of the salt.)

(ii) Are soluble in water.

(iii) Often react with metals, one of the products of the chemical reaction being the gas hydrogen.

Other details (such as electrical characteristics) necessary to give a complete definition of an acid are not included in this Unit.

A base is a substance which can neutralize an acid.

An acid–base reaction is one in which the combination of an acid and a base produce a salt and water only. This type of reaction is frequently called a neutralization reaction.

All neutralizations can be summarized by the chemical equation:

Base + Acid → Salt + Water

A typical example of this type of chemical reaction is:

Copper oxide + Sulphuric acid → Copper sulphate + Water

$$CuO + H_2SO_4 \rightarrow CuSO_4 + H_2O$$

If a base is soluble in water it is called an *alkali*. Most bases are insoluble in water, but a small number are fairly soluble. These produce alkaline solutions that typically feel slippery or greasy.

It is frequently necessary to know whether substances are alkaline or acid in nature. This is particularly important in biological systems; our health depends upon our body system keeping acid things which need to be acid, and keeping alkaline things which need to be alkaline.

Simple indicators like litmus (a mixture of dyes extracted from lichens) or screened methyl orange will, by changing colour, tell whether a substance is acid or alkaline. They do not, however, give any idea of the degree of acidity or alkalinity.

Universal indicator, a mixture of several indicators, changes to a different colour according to the acidity or alkalinity of the solution to which it is added. This has helped to create what is known as the pH scale; on this scale the degrees of acidity and alkalinity are represented by numbers. The Danish biochemist S. P. Sorensen proposed this scale, and the letter pH mean 'the potency of hydrogen'. Figure 198 shows how the Universal indicator gives a measure of acidity and alkalinity. The range of numbers 1 to 14 is accepted by all chemists.

	pH	colour of universal indicator
strongly acidic	1	
	2	red
	3	
	4	orange
	5	
weakly acidic	6	yellow
	7	green
	8	torquoise
weakly alkaline	9	blue
	10	
	11	purple
	12	
strongly alkaline	13	
	14	

Figure 198 *Universal indicator and pH values*

The lower the pH value, the more acidic is the solution. A solution like pure water which is neither acid nor alkaline is said to be neutral and has a pH value of 7. pH values greater than 7 indicate an alkaline solution.

A knowledge of the pH of solutions is needed frequently by doctors. The human body keeps the blood alkaline by automatic adjustment. If there is an increase in the acidity of the blood, doctors can use this information to find the cause of the illness. The pH value of the blood for good health is about 7.4. Gastric juices, on the other hand, are operating most efficiently when their pH value is as low as 1.5.

pH values are needed by farmers; crops are known to grow best on soils having particular acid or alkaline characteristics.

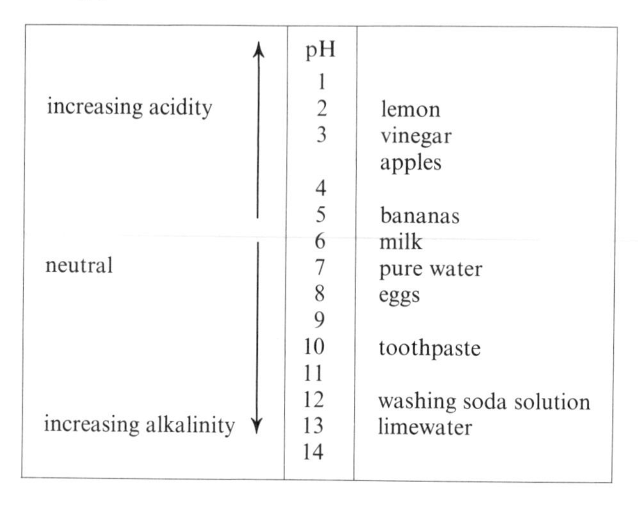

	pH	
	1	
increasing acidity ↑	2	lemon
	3	vinegar
		apples
	4	
	5	bananas
	6	milk
neutral	7	pure water
	8	eggs
	9	
	10	toothpaste
	11	
	12	washing soda solution
increasing alkalinity ↓	13	limewater
	14	

Figure 199 *pH values of common materials*

Figure 199 shows the pH values of familiar substances.

Indicators give a fair indication of the acidity or alkalinity of a solution; whenever really accurate pH values are required, a pH meter must be used. This instrument makes use of the electrical characteristics of the solution.

Self-assessment questions

20 Rust on a steel component is an example of:

(*a*) a solution
(*b*) a mixture
(*c*) a compound
(*d*) an element

Select the correct alternative.

21 When copper is oxidized as a result of being heated in air, the resulting copper oxide is

(*a*) lighter than the original metal
(*b*) the same mass as the original value
(*c*) heavier than the original metal

Select the correct alternative.

22 When a large jar is placed over a burning candle, the candle burns for a while before going out. Explain this in terms of chemical reactions.

23 Complete the table below, indicating which are physical changes and which are chemical changes.

	physical change	chemical change
rusting of iron		
mixing of sand and stones		
baking of a cake		
melting of cooking fat		
burning of coal		
magnetizing a piece of steel		
explosion of a firework		
boiling of water		

24 Name (*a*) (*b*) (*c*) as formula, symbol, or chemical equation:

(*a*) $C+O_2 \rightarrow CO_2$
(*b*) C
(*c*) O_2

25 Define, in words, a chemical equation.

26 Name two important characteristics of acids.

27 (*a*) A base neutralizes an acid.

TRUE/FALSE

(*b*) Alkalis are soluble bases.

TRUE/FALSE

28 The following table shows a number of solutions and their pH value.

solution	I	II	III	IV	V	VI
pH	7	4	9	5	6	2

(*a*) Which solution is most alkaline?
(*b*) Which solution is most acidic?
(*c*) If equal volumes of I and III are mixed, is the resulting solution acid, neutral, or alkaline?

29 Pure water has a pH value of

(*a*) 14
(*b*) more than 7

(*c*) 7
(*d*) less than 7
(*e*) 1
Select the correct alternative.

30 The pH value of a solution of hydrochloric acid is always
(*a*) 14
(*b*) greater than 7
(*c*) 7
(*d*) less than 7
(*e*) 1
Select the correct alternative.

Solutions to self-assessment questions

20 (*c*) a compound

21 (*c*) heavier than the original metal

22 When the candle is burning, the oxygen in the air is being consumed. When the oxygen has all been burned, no further chemical reaction can take place, and the candle goes out.

23

	physical change	chemical change
rusting of iron		√
mixing of sand and stones	√	
baking of a cake		√
melting of cooking fat	√	
burning of coal		√
magnetizing a piece of steel	√	
explosion of a firework		√
boiling of water	√	

24 (*a*) chemical equation
(*b*) symbol
(*c*) formula

25 A chemical equation is a representation of a chemical reaction using formulae and symbols.

26 Acids:
(i) React with bases to form a salt and water only,
(ii) Are soluble in water,
(iii) Often react with metals to release the gas hydrogen.

27 (*a*) True.
(*b*) True.

28 (*a*) III. (*b*) VI. (*c*) alkaline.

29 (*c*) 7.

30 (*d*) less than 7.

Topic area: Rays

After reading the following material, the reader shall:

1 Be aware of some of the simple properties of light.
1.1 Identify some well known properties of light.

The topic of light is closely related to many parts of physical science. Light is a form of energy, and energy interchanges are basic to the study of many topics. Light has wave properties, and wave properties have been discussed in the work on Topic area Waves. Light absorption and production involve energy changes within the atom, and these depend upon electrical effects.

It is not easy to give simple precise answers to the questions, 'What is energy?' and, 'What is electricity?' In a similar fashion, there is no easy explanation as to the exact nature of light. Uncertainties about the properties of light have puzzled scientists since the original ideas were put forward by Isaac Newton in the seventeenth century. Early in the present century light was discovered to have the properties both of waves and of particles. The complex nature of light fortunately does not make difficult a study of its definite and familiar properties.

Everyday experiences indicate that
(i) Mirrors reflect light.
(ii) Lenses change the direction of light.
(iii) Light must move.
(iv) Light must travel in straight lines.

The term 'electromagnetic spectrum' takes in all vibrations with the same general characteristics of light. Radio waves have the lowest frequency, next come radiant heat waves, and then comes 'visible' light. The normal human eye can detect vibrations having frequencies between 400×10^{12} Hz (red light) and 750×10^{12} Hz (violet light). The visible spectrum is a very small part of the electromagnetic spectrum. Beyond visible light are radiations of greater and greater frequency. These radiations include ultra-violet rays, X-rays and gamma rays.

It has been known for over 300 years that light travels through space at about 300×10^6 m/s. Light does not however travel at this same enormous speed through substances such as water or glass; light slows down when it enters a substance which has an 'optical density' greater than that of empty space.

The term 'medium' is used for any substance through which light may pass. If the luminous object can be seen clearly, then the medium is described as 'transparent'. Ordinary glass and air are two of the commonest transparent media. If the luminous object can be detected through the medium, but only roughly defined, then the medium is said to be 'translucent'. Ground glass is a typical example.

Before studying reflection and refraction, it is useful to be reminded that light, when moving in one medium, travels in straight lines. A ray of light is the straight line between two points along which the light is travelling.

After reading the following material, the reader shall:

1.2 State the laws of reflection for plane mirrors.
1.3 Draw diagrams for a plane mirror using the laws of reflection.

One of the most generally recognized properties of light is that it can be reflected. Objects that do not emit light are only visible because they reflect light. This means that the great majority of objects are seen solely by reflected light. Everyday experience demonstrates that a shiny or polished surface is a better reflector of light than a dull or matt surface.

There are mirrors of many shapes: a plane mirror is one in which the reflecting surface lies in one plane. It is a flat sheet of polished glass coated on the back with some shiny metal, commonly silver. The light passes through the glass before being reflected by the metal surface.

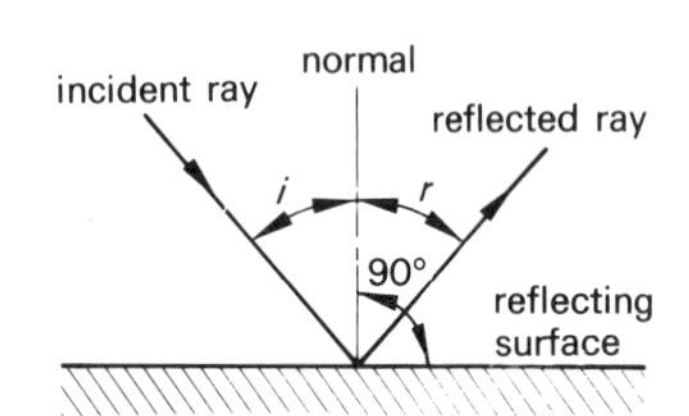

Figure 200 *Reflection of a ray of light*

A ray of light falling on a reflecting surface is called an incident ray, and the ray that 'bounces' off the shiny material is called the reflected ray. The direction of a ray is measured by the angle it makes with the normal (see Figure 200).

The two laws of reflection state that

(i) *When light is reflected from a plane surface, the angles of incidence and reflection are equal.*
In Figure 200 this means that $i = r$.

(ii) *The normal to the mirror, the incident ray and the reflected ray all lie in the same plane.*

The laws help to explain how images are seen in mirrors.

If a person looks into a plane mirror he can 'see' himself. If the mirror is moved to one side, objects behind the observer can be 'seen'. What is 'seen' in the mirror is referred to as the 'image' of the object.

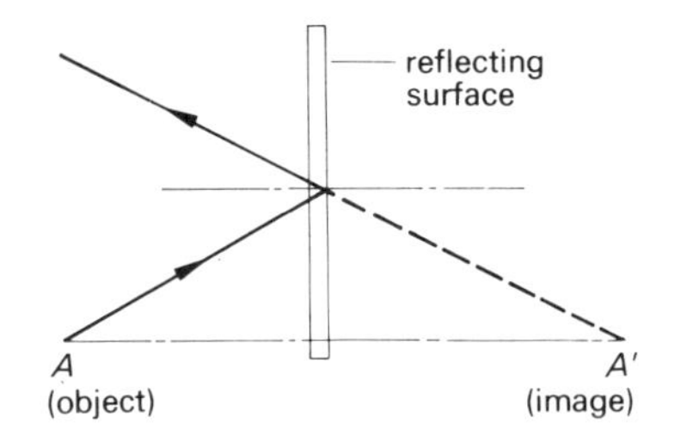

Figure 201 *Single ray reflected: position of the image*

Figure 201 shows what happens to a ray of light from a point A.

By looking at the symmetry of Figure 201 it can be seen that the image A^1 is as far behind the mirror as the object A is in front of the mirror.

The image formed by a plane mirror lies as far behind the mirror as the object is in front of the mirror. The image also lies on the normal from the object to the mirror.

Consider what happens when light from an object is reflected in a plane mirror. In Figure 202 an object AB is reflected in a plane mirror. The paths of the rays of light from the ends of the object are shown. The observer at 0 will 'see' the image of the object, and the object, to the observer, will appear to be at position A^1B^1.

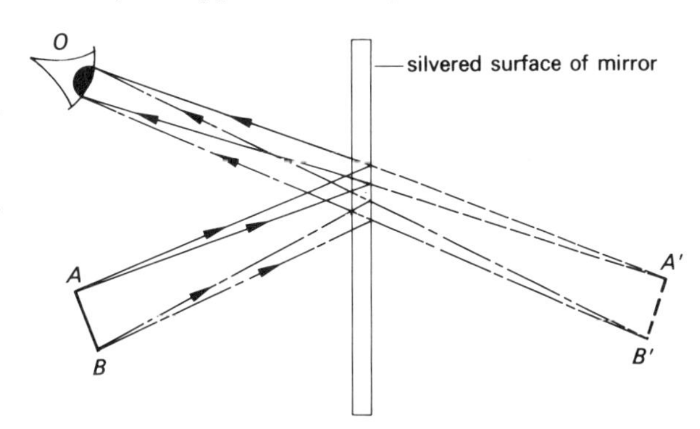

Figure 202 *The 'seeing' of an image in a plane mirror*

Self-assessment questions

1 (*a*) Light is a form of energy.

TRUE/FALSE

(*b*) The human eye cannot detect gamma rays.

TRUE/FALSE

(*c*) Pure water is translucent.

TRUE/FALSE

2 State the two laws of reflection, and mark the angles of incidence and reflection on a diagram.

3 (i) If a ray of light strikes a mirror at an angle of incidence of 30°, indicate on a diagram (drawn to show angles correctly) the angle between the incident and reflected rays.

(ii) If the angle of incidence is now increased by rotating the mirror through 20°, what is the new angle between the incident and reflected rays?

After reading the following material, the reader shall:

2 Understand the refraction of light rays.
2.1 Define refraction.
2.2 Name and sketch the shapes of biconvex and biconcave lenses.
2.3 Sketch the path of a parallel beam of light passing through the lens parallel to the axis.
2.4 Define focal length.
2.5 Explain the formation of virtual and real images in terms of rays.

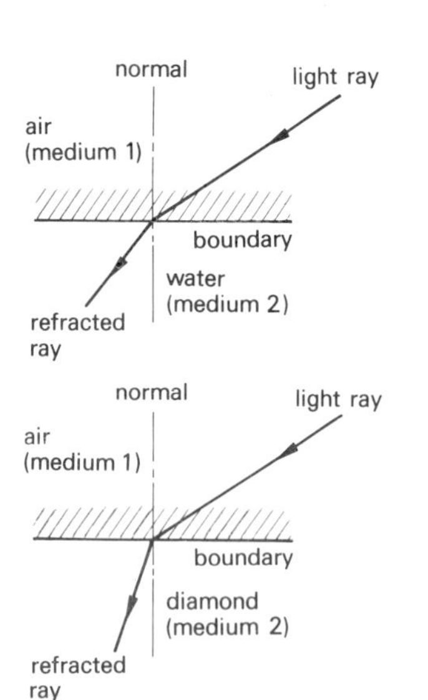

Figure 203 *Refraction of light rays*

Although mirrors of all types are useful common articles, they are not so important to general well-being as are the lenses used in spectacles, microscopes and optical instruments.

When a ray of light meets a boundary between two transparent media, some of the light is reflected, whilst some is transmitted into the second medium. The transmitted ray undergoes a sharp change in direction. (The only exception to this rule is when the ray of light is travelling at right angles to the boundary surface.) Figure 203 shows rays of light travelling through air and meeting surfaces of water and of diamond. The deviation of the ray in the diamond is greater than the deviation in the water. Diamond is said to be optically denser than water.

When a ray of light passes from one medium into another medium, the transmitted light energy undergoes a direction change. This deviation of the transmitted ray is known as refraction. Refraction is caused by the speed of movement of the light changing as the light crosses the boundary between the media.

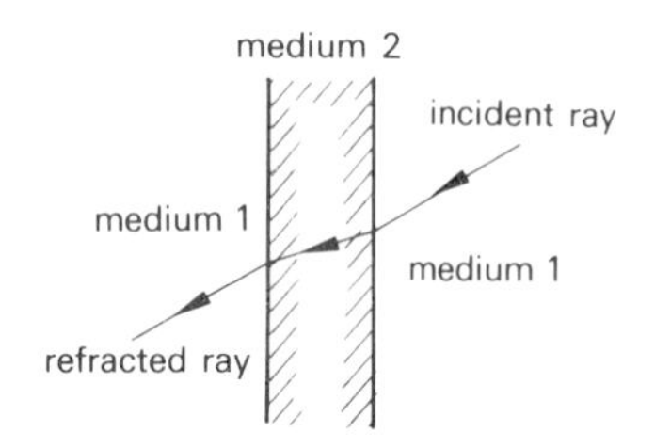

Figure 204 *Passage of a ray of light through a flat plate*

So far only one boundary surface has been mentioned. Generally, however, light that enters a material leaves it at another surface. When this happens, the bending of the light ray is usually in the opposite direction from what it was on entering. Figure 204 shows a beam of light which leaves a flat plate parallel to its original direction but displaced to one side when the surrounding media are the same on both sides of the plate.

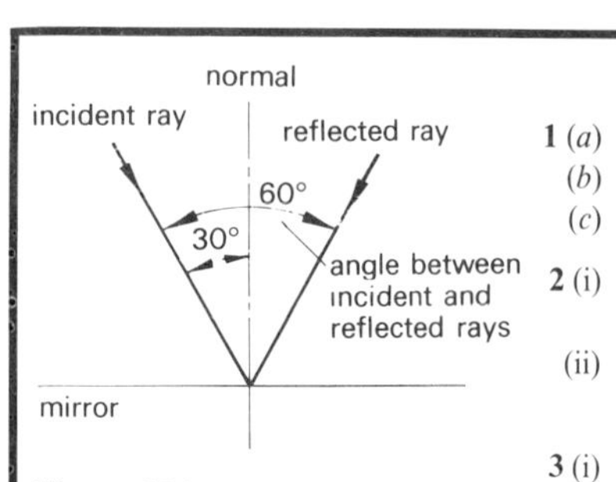

Figure 205 *Reflection in a plane mirror*

Solutions to self-assessment questions

1 (*a*) True.
(*b*) True.
(*c*) False. Pure water is transparent.

2 (i) When light is reflected from a plane surface, the angles of incidence and reflection are equal.
(ii) The normal, the incident ray and the reflected ray all lie in the same plane.
Refer to Figure 200 and check that angle i = angle r.

3 (i) See Figure 205.
(ii) The angle of incidence is now 50°. This means that the reflected ray is turned through 100° from the incident ray.

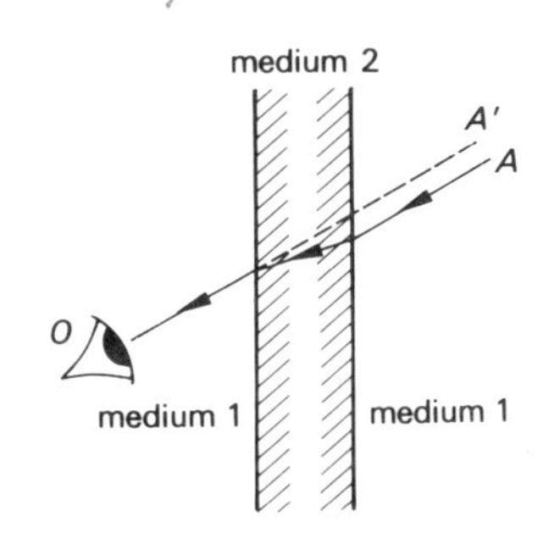

Figure 206 *'Shift' of an object caused by refraction*

This effect explains the apparent displacement of objects looked at through a pane of glass (e.g. a window). The observer at *O* 'sees' the object as though it were at point A^1 although the object is really situated at point *A* (Figure 206).

The 'shallowing' effect noticed when objects are observed below the surface of water is explained by the idea of refraction. Figure 207 shows how an object *A* on the bottom of a pond appears to be in a different position to an observer *O*. The more obliquely the observer views the top surface, the more the object appears to move upwards from its true position. Bathers need to be aware of this effect of the refraction of light; non-swimmers may enter casually into water they have mistakenly taken to be shallow.

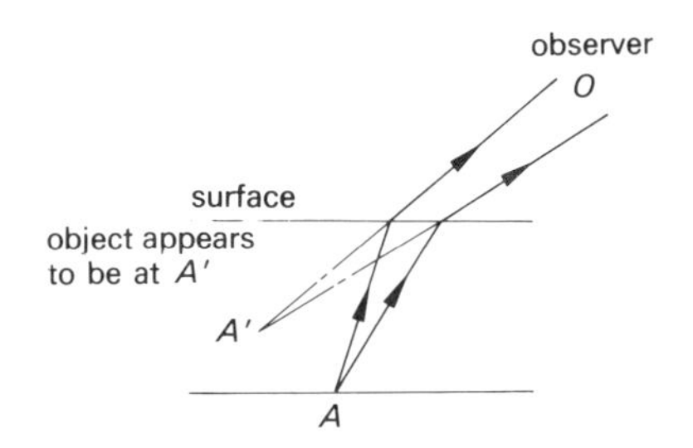

Figure 207 *Apparent position of an object viewed through water*

Refraction at plane surfaces is worthy of note, but refraction through lenses has many important practical applications.

A lens consists of transparent material bounded (usually) by two curved surfaces through which the light passes in turn. Figure 208 shows two groups of common lenses. In Group I the lenses are called 'converging', since they tend to make a parallel beam of light converge (come to a point). In Group II the lenses are called 'diverging', since they tend to make a parallel beam of light diverge (spread out). In general, if a lens is thicker in the middle than at the circumference, it causes light rays to converge. If it is thinner at the centre, it causes them to diverge.

Group I diverging

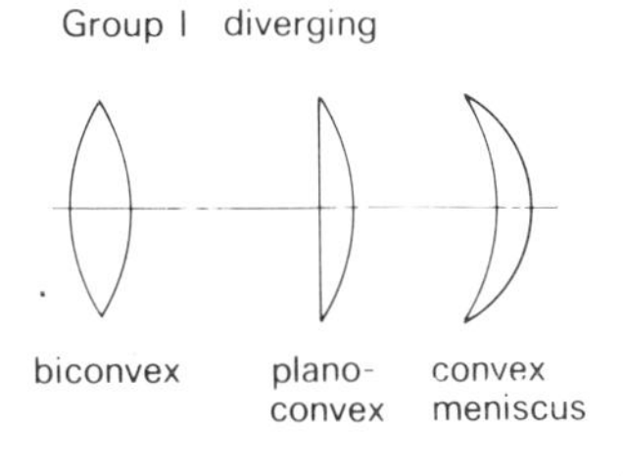

Group II converging

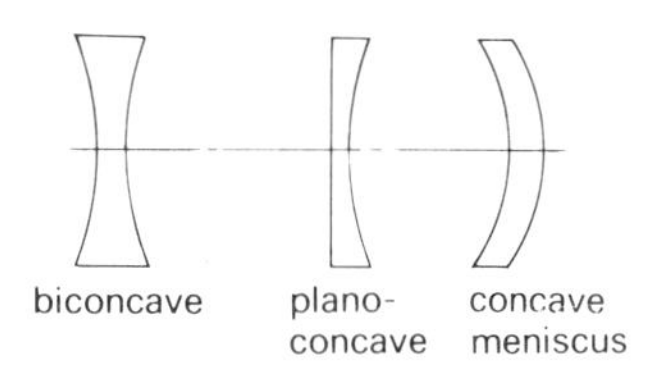

Figure 208 *Diverging and converging lenses*

Most lenses are ground with their surfaces spherical. The majority of lenses used in optical instruments are called 'thin'. This means that the thickness of the lens is very small compared with the radius of the curved surface.

The camera of the TV or film producer, the telescope of the surveyor, the sextant of the sailor, and the microscope of the biologist are just a few of the optical instruments which have helped to create the complex technological world of today. Each of these, and of many other optical devices, has been developed using the ideas of reflection and refraction. Much of our present-day comfort and safety depends on devices where light is refracted and reflected.

It is possible to describe the action of a lens by considering how light bends as it passes through the surfaces. Consider the important cases of a parallel ray of light striking (i) a biconvex lens (ii) a biconcave lens.

Readers will understand that the light actually changes direction at each surface of the lens. Ray diagrams are easier to draw if the bending is shown taking place overall on a line drawn through the optical centre and at right angles to the principal axis.

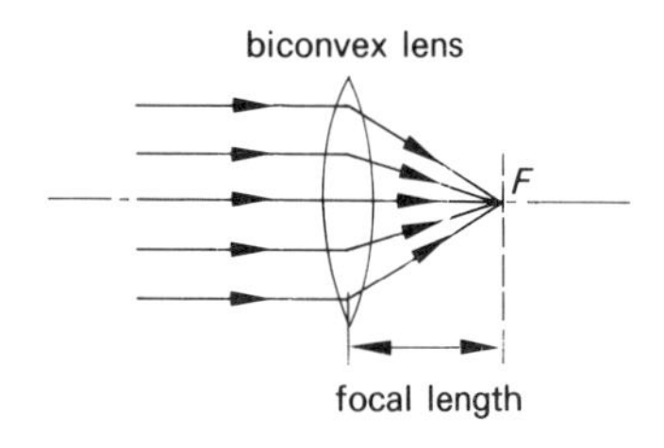

Figure 209 *Parallel rays brought into focus by a biconvex lens*

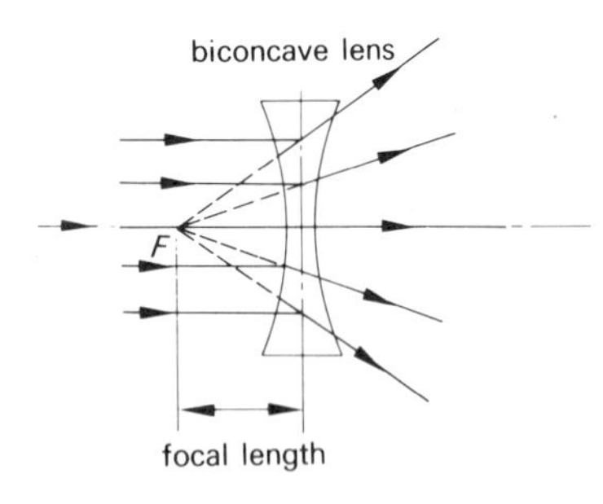

Figure 210 *Parallel rays meeting a biconcave lens*

In Figure 209 parallel rays of light strike a biconvex lens. The curved surfaces are shaped in such a way that the beam is brought to a definite focus at *F*. The distance between the lens and this focus is called the *focal length of the lens*.

A biconcave lens behaves as shown in Figure 210. An observer would see the rays coming to a virtual focus at *F*. The distance from the lens to *F* is called the focal length of the lens.

Before considering how images are formed by lenses, some of the terms used in connection with lenses need some explanation. The optical centre of a lens, for all practical purposes, lies on the geometrical centre of the lens. Light passes through the optical centre without change of direction. The three principal rays which are convenient for drawing the object–image relationship of a lens are shown in Figure 211. The rays parallel to the principal axis are refracted through a focal point; the rays passing through a focal point and then entering the lens emerge parallel to the principal axis.

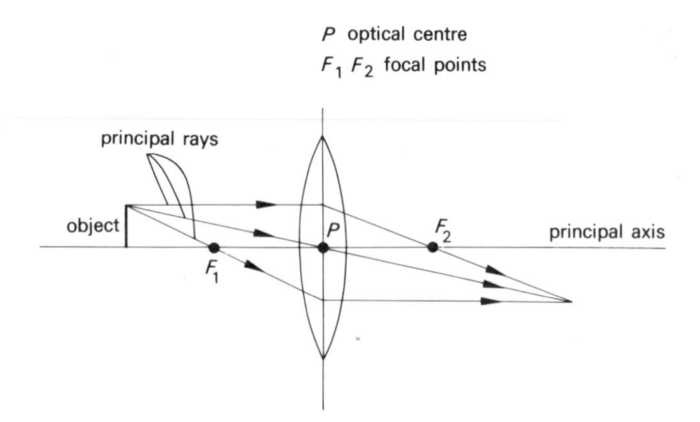

Figure 211 *Main terms in lens diagrams*

Lenses are of importance because they form images. The position and size of the image can be found by drawing a diagram of the light rays; this diagram also indicates whether the image is upright or inverted.

Four important cases of locating images will be reviewed. Other cases can be solved by using the principles laid down here.

Case 1

When the object to be viewed is at least twice the focal length from a biconvex lens, then the ray diagram is as shown in Figure 212. The principal rays are numbered 1, 2, 3.

In Figure 212 the image of object *AB* is formed just beyond the focal point at A^1B^1. It is inverted, and the image is clearly smaller than the object. This type of image is called a *real image*, because it is formed by the meeting of real light rays.

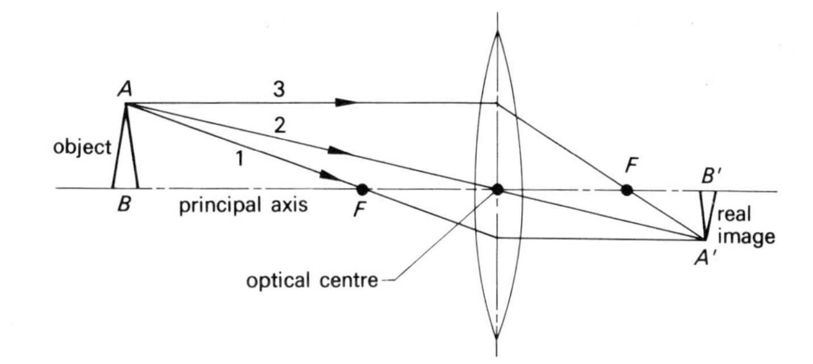

Figure 212 *Image seen through biconvex lens when object is more than twice the focal length from the lens*

Case 2

When the object to be viewed is placed slightly more than the focal length from a biconvex lens, then the ray diagram is as shown in Figure 213.

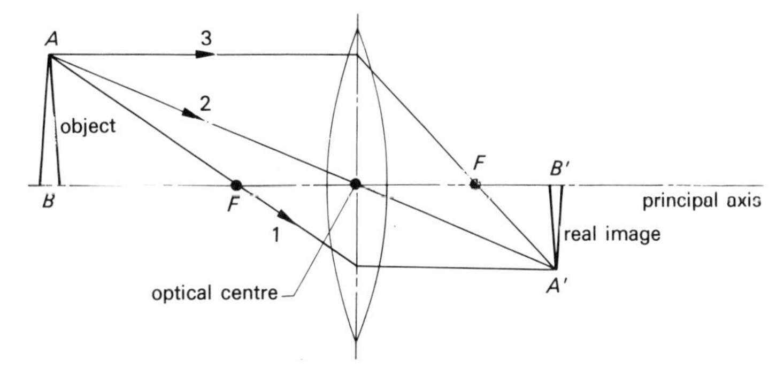

Figure 213 *Image seen through biconvex lens when object is just more than the focal length from the lens*

The image is formed well away from the lens: it is inverted, and as is shown in the diagram, considerably larger than the object. This image is also 'real', being formed by the meeting of real light rays.

Case 3

When the object to be viewed is placed at a distance less than the focal length from a biconvex lens, then the ray diagram is as shown in Figure 214.

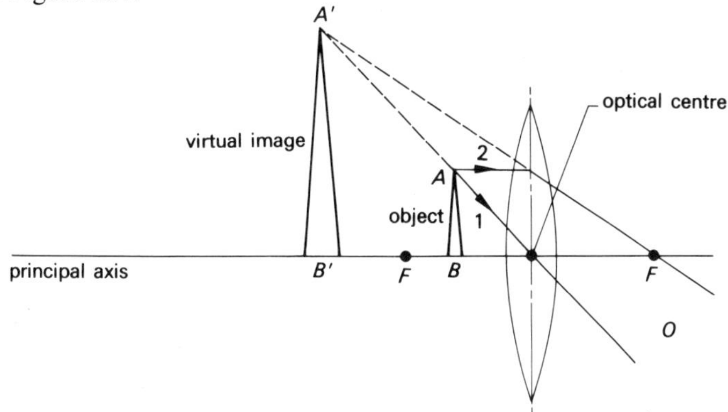

Figure 214 *Image seen through biconvex lens when object is less than the focal length from the lens*

The rays of light which enter the eye of the observer at *O* have been bent in such a way that the image appears to be much larger than the object. The image is formed on the same side of the lens as the object. The image is enlarged and upright. It is a *virtual* image because the rays of a light only *appear* to come from it. It should be noted that in this case only two principal rays can be usefully drawn. The ray from *A* through the focal point to the left of the lens does not pass through the lens.

The most obvious application of Figure 214 is a magnifying glass. A small object placed close to the lens appears much larger, and the image is the same way up as the object (i.e. it is upright).

Case 4

When the object to be viewed is placed in front of a biconcave lens, then the ray diagram is as shown in Figure 215.

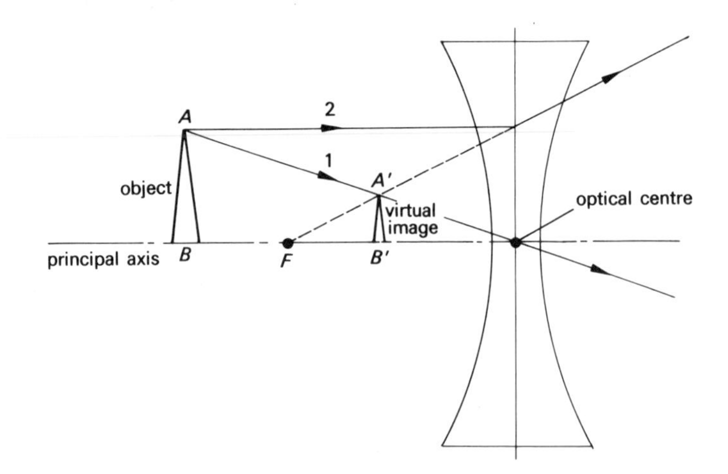

Figure 215 *Image seen through biconcave lens*

The rays of light are bent in such a manner that the image appears to be much smaller than the object. Once again the ray diagram only contains two principal rays. The image and the object are formed on the same side of the lens. The image is a virtual image since the rays only appear to come from it.

Self-assessment questions

4 By the use of a diagram explain what is meant by refraction.

5 Sketch (i) a biconcave lens and (ii) a biconvex lens. Which of these lenses would be used in a magnifying glass?

6 A beam of light originally parallel to the principal axis passes through a biconvex lens. Draw a ray diagram to show the path of the beam after it has passed through the lens. Indicate on the diagram the focal length of the lens.

7 A tribesman is shooting a fish with a bow and arrow. Assuming that he is not directly above the fish, should he aim below, at, or above the fish? Illustrate your answer by means of a sketch.

8 An object 20 mm high stands on the principal axis of a biconvex lens. The object is 50 mm from the lens. The focal length of the lens is 30 mm.

By means of a ray diagram, find the height of the image. Is the image real or virtual?

9 An object 30 mm high stands on the principal axis of a biconcave lens. The object is 90 mm from the lens. The focal length of the lens is 60 mm.

By means of a ray diagram find the height of the image. Is the image upright? Is the image real or virtual?

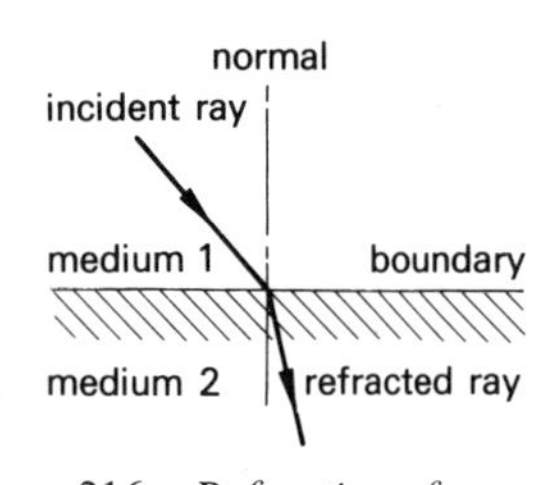

Figure 216 *Refraction of a light ray*

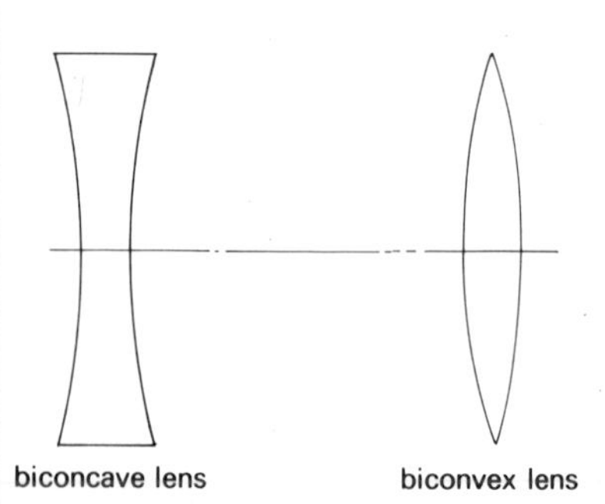

Figure 217 *Biconcave and biconvex lenses*

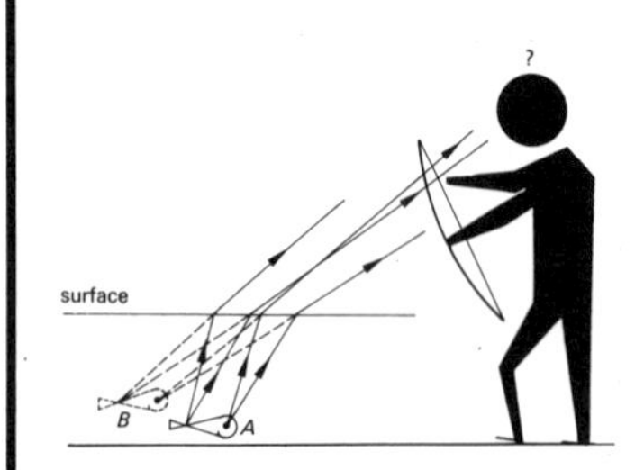

Figure 219 *Apparent position of a fish*

Solutions to self-assessment questions

4 When a ray of light passes from one medium into another of different optical density, the light ray undergoes a sharp change in direction. This is known as refraction; see Figure 216.

5 See Figure 217. The bi-convex lens could be used as a magnifying glass.

6 See Figure 218.

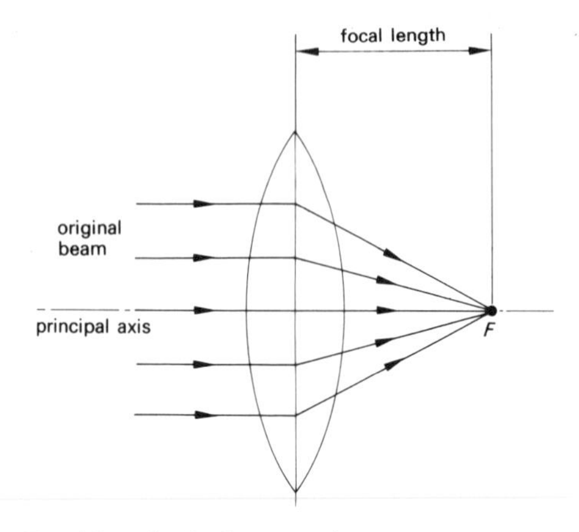

Figure 218 *Focal length of a biconvex lens*

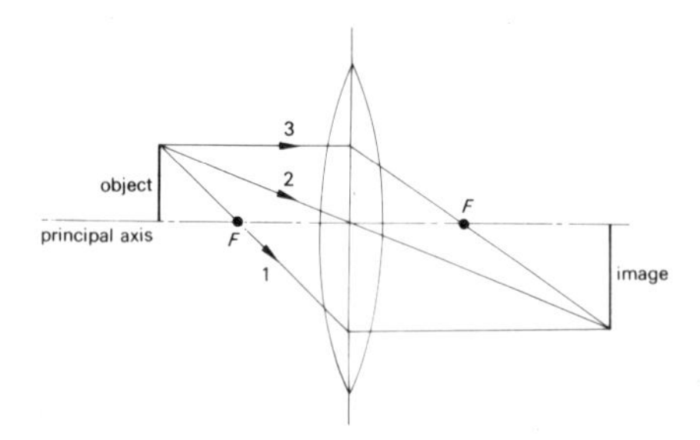

Figure 220 *Problem on biconvex lens*

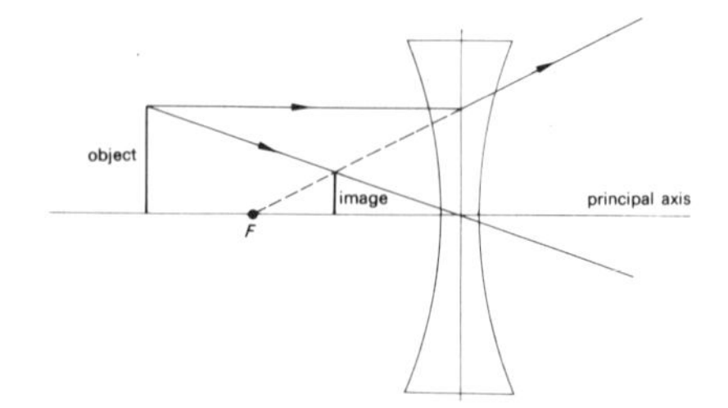

Figure 221 *Problem on biconcave lens*

7 The fish appears to be in position *B* when it is actually in position *A* (see Figure 219). The tribesman needs to aim below where the fish appears to be.

8 By measurement, the height of the image is 28 mm. The image is real. See Figure 220.

9 The height of the image is 12 mm. The image is upright and virtual.

Topic area: Further electricity

After reading the following material, the reader shall:

1 Know the effect of temperature on the resistance of a conductor and an insulator.

1.1 Define resistivity.

1.2 Solve simple problems involving resistivity.

1.3 Use resistivity values to classify materials as either conductors or insulators.

1.4 Define the temperature coefficient of resistance.

1.5 Solve simple problems involving the temperature coefficient of resistance.

1.6 Select, from a list, examples of conducting and insulating materials in common use.

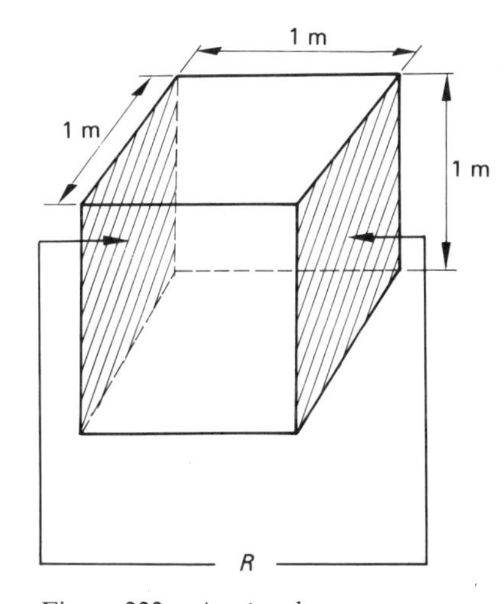

Figure 222 *A unit cube*

An electric current is a drift of electrons through a material. The material may be solid, liquid or gas. In some materials the atomic structure offers little opposition to the flow of current. In other materials, there is considerable opposition to current flow. The opposition to a flow of current is called the resistance of the material.

In order to compare the resistances of materials, conditions need to be standardized. When comparing the resistances of different solid materials the shape, size and the temperature of each sample must be the same.

The shape used is a cube with a side length of 1 m, the resistance being measured between opposite faces of the cube. This resistance is called the *resistivity* of the material. The resistivity of the material is defined as the resistance measured between opposite faces of a unit cube of the material at a specified temperature, which is usually 0 °C or 20 °C. The symbol for resistivity is ρ (rho), and its units can be found from the equation

$$R = \frac{\rho l}{A}$$

In this equation
R is the resistance of the material in ohms.
ρ is the resistivity of the material at a stated temperature,
l is the length of the material in metres,
A is the cross-sectional area of the material in square metres.

Thus the units of resistivity are

$$\frac{\text{ohms} \times \text{m}^2}{\text{m}}, \text{ i.e. ohm metre}$$

Hence the fundamental unit of resistivity is composed of two quantities:

(i) the unit of resistance
(ii) the length of the side of a unit cube

Many variations of the resistance unit and the length of the side of the cube are in use. It is common to find the resistance units in multiples and sub-multiples of the ohm, and the length of the side of the cube in multiples and sub-multiples of the metre.

When solving problems which include resistivity values, it is important that all the quantities in the problem are expressed in consistent units as shown in the following example.

Example 1

Find the resistance of a conductor having a diameter of 2.5 mm and a length of 100 m if the conductor has a resistivity of 17.2 μΩmm at 0 °C.

The resistance of the conductor in ohms

$$R = \frac{\rho l}{A}$$

ρ is the resistivity in ohm metres,
l is the length of the conductor in metres,
A is the cross-sectional area of the conductor in square metres.

$$\text{Conductor area, } A = \frac{\pi}{4} \times 2.5^2 \text{ mm}^2$$

$$\text{Conductor area in square metres, } A = \frac{\pi}{4} \times \frac{2.5^2}{10^3 \times 10^3} \text{ m}^2$$

$$\therefore A = 4.91 \times 10^{-6} \text{ m}^2$$

Resistivity in microhm millimetre
$= 17.2$ μΩmm

Resistivity in ohm millimetre
$= 17.2 \times 10^{6}$ Ωmm

Resistivity in ohm metre
$= 17.2 \times 10^{-6} \times 10^{-3}$ Ωm
$= 17.2 \times 10^{-9}$ Ωm

Conductor length
$= 100$ m

The conductor resistance in ohms is

$$R = \frac{\rho l}{A}$$

$$= \frac{17.2 \times 10^{-9} \times 100}{4.91 \times 10^{-6}} \left[\frac{\Omega m \times m}{m^2}\right]$$

$$\therefore R = 0.35\ \Omega$$

The resistance of the conductor at 0 °C is 0.35 Ω.

The following table gives typical resistivity values for some materials which are used in electrical work. The values are approximate as the resistivity of each material depends upon the method of production and the purity of the materials.

material	resistivity at 0 °C
copper	156 μΩ m
aluminium	245 μΩ m
platinum	981 μΩ m
PVC	300 MΩ m
glass	between 10^2 and 10^6 MΩ m
mica	between 10^4 and 10^7 MΩ m

Note the very considerable difference between the resistivity of the metals and the other materials. Metals have low resistivities and are used as conductors. Non-metals such as PVC, glass, mica, slate, rubber, wood, paper, fibre, polystyrene, polythene, etc. are used as insulators. The resistivity of metals is always low, i.e. in microhms millimetre, and the resistivity of the non-metals is always high, i.e. in megohms millimetre.

When the temperature of any solid material is increased, the rate at which the atoms within the material vibrate about their mean position increases. This increase in activity releases more electrons from their parent atoms, and these electrons move in a random fashion. The increase in the activity within a material caused by a rise in temperature changes the resistance of the material.

The resistance of materials commonly used as conductors (e.g. copper or aluminium) increases when the temperature of the conductor increases. Atomic theory suggests that the increase in the resistance of conductors produced by a temperature rise is due to the increase in the number of collisions between the electrons in random motion, and the electrons which constitute a flow of current.

When values of resistance are plotted against temperature, it is found that for metals used as conductors, the relationship between the resistance of the conductor and a limited change in temperature

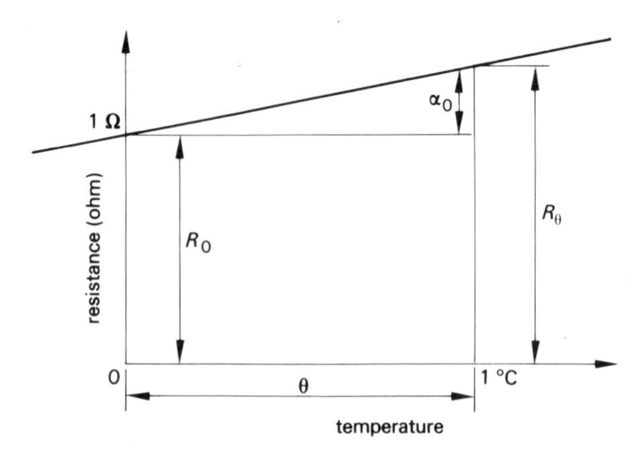

Figure 223 *Variation of the resistance of metal conductors with temperature*

is linear. The graph in Figure 223 shows the change in the resistance of a copper conductor caused by a rise in temperature from 0 °C to 1 °C. The resistance of the copper conductor at 0 °C, (R_0) is 1 Ω. The resistance of the copper conductor at 1 °C (R_θ) is 1.004 264 Ω. The change in the resistance of the conductor caused by an increase in the temperature of the conductor from 0 °C to 1 °C is referred to as the temperature coefficient of resistance or resistance-temperature coefficient at 0 °C. The symbol for the resistance-temperature coefficient is α (alpha) and at 0 °C is denoted by α_0.

The resistance-temperature coefficient of copper at 0 °C is written as $\alpha_0 = 0.004\,264$ Ω/Ω °C. This means that each ohm of the resistance at 0 °C increases by 0.004 264 Ω when the temperature rises to 1 °C. If the temperature rises above 1 °C, the rate at which the atoms within the material vibrate changes. This change alters the resistivity of the material and the resistance-temperature coefficient. The resistance-temperature coefficient is always related to a specific temperature. It may be defined as *the change in each ohm of resistance produced by a 1 °C rise from the specified temperature.* The table shows the values of the resistance-temperature coefficient of standard annealed copper for several temperatures.

temperature (°C)	resistance temperature coefficient (α)
0 °C	0.004 26 Ω/Ω °C
10 °C	0.004 09 Ω/Ω °C
20 °C	0.003 92 Ω/Ω °C
50 °C	0.003 51 Ω/Ω °C

Inspection of these values shows that, corrected to three decimal places, the value of the temperature coefficient is 0.004 Ω/Ω °C. Thus the resistance-temperature coefficient for copper, used as a conductor, can be regarded as constant if the operating temperature of the copper is below 50 °C. In these cases if a conductor has a resistance R_0 ohms at 0 °C, and a resistance-temperature coefficient of α, the increase in resistance for a 1° rise in temperature is $R_0\alpha$ ohms. At a temperature of θ °C, which is within the range of temperature rise in which the resistance-temperature coefficient is regarded as a constant value, the increase of resistance is $R_0\alpha\theta$ ohms. Hence for practical purposes,

$$R_\theta = R_0(1+\alpha\theta)$$

where R_θ is the resistance of the conductor at a temperature
θ °C between the limits in which
α is regarded as having a constant value,
R_0 is the resistance of the conductor at 0 °C,
α is the resistance temperature coefficient from and above 0 °C, to a known limit,
θ is the temperature in °C.

In other cases in which the value of the resistance-temperature coefficient is specific to a particular temperature it is denoted by a subscript, e.g. α_{20}. This means the resistance-temperature coefficient from and above 20 °C. For example, the resistance of a conductor at 20 °C is given by

$$R_{20} = R_0(1+\alpha_{20}.20).$$

Example 2
The resistance of a conductor at 0 °C is 100 Ω. The resistance-temperature coefficient from and above 0 °C is 0.004 Ω/Ω °C. Calculate the resistance of the conductor at 10 °C. Assume a constant value for the resistance-temperature coefficient over this range of temperature.

$$R_\theta = R_0(1+\alpha\theta)$$

When θ is 10 °C

$$R_{10} = 100\,[1+(0.004\times 10)]$$
$$= 100\,[1.04]$$

Resistance, $R_{10} = 104\ \Omega$

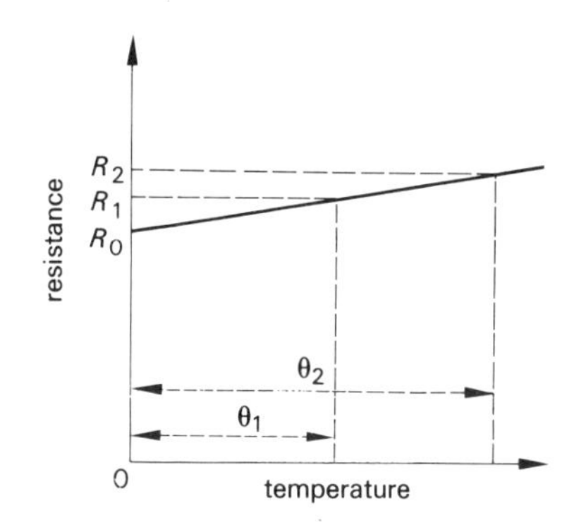

Figure 224 *Resistance–temperature graph for a resistor*

In practical situations the resistance of a conductor at 0 °C is difficult to obtain. Suppose a conductor has a resistance of R_1 at a temperature θ_1, and a resistance of R_2 at a higher temperature, θ_2. The relationship between the resistance at 0 °C and R_1 and R_2 is illustrated in Figure 224. Over the temperature range 0 °C to θ_2 °C the resistance-temperature coefficient is regarded as having a constant value.

In this case

$$R_1 = R_0(1+\alpha\theta_1)$$
$$\text{and } R_2 = R_0(1+\alpha\theta_2)$$

Dividing the two equations,

$$\frac{R_1}{R_2} = \frac{(1+\alpha\theta_1)}{(1+\alpha\theta_2)}$$

The use of this equation is illustrated in the following two examples.

Example 3
The resistance of a 100 W, 240 V filament lamp when it is connected to a 240 V supply is calculated to be 576 ohms. The resistance of the lamp when disconnected from the supply is measured, and found to be 50 ohms at 20 °C. The filament is made from a material with a resistance-temperature coefficient with a constant value of 0.005 Ω/Ω °C. Calculate the working temperature of the filament.

R_1 is 50 Ω
θ_1 is 20 °C
R_2 is 576 Ω
θ_2 is not known
α is 0.005 Ω/Ω °C

$$\text{as } \frac{R_1}{R_2} = \frac{1+\alpha\theta_1}{1+\alpha\theta_2}$$

$$1+\alpha\theta_2 = R_2 \times \frac{(1+\alpha\theta_1)}{R_1}$$

$$= \frac{576 \times [1+(0.005 \times 20)]}{50}$$

$$= \frac{633.6}{50} = 12.67$$

$$\therefore \theta_2 = \frac{12.67-1}{0.005}\ ^{\circ}\text{C}$$

$$\text{Temperature, } \theta_2 = 2\,334\ ^{\circ}\text{C}$$

Example 4
The resistance of a coil of wire is measured at 20 °C and is found to be 100 ohms. The coil is connected to a supply, and after some time, the temperature of the coil is found to be 80 °C. The resistance-temperature coefficient of the wire is regarded as having a constant value of 0.004 Ω/Ω °C over this range of temperature. Calculate the resistance of the coil at 80 °C.

R_1 is 100 Ω
θ_1 is 20 °C
R_2 is not known
θ_2 is 80 °C
α is 0.004 Ω/Ω °C

$$\text{as } \frac{R_1}{R_2} = \frac{1+\alpha\theta_1}{1+\alpha\theta_2}$$

$$R_2 = R_1 \times \left[\frac{1+\alpha\theta_2}{1+\alpha\theta_1}\right]$$

$$= 100 \times \left[\frac{1+(0.004\times 80)}{1+(0.004\times 20)}\right]$$

$$= 100 \times \frac{1.32}{1.08}$$

$$= 122\ \Omega$$

The resistance of the coil at 80 °C is 122 Ω.

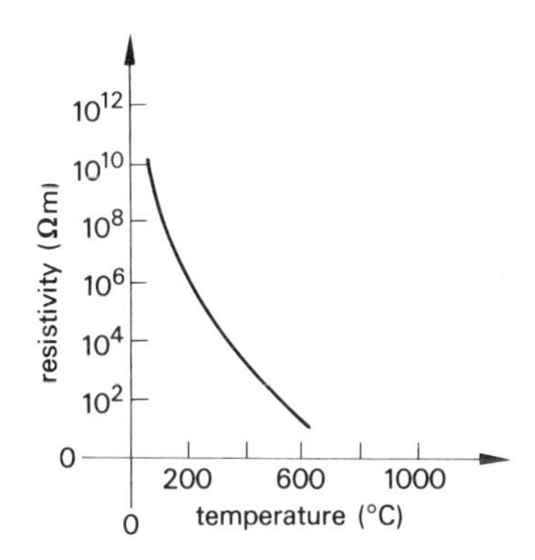

Figure 225 *Variation of resistivity with temperature of an insulating material*

In each of the examples the resistance-temperature coefficient is positive, i.e. as the temperature of the conductor has increased the resistance has also increased. In each case the relationship between the resistance and the temperature of the conductor is a linear relationship.

In the case of materials which are used as insulators, the relationship between the resistance of the material and the temperature is more complex. Firstly, the resistance temperature coefficient is negative. This means that as the temperature increases the resistance of the material decreases. Secondly, the relationship between temperature and resistance is non-linear.

The curve in Figure 225 shows how the resistance of a typical ceramic insulator decreases with increasing temperature.

Self-assessment questions

1 Each of the following statements refer to either a pure metallic conductor or an insulating material. Classify each statement as referring to a conductor or an insulator:

(*a*) Resistance increases with increase in temperature.
CONDUCTOR/INSULATOR

(*b*) Resistances decrease with increase in temperature.
CONDUCTOR/INSULATOR

(*c*) Positive resistance-temperature coefficient.
CONDUCTOR/INSULATOR

(*d*) The resistance-temperature relationship is linear.
CONDUCTOR/INSULATOR

(*e*) Negative resistance-temperature coefficient.
CONDUCTOR/INSULATOR

(*f*) The resistance-temperature relationship is non-linear.
CONDUCTOR/INSULATOR

2 Classify each of the following materials as either a conductor or an insulator:

material	classification
wood	
aluminium	
slate	
copper	
paper	
mica	
rubber	
lead	
zinc	
glass	
PVC	
silver	

3 Classify each of the following values of resistivity as referring to either a conductor or an insulator:

resistivity at 0 °C	classification
1.58 μΩ cm	
22 μΩ cm	
300 MΩ m	
480 μΩ cm	
40 MΩ cm	
17.2 μΩ mm	

4 Complete the following two statements:

(i) Resistivity is defined as ______.

(ii) The resistance-temperature coefficient of a material is defined as ______.

5 Calculate the resistivity at 20 °C of a conductor which is 100 m long, has a cross-sectional area of 2×10^{-6} m^2 and a resistance of 0.85 Ω at 20 °C.

6 The resistance of a conductor at 0 °C is 2.4 Ω. The resistance-temperature coefficient of the conductor has a constant value of 0.005 Ω/Ω °C. Calculate the resistance of the conductor at 110 °C.

7 A coil has a resistance of 55 Ω at 20 °C. The coil is energized, and after some time the resistance is found to be 65 Ω. The resistance-temperature coefficient of the coil has a constant value of 0.004 3 Ω/Ω °C. Calculate the temperature when the resistance of the coil is 65 Ω.

After reading the following material, the reader shall:

2 Know the concepts of e.m.f. and internal resistance.

2.1 Describe the potential difference (voltage) of a source on no load as the e.m.f.

2.2 Explain the reasons for a voltmeter having a high resistance.

2.3 Define internal resistance.

2.4 State the effect of load current on terminal p.d.

2.5 Calculate the internal resistance of a source of e.m.f.

An electromotive force is that which tends to produce an electric current in a circuit. The unit of e.m.f. (E) is the volt (V). The most utilized sources of e.m.f. are those produced by

(i) variations of magnetic flux interlinked with conductors, e.g. in alternators.

(ii) the chemical action which takes place when dissimilar conductors are placed in an electrolyte, e.g. in batteries.

These and all other sources of e.m.f. have internal resistance. The internal resistance of a source of e.m.f. is the opposition offered by the materials within the source of the e.m.f. to a flow of current.

The maximum current that a given e.m.f. can supply is determined by Ohm's law.

$$\text{Maximum } I = \frac{E}{r} \text{ ampere,}$$

when r is the internal resistance in ohms of the source of e.m.f.

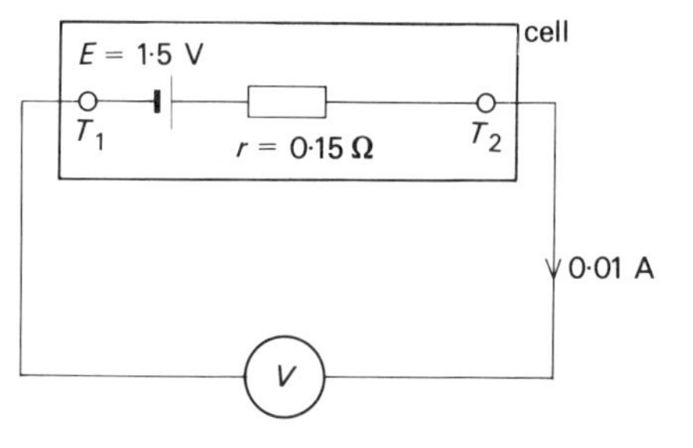

Figure 226 *The measurement of the terminal voltage of a cell*

The circuit diagram in Figure 226 shows a single cell with an e.m.f. of 1.5 V. The internal resistance is 0.15 Ω. The terminals of the cell (T_1 and T_2) are connected to a voltmeter which will measure the p.d. across the cell. The voltmeter is in this case a moving coil instrument. A moving coil instrument draws a current, which in Figure 226 is 0.01 A. This current flows through the internal resistance of the cell and produces a p.d. across the internal resistance. As the voltmeter draws a current of 0.01 A, and the internal resistance is 0.15 Ω, the p.d. across the internal resistance

$= I \times r$ volts

$= 0.01 \times 0.15$

$= 0.001\,5$ V

Solutions to self-assessment questions

1 (*a*) conductor
(*b*) insulator
(*c*) conductor
(*d*) conductor
(*e*) insulator
(*f*) insulator

2 wood – insulator
aluminium – conductor
slate – insulator
copper – conductor
paper – insulator
mica – insulator
rubber – insulator
lead – conductor
zinc – conductor
glass – insulator
PVC – insulator
silver – conductor

3

resistivity at 0 °C	classification
1.58 μΩ cm	conductor
22 μΩ cm	conductor
300 MΩ cm	insulator
480 μΩ cm	conductor
40 MΩ cm	insulator
17.2 μΩ cm	conductor

4 (i) Resistivity is defined as *the resistance measured between opposite faces of a unit cube of a material at a stated temperature.*
(ii) The resistance-temperature coefficient of a material is defined as *the change in each ohm of resistance produced by a 1 °C rise from a specified temperature.*

5 Since $R = \frac{\rho l}{A}$

$$\rho = \frac{AR}{l}$$

when A is 2×10^{-6} m²
R is 0.85 Ω
l is 100 m

$$\rho = \frac{2 \times 10^{-6} \times 0.85}{100} \left[\frac{m^2\Omega}{m}\right]$$
$$= 0.017 \times 10^{-6}\ \Omega\ m$$

∴ Resistance-temperature coefficient, $P = 0.017\ \mu\Omega$ m

6 Since $R = R_0(1+\alpha\theta)$
As R is the resistance at 110 °C = R_{110}
R_0 is the resistance at 0 °C = 2.4 Ω
α is the resistance-temperature coefficient of the material = 0.005 Ω/Ω °C

$$R_{110} = 2.4\,[1 + (0.005 \times 110)]$$
$$= 2.4 \times 1.55$$

∴ Resistance, $R_{110} = 3.72\ \Omega$

It follows that the p.d. measured by the voltmeter

$= E - I \times r$

$= 1.5 - 0.0015 \simeq 1.499\ \text{V}$

The p.d. measured when a current is being supplied by a source is referred to as a terminal voltage, and is given the symbol V. Therefore

$$V = E - Ir$$

when V is the terminal voltage in volts
E is the e.m.f. of the source in volts
I is the current in the circuit in amperes
r is the internal resistance of the source of e.m.f. in ohms

In all instances when a p.d. is measured with an instrument which takes a current from the source of e.m.f., the p.d. measured is a terminal voltage. This p.d. is always less than the p.d. of the source of e.m.f. because of the effect of the instrument current flowing through the internal resistance of the source.

Many voltmeters include a moving coil instrument. The moving coil instrument pointer is deflected by a force generated from the interaction of the magnetic field of a permanent magnet and the current in the coil. The magnitude of the current required to cause a full-scale deflection is known, and is usually shown on the instrument. The lower the current required to cause a full scale deflection, the higher is the resistance of the instrument. When an instrument with a high resistance is used to measure the p.d. of a source of e.m.f. which is supplying only the current required by the instrument, the difference between the terminal voltage shown on the instrument and the e.m.f. of the source is often negligible.

7

Since $\dfrac{R_1}{R_2} = \dfrac{1+\alpha\theta_1}{1+\alpha\theta_2}$

As $R_1 = 55\ \Omega$ and is the resistance at 20 °C

$R_2 = 65\ \Omega$ and is the resistance at the final temperature

$\theta_1 = 20$ °C and is the initial temperature

$\theta_2 =$ final temperature

$\alpha = 0.0043\ \Omega/\Omega$ °C and is the resistance-temperature coefficient at 0 °C

Then $1+\theta_2 = R_2 \times \left[\dfrac{1+\alpha\theta_1}{R_1}\right]$

$= 65 \times \left[\dfrac{1+(0.0043 \times 20)}{55}\right]$

$\therefore\ 1+\alpha\theta_2 = 1.283$

$\therefore\ \theta_2 = \dfrac{1.283-1}{0.0043}$

Temperature, $\theta_2 = 65.8$ °C

Example 5
A battery has an e.m.f. of 1.5 V and an internal resistance of 0.15 Ω. The terminal voltage is measured with a 0–10 V moving coil instrument which requires a current of 15 mA to cause a full scale deflection. Calculate the difference between the terminal voltage measured and the e.m.f. of the cell.

The p.d. required to cause a full-scale deflection is 10 V. The current required to cause a full-scale deflection is 15 mA $= 15 \times 10^{-3}$ A.

From Ohm's law, the resistance of the instrument

$$= \frac{V}{I}$$

$$= \tfrac{10}{15} \times 10^{-3}$$

$$= 666.67\ \Omega$$

The current taken by the voltmeter when connected to an e.m.f. of 1.5 V

$$= \frac{E}{R}\ \text{ampere}$$

$$= \frac{1.5}{666.67} = 0.002\,25\ \text{A}$$

The voltage drop produced by this current in the internal resistance

$$= I \times r$$

$$= 0.002\,25 \times 0.15 = 0.000\,34\ \text{V}$$

The terminal voltage, i.e. the voltage reading on the instrument is

$$\begin{aligned} V &= E - Ir \\ &= 1.5 - (0.15 \times 0.002\,25) \\ &= 1.5 - 0.000\,34 \\ &= 1.499\,7\ \text{V} \end{aligned}$$

The difference between the e.m.f. and the measured terminal voltage

$$= 1.5 - 1.499\,7 = 0.000\,3\ \text{V}$$

This terminal voltage is very close to the p.d. of the source of e.m.f., and can be regarded for most practical purposes as the e.m.f. of the source. From the equation $V = E - Ir$ it can be seen that the lower the current drawn by the voltmeter, the nearer to the e.m.f. is the reading on the voltmeter. The internal resistance of an e.m.f. can be calculated as in the following example.

Example 6
An e.m.f. of 14 V is connected to a circuit with a resistor of 7 Ω. The p.d. across the resistor is 12 V. Calculate the internal resistance of the source of e.m.f.

From Ohm's law, $I = \frac{E}{R}$

$$= \tfrac{14}{7} = 2\ \text{A}$$

$$\text{From } V = E - Ir$$

$$r = \frac{E-V}{I}$$

$$\text{Internal resistance, } r = \frac{14-12}{2} = 1\ \Omega$$

If the e.m.f. and the internal resistance of the source in example 6 are maintained constant, but the current is increased to 4 A, the terminal voltage decreases.

$$\begin{aligned}\text{In these conditions } V &= E - Ir \\ &= 14-(4\times 1) \\ &= 10\text{ V}\end{aligned}$$

If the current is increased to 8 A the terminal voltage

$$\begin{aligned}V &= E - Ir \\ &= 14-(8\times 1) \\ &= 6\text{ V}\end{aligned}$$

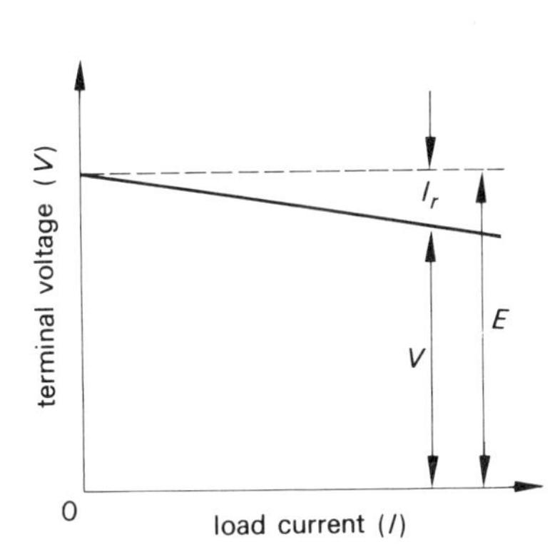

Figure 227 *Variation of terminal voltage with load current*

The graph in Figure 227 shows the effect of an increase in load current on the terminal voltage, when the e.m.f. and internal resistance are maintained constant.

From the graph it can be seen that when the e.m.f. and the internal resistance of a source are maintained constant, an increase in the current in the circuit reduces the p.d. across the circuit.

Self-assessment questions

8 Complete the following four statements:

(i) The p.d. between the terminals of a source of electrical energy which is not supplying current is referred to as the ______.

(ii) The p.d. between the terminals of a source of electrical energy which is supplying a current is referred to as the ______.

(iii) The internal resistance of a source is defined as the ______.

(iv) In conditions in which the e.m.f. and the internal resistance are constant, the terminal voltage decreases when the current ______.

9 An e.m.f. of 10 V with a constant internal resistance is to be measured.

(i) Select from the following instruments the one which will provide the best approximation of the e.m.f.

(ii) Explain the choice of instrument.

Instrument 1: 0–10 V which requires 20 mA to cause a full scale deflection.

Instrument 2: 0–15 V which requires 20 mA to cause a full scale deflection.

Instrument 3: 0–12 V which requires 15 mA to cause a full scale deflection.

10 A torch battery with an e.m.f. of 3 V is supplying a load which takes 0.3 A. The p.d. across the load measured with a high resistance voltmeter is 2.88 V. Calculate the internal resistance of the battery.

After reading the following material, the reader shall:

3 Know the chemical action in, and the major parts of, cells.
3.1 Explain the difference between primary and secondary cells.
3.2 Describe the charging and discharging of a simple lead–acid cell.
3.3 Label, given diagram, the main parts of:
(*a*) lead–acid cells,
(*b*) alkaline cells,
(*c*) mercury cells.

Electric cells

Electric cells, or batteries, are grouped into two classes:
(i) primary cells
(ii) secondary cells
Both classes of electric cell contain two conductors, called *plates*, immersed in a conducting solution, called an *electrolyte*. The plates are made from dissimilar materials, e.g. zinc and carbon, zinc and copper, lead and lead peroxide. One plate has a greater tendency to lose electrons than the other. This plate is referred to as the negative plate. The other is called the positive plate. Between the plates is a difference in their state of charge called an e.m.f. This is produced as a direct result of the chemical action which takes place between the two plates when they are immersed in an electrolyte. The chemical activity in both types of cell is increased when the cells are supplying electrical energy to an external circuit. The electrical energy is produced as a direct result of chemical changes in some of the chemically active substances within the cells. When the cells have supplied electrical energy for some time, the chemical changes produce an increase in the internal resistance of the cells which reduces the magnitude of the e.m.f., and therefore the quantity of electrical energy supplied to the external circuit. When the e.m.f. has decreased to a level such that the energy supplied to the external circuit is reduced below that required for the efficient operation of the external circuit, the cells are said to be discharged.

In a discharged primary cell the chemically active material in the negative plate has been dissolved into the electrolyte. In order to restore a flow of current in an external circuit, both the electrolyte, and the active material in the negative plate must be replaced. Modern primary cells (e.g. those use in torches, calculators, transistor radios, etc.) are sealed; when they are discharged, it is impossible to replace the chemically active substances. They are therefore discarded when they do not function efficiently. When a secondary cell is discharged, the chemical changes produced during the discharge can be reversed by connecting the cell to a supply of electrical energy. The process is called charging or re-charging. The charge–discharge–charge cycle can be repeated many times with a secondary cell.

Hence the major difference between primary and secondary cells is that secondary cells can be recharged, whilst primary cells cannot be recharged.

The mercury cell

As soon as most primary cells are manufactured, the negative plate starts to dissolve into the electrolyte. This means that most types of primary cell have a limited life. Also, when most types of primary cell are supplying electrical energy to external circuits, the increase in chemical activity produces a layer of bubbles on the positive plate. This effect is called *polarization*. Polarization increases the resistance between the positive plate and the electrolyte, and also reduces the e.m.f. between the positive plate and the negative plate. The mercury cell was developed to increase the life of primary cells, and to reduce the effects of polarization.

The mercury cell consists of seven main parts, illustrated in Figure 228.

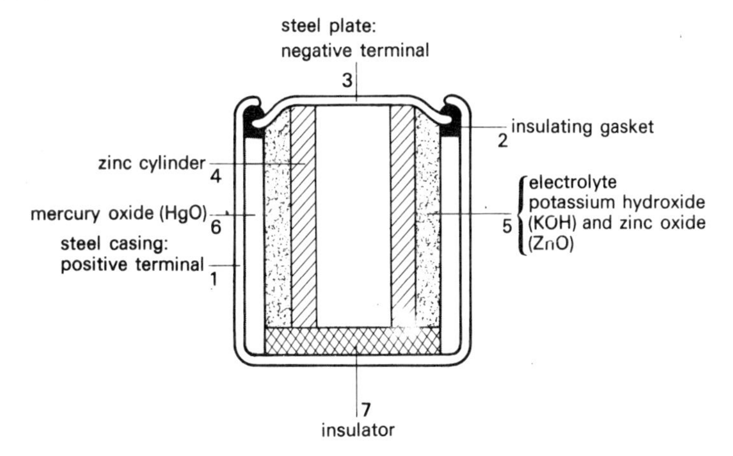

Figure 228 *A mercury cell*

These are

1 The steel casing of the cell, which forms a part of the positive plate, and is the positive terminal.

2 An insulating gasket which supports the negative terminal and seals the cell.

3 A steel plate which is the negative terminal, and is a part of the negative plate.
4 A cylinder, made of zinc, which forms the second part of the negative plate.
5 Potassium hydroxide and zinc oxide electrolyte between the positive and negative plates.
6 Mercury oxide, which in conjunction with the steel case, forms the positive plate.
7 An insulator separating the positive and negative plates.

The mercury cell can maintain a reasonably constant e.m.f. over a much longer period of time than most other forms of primary cell. The interactions between the chemically active materials when the cell is supplying a current ensure that no gas is produced; there is therefore no polarization. This cell has considerable advantages over other types of primary cell, but when it is discharged, it also must be replaced by a new cell.

Solutions to self-assessment questions

8 (i) The p.d. between the terminals of a source of electrical energy which is not supplying current is referred to as the *e.m.f.*

(ii) The p.d. between the terminals of a source of electrical energy which is supplying a current is referred to as the *terminal voltage*.

(iii) The internal resistance of a source is defined as the *opposition to a flow of current offered by the materials within the source*.

(iv) In conditions in which the e.m.f. and the internal resistance are constant, the terminal voltage decreases when the current *increases*.

9 (i) The instrument which will provide the best approximation of the e.m.f. of the source is instrument 3.

(ii) The resistance of instrument 3

$$= \frac{V}{I} = \frac{12}{15} \times 10^{-3} = 800\ \Omega$$

The resistance of instrument 2

$$= \frac{V}{I} = \frac{15}{20} \times 10^{-3} = 750\ \Omega$$

The resistance of instrument 1

$$= \frac{V}{I} = \frac{10}{20} \times 10^{-3} = 500\ \Omega$$

The instrument with the higher resistance draws the minimum current, and therefore causes the minimum voltage drop across the internal resistance. Therefore the instrument with the highest resistance provides the best approximation of the e.m.f.

10

$$\text{As } V = E - Ir$$

$$r = \frac{E - V}{I}\ \text{ohm}$$

$$= \frac{3 - 2.88}{0.3} = \frac{0.12}{0.3}$$

$\therefore$ Internal resistance, $r = 0.4\ \Omega$

The lead–acid cell

An example of a secondary cell, in which the chemical changes produced during discharge can be reversed, is the lead–acid cell. In its simplest form a lead–acid cell consists of:

(i) A container made from an insulating material (e.g. glass).
(ii) An electrolyte, consisting of sulphuric acid and water.
(iii) Two conductors (plates) immersed in the electrolyte.

Discharge of a lead–acid cell
When sulphuric acid (H_2SO_4) and water (H_2O) are mixed to produce the electrolyte, the molecules of sulphuric acid and water produce positively charged hydrogen ions and negatively charged sulphate ions. These charged particles obey the law that *like charges repel, unlike charges attract*.

One of the plates in a charged lead–acid cell contains lead (Pb), and is a slate-grey colour. The second plate contains lead peroxide (PbO_2), and is a reddish-brown colour. When the plates are immersed in the electrolyte and joined by an external circuit, chemical activity takes place. This causes a drift of electrons from the lead plate through the external circuit to the lead peroxide plate, and into the electrolyte.

The electrons which drift from the lead plate leave this plate with a positive charge, which attracts negative ions from within the electrolyte. These ions combine with the lead plate to produce lead sulphate ($PbSO_4$). The lead sulphate covers the plate, and the resistance between the plate and the electrolyte increases as the discharge continues. Lead sulphate appears on the plate as a white salt.

The electrons which have drifted through the external circuit, from the lead plate to the lead peroxide plate, combine with positive ions from the electrolyte (i) to form a gas which is released to the atmosphere and (ii) to produce water molecules in the electrolyte. The increase in the quantity of water in the electrolyte decreases the relative density which gradually increases the resistance of the electrolyte as the discharge continues.

The lead peroxide plate has a positive charge when compared with the lead plate; it therefore attracts negative ions from the electrolyte. The negative ions combine with the lead peroxide to produce lead sulphate as on the lead plate. The lead sulphate covers the plate and, as discharge continues, increases the resistance between the plate and the electrolyte.

As the discharge of a lead–acid cell continues, the formation of water in the electrolyte and lead sulphate on the plates increases the internal resistance of the cell, and therefore increases the p.d. caused by the discharge current through the internal resistance. From the equation $V = E - Ir$ it can be seen that the terminal voltage will gradually decrease during discharge as the internal resistance increases.

Re-charging of a lead–acid cell
The chemical changes produced in the cell during discharge can be reversed by connecting the plates to a source of electrical energy. The cell is connected to the source so that electrons drift into the lead plate. The drift of electrons through the cell restores the lead sulphate formed on both plates to the electrolyte. This reduces the resistance of the electrolyte, and decreases the resistance between the plate and electrolyte.

Re-charging is continued until the plates and the electrolyte are restored to their charged condition. This condition is reached when the relative density of the electrolyte is 1.230, and the plates are respectively grey and reddish-brown as in Figure 229. The nominal e.m.f. of a lead–acid cell is 2 V.

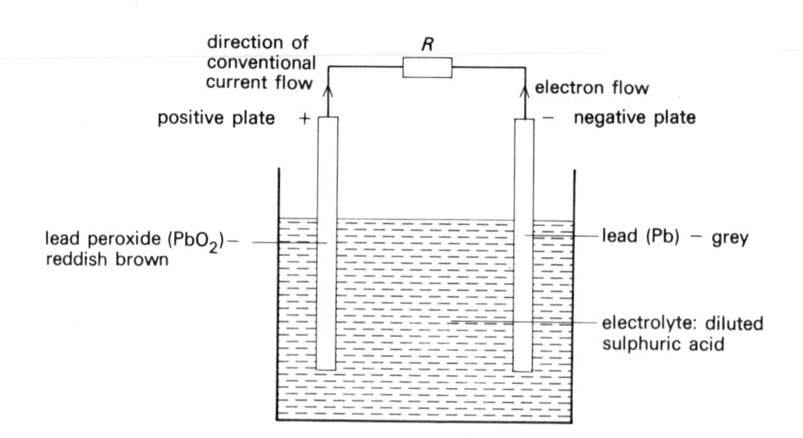

Figure 229 *A charged simple lead–acid cell*

Sulphation
If a lead–acid cell is left in a discharged condition the lead sulphate produced during the discharge hardens and does not dissolve during re-charging. This condition is known as *sulphation.* A lead–acid cell in this condition cannot be re-charged and has to be discarded. The condition is avoided by maintaining a charge–discharge–charge cycle.

The alkaline cell

A second form of secondary cell is the alkaline cell. The alkaline cell has a nominal e.m.f. of 1.2 V. Its main advantage when compared with a lead–acid cell is that it may be left in a discharged condition

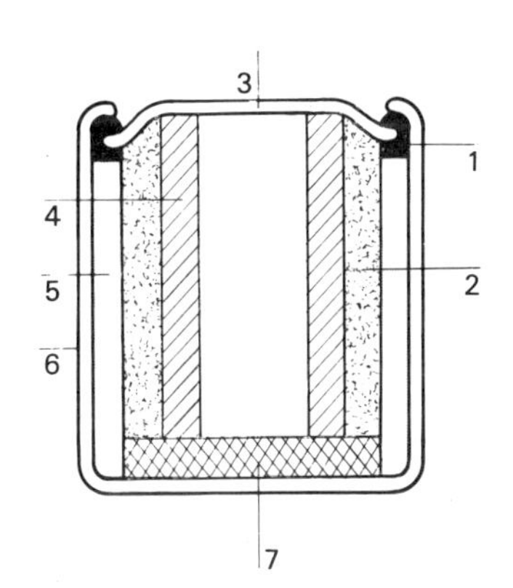

Figure 231 *A mercury cell*

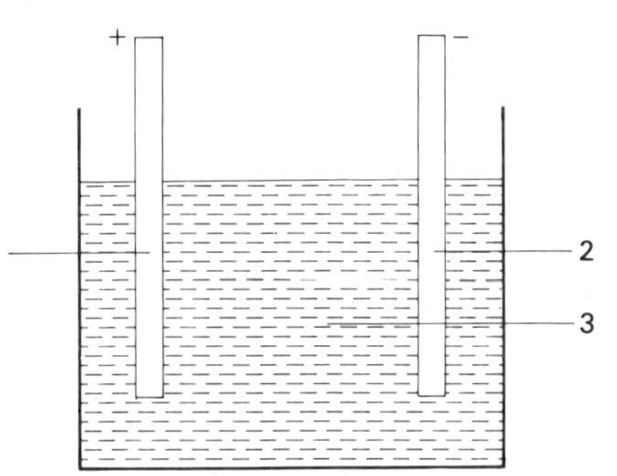

Figure 232 *Charged simple lead–acid cell*

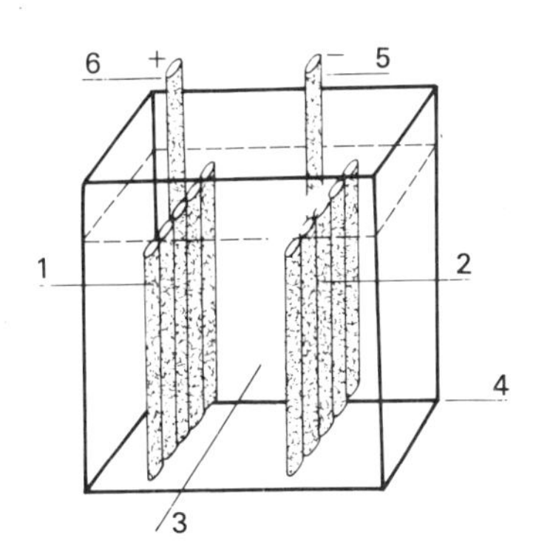

Figure 233 *Charged alkaline cell (wooden case not shown)*

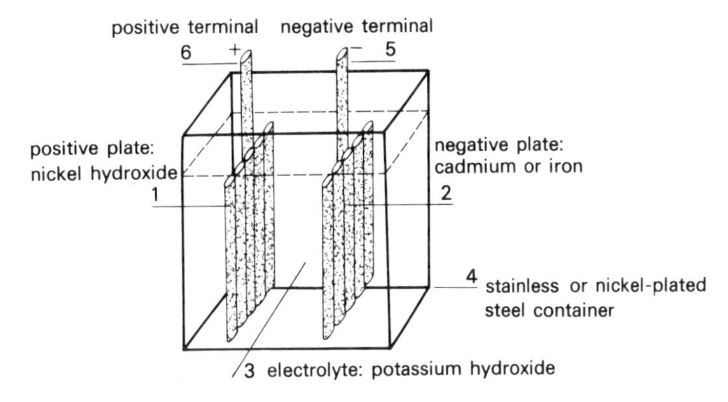

Figure 230 *Charged alkaline cell (wooden case not shown)*

for a considerable time, and still be capable of being re-charged. The main parts of an alkaline cell are shown in Figure 230.

In the alkaline cell the electrolyte is potassium hydroxide (KOH). The plates are constructed from perforated steel tubes joined as shown in Figure 230. The steel tubes in the positive plate hold nickel hydroxide ($Ni(OH)_3$). The tubes in the negative plate contain either cadmium (Cd) or iron (Fe). Two types of alkaline cell are manufactured, the nickel–cadmium cell and the nickel–iron cell. The plates and the electrolyte are enclosed in a stainless or nickel-plated steel container. The steel container is mounted in a wooden case.

Self-assessment questions

11 State the major difference between a discharged primary cell and a discharged secondary cell.

12 Name and describe as appropriate the numbered items in the sketch of a mercury cell shown in Figure 231.

13 Name and describe as appropriate the numbered items in the sketch of a charged simple lead–acid cell shown in Figure 232.

14 Name and describe as appropriate the numbered items in the sketch of a charged alkaline cell shown in Figure 233.

15 Explain the effect on the output of a lead–acid cell of the formation of lead sulphate on the plates, and the production of additional water in the electrolyte during a discharge.

16 Explain the effect on the internal resistance of a lead–acid cell of a re-charge.

17 Explain why primary cells cannot be re-charged.

Solutions to self-assessment questions

11 The major difference between a discharged primary cell and a discharged secondary cell is that the secondary cell can be re-charged.

12 (i) Insulating gasket.
(ii) Electrolyte consisting of potassium hydroxide and zinc oxide.
(iii) A steel plate which is the negative terminal, and is a part of the negative plate.
(iv) A zinc cylinder which is part of the negative plate.
(v) Mercury oxide which is a part of the positive plate.
(vi) Steel case which is the positive terminal, and a part of the positive plate.
(vii) An insulator separating the positive and negative plates.

13 The names and description of the numbered items in a charged simple lead–acid cell are shown in Figure 234.

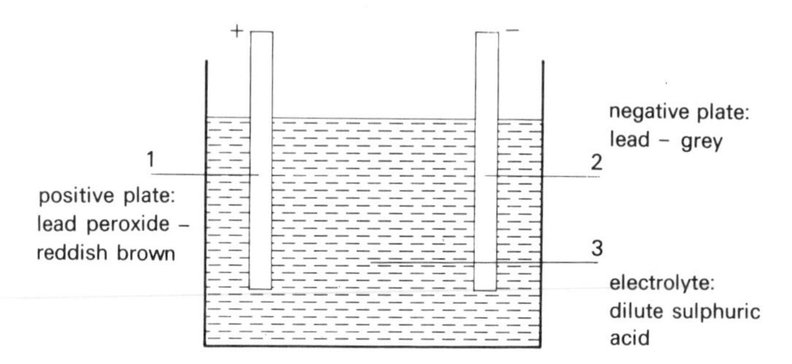

Figure 234 *Charged simple lead–acid cell*

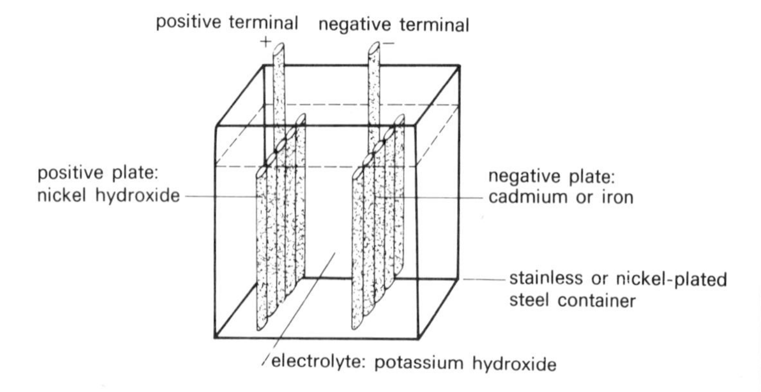

Figure 235 *Charged alkaline cell (wooden case not shown)*

14 The names and descriptions of the numbered items in a charged alkaline cell are shown in Figure 235.

15 The effect on the output of the cell caused by the formation of lead sulphate on the plates and additional water in the electrolyte, is to increase the internal resistance of the cell, and therefore cause a reduction in the e.m.f. produced by the cell.

16 The effect of a re-charge on the internal resistance of a lead–acid cell is to decrease the internal resistance. This is accomplished by restoring to the electrolyte the lead sulphate which was produced during discharge. The removal of the lead sulphate from the plates also decreases the resistance between the plates and the electrolyte.

17 Primary cells cannot be re-charged because the chemically active substances in the negative plate are dissolved into the electrolyte, and the process cannot be reversed by reversing the flow of current, i.e. by re-charging.

After reading the following material, the reader shall:

4 Know methods of connecting cells.

4.1 Calculate the internal resistance and e.m.f. of cells connected (i) in series (ii) in parallel.

The series connection of cells

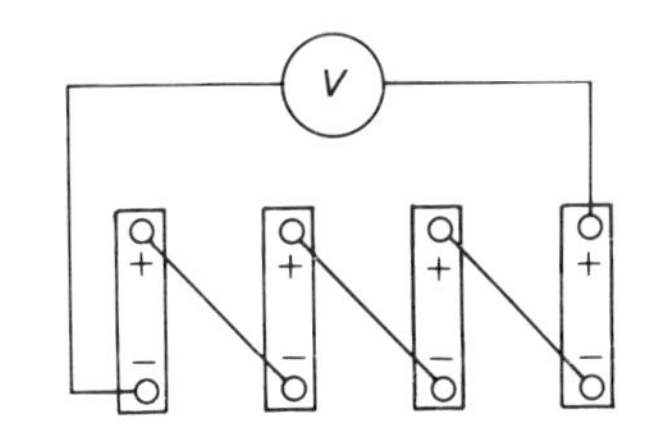

Figure 236 *Cells connected in series*

The discussion on cells has been limited to single cells. The nominal e.m.f. of many single cells is about 2 V. There are many applications in science and engineering which require a p.d. greater than that provided by a single cell. This is achieved by connecting cells in series to form batteries. The diagram in Figure 236 shows four cells connected in series. When cells are connected in series, the e.m.f. of the battery is the sum of the e.m.f. of each cell.

$E = e \times n$ volts

when E is the e.m.f. of the battery
e is the e.m.f. of a cell
n is the number of cells

The internal resistance of a battery consisting of cells connected in series is the sum of the internal resistance of each cell in the battery.

$R = r \times n$ ohms

when R is the internal resistance of the battery
r is the internal resistance of a cell
n is the number of cells

Example 7

Ten cells, each with an internal resistance of 0.2 Ω and an e.m.f. of 2 V, are connected in series. Calculate the internal resistance and the e.m.f. of the battery.

From $R = n \times r$ ohms
$R = 10 \times 0.2$
$R = 2$ ohms

The internal resistance of the battery is 20 ohms.

From $E = n \times e$ volts
$E = 10 \times 2$
$E = 20$ volts

The e.m.f. of the battery is 20 volts.

The parallel connection of cells

There are uses of electrical energy in engineering and science which require an e.m.f. of the magnitude that can be provided by a single

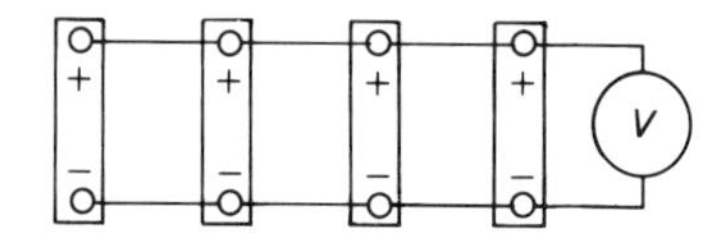

Figure 237 *Cells connected in parallel*

cell, but which also require a flow of current to be maintained for long periods of time. In these cases cells may be connected in parallel as shown in Figure 237.

When cells are connected in parallel the e.m.f. of the battery is the e.m.f. of an individual cell.

$$E = e$$

when E is the e.m.f. of the battery
e is the e.m.f. of a single cell

The internal resistance of the battery is calculated from

$$\frac{1}{R} = \frac{1}{r_1} + \frac{1}{r_2}, \text{ etc.}$$

when R is the internal resistance of the battery
r is the internal resistance of a single cell

When similar cells are connected in parallel, the internal resistance of the battery

$R = \frac{1}{n}$th of the internal resistance of one cell.

Example 8

Ten cells each with an internal resistance of 0.2 Ω and an e.m.f. of 2 V are connected in parallel. Calculate the internal resistance and the e.m.f. of the battery.

Internal resistance of the battery

$R = \frac{1}{n}$th the internal resistance of a cell

when n is the number of cells
r is the internal resistance of one cell

$$R = \frac{r}{n} = \frac{0.2}{10} = 0.02\ \Omega$$

The e.m.f. of the battery

$$E = e$$

when e = e.m.f. of one cell

$$E = 2\ \text{V}$$

When cells are connected in parallel, each cell should be in a similar state of charge or currents will flow between cells.

Self-assessment question

18 A battery consists of fifteen cells each with an e.m.f. of 1.5 V and an internal resistance of 0.4 Ω. Calculate the e.m.f. and internal resistance of the battery when the cells are connected

(*a*) in series

(*b*) in parallel

Solution to self-assessment question

18 (*a*) series connection

The e.m.f. of the battery

$E = n \times e$ volts

$= 15 \times 1.5$

$= 22.5$ V

The internal resistance of the battery

$R = n \times r$ ohms

$= 15 \times 0.4$

$= 6\ \Omega$

(*b*) parallel connection

The e.m.f. of the battery

$E =$ e.m.f. of one cell

$\therefore E = 1.5$ V

The internal resistance of the battery

$R = \frac{1}{n}$th the internal resistance of one cell when n is the number of cells.

$\therefore R = \frac{r}{n}$ ohms

$= \frac{0.4}{15} = 0.027\ \Omega$